1 MONTH OF
FREE
READING

at

www.ForgottenBooks.com

---◆---

By purchasing this book you are eligible for one month membership to ForgottenBooks.com, giving you unlimited access to our entire collection of over 1,000,000 titles via our web site and mobile apps.

To claim your free month visit: www.forgottenbooks.com/free78937

ISBN 978-0-265-28883-2
PIBN 10078937

This book is a reproduction of an important historical work. Forgotten Books uses
state-of-the-art technology to digitally reconstruct the work, preserving the original format
whilst repairing imperfections present in the aged copy. In rare cases, an imperfection in
the original, such as a blemish or missing page, may be replicated in our edition. We do,
however, repair the vast majority of imperfections successfully; any imperfections that
remain are intentionally left to preserve the state of such historical works.

BUREAU OF PROVINCIAL INFORMATION.

BULLETIN No. 10.

LAND AND AGRICULTURE

—IN—

BRITISH COLUMBIA.

THE GOVERNMENT OF
THE PROVINCE OF BRITISH COLUMBIA

(FIFTH EDITION.)

Printed by authority of the Legislative Assembly.

VICTORIA, B. C.:
Printed by RICHARD WOLFENDEN, V.D., I.S.O., Printer to the King's Most Excellent Majesty.
1906.

Captain the Honourable R. G. Tatlow,

    Minister of Finance and Agriculture,

        Victoria, B. C.:

SIR,—I have the honour to present for your approval a new edition of Bulletin No. 10, on Land and Agriculture, four large editions of which have been printed and distributed.

This Bulletin is compiled from the latest available authorities, and contains information which should prove invaluable to the thousands of prospective settlers in British Columbia, from whom letters of inquiry are being received.

I have the honour to be,

Sir,

Your obedient servant,

R. M. PALMER,

Secretary Bureau of Provincial Information.

Victoria, B. C., July, 1906.

ROYAL HORTICULTURAL SOCIETY

ESTABLISHED A.D. 1804.   INCORPORATED A.D. 1809.

GOLD MEDAL.

Colonial Fruit Show, London, Dec. 5 & 6th 1905

Awarded to The Province of British Columbia.

For   Collection of Apples

AWARDED EXHIBITS OF BRITISH COLUMBIA APPLES.
December 1905.

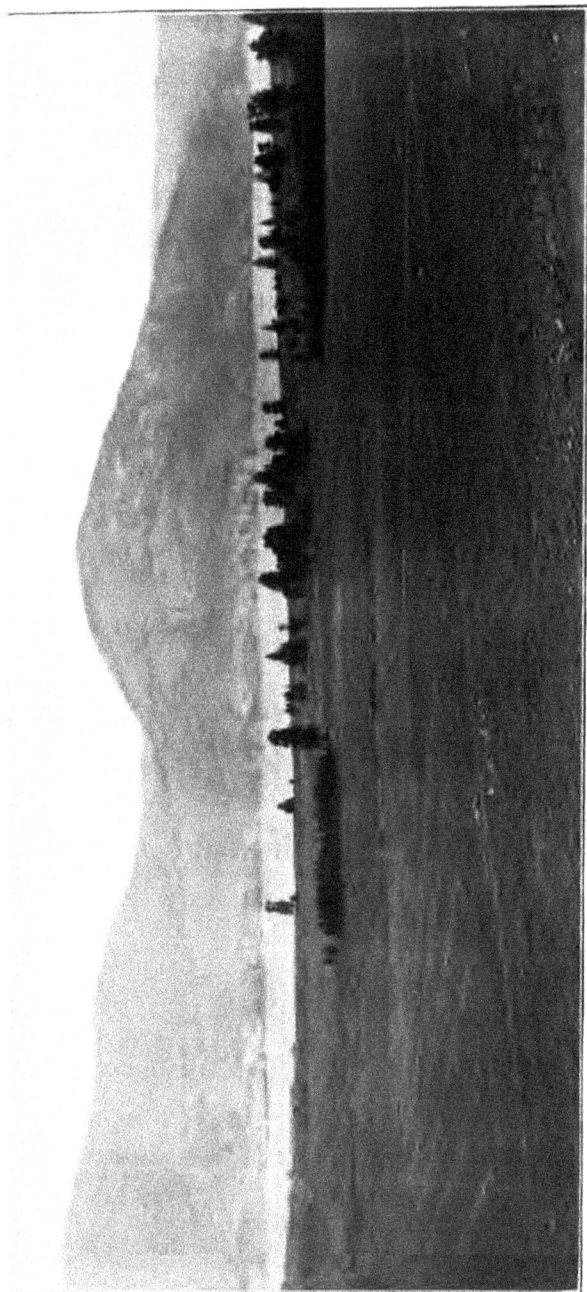

BENCH LANDS, PENTICTON.—SHOWING LAND READY FOR PLANTING TREES.

LAND AND AGRICULTURE.

————:o:————

A YEAR OF PROGRESS.

DURING the past year, or since the publication of the fourth edition of Bulletin No. 10, Agriculture in British Columbia, conditions have materially improved in all branches of this important industry. In 1905 the importations of agricultural products decreased by about $500,000, and as the population increased considerably during the year, it can safely be inferred that this falling off in imports is far more than balanced by increased home production.

One of the important events of the year was the demonstration of the practicability of supplying the British market with British Columbia fruit in prime condition.

The Provincial Department of Agriculture has pursued the policy of collecting all available data as to local progress and conditions from every part of the Province, and its officials have done excellent work in protecting the farms and orchards from insect pests, through a rigorous inspection of all imported seeds and nursery stock, and by practical demonstrations on the ground of the best methods of destroying those insect foes which have succeeded in gaining an entrance. As a result of the untiring efforts of these officers British Columbia is singularly free from the injurious insects and plant diseases which work such havoc in other countries.

The efficiency of the system in vogue in British Columbia is recognised abroad, no less an authority than the London "Times" commending it as follows:—

"The insect plagues which worked such mischief in California have been excluded hitherto, and this immunity gives B. C. an enormous advantage. The expense and risk of insecticides are avoided."

Mr. McNeill, of the Dominion Department of Agriculture, says:—"The one thing which impressed me was the fact that as far as the fruit industry is concerned, the growers commence in this country with a clean slate."

The Dominion Department of Agriculture has rendered valuable assistance in this work of prevention, as well as supplying lecturers on agricultural subjects, who visit the Province and address the Farmers' Institutes.

Arrangements have been perfected with the Dominion Government whereby sales of pure-bred live stock are held at convenient times, thus enabling stockmen and farmers generally to improve their herds.

An Inspector of live stock and dairies has been appointed by the Provincial Government, whose work, although as yet in the organisation stage, is already showing good results. Although general farming has made gratifying progress, in the way of an increase in the acreage of cultivated land and a consequent addition to the total of agricultural products, the most noticeable advance has been made in fruit-growing, the acreage under fruit having increased by about 20,000 acres, representing an aggregate of over a million trees planted, besides which about 500 additional acres were set out in strawberries and other small fruits.

There has also been a marked improvement in dairying, the co-operative creamery taking over the work formerly done by the farmers' wives and daughters. The output of the creameries doubled in 1905, producing about 1,400,000 pounds, while the dairy butter aggregated about 400,000 pounds. While these figures are gratifying, and indications are that they will be greatly surpassed this year, there is ample room for further expansion in the dairying industry, for large imports of butter, exceeding the total local production, are still necessary to supply the provincial markets. The establishment of a cheese factory at Langley marks the initiation of a new industry, which, under good management, should prove highly profitable.

Considerable progress has been made in the live stock industry, a general movement being in the direction of getting rid of scrubs and replacing them with the best obtainable strains of beef and dairy cattle. Sheep and hog-raising are also securing more attention, and give promise of growing to important proportions in the future.

One of the most promising signs of the times is the breaking up of many of the large cattle ranches into small farms and orchards which are being eagerly bought by actual settlers, a majority of whom are taking up mixed farming and fruit-growing.

DESCRIPTIVE.

———o———

PHYSICAL CHARACTERISTICS.

BRITISH Columbia, the Pacific Maritime Province of Canada, is a great quadrangle of territory, 700 miles long and averaging 400 in width, lying between the Rocky Mountains and the Pacific Ocean. Extending north from the 49th parallel of latitude, it has a coast-line of 450 miles on the Pacific Ocean, the northern portion being cut off from the sea by a narrow strip of Alaskan Territory. The boundary on the north is the 60th parallel.

The Province may be divided roughly into three areas, each having its special characteristics, viz.:—(1) The islands adjacent to the Coast; (2) the great interior plateau, flanked by mountains on east and west, and forming the southern half of the Mainland; (3) the northern half separated from the plateau by various cross mountain chains from whence spring the headwaters of the Peace River.

The first area comprises Vancouver Island, the Queen Charlotte group, and the innumerable islands of various sizes that dot the coast-line. Washed by the waters of the Japanese current, the climate is mild and moist, and the same may be said of the narrow strip of territory intervening between the Coast Range and the sea-shore. This influence also affects to some extent the estuaries of the rivers flowing into the Pacific.

The great interior plateau, elevated some 3,500 feet above sea-level, has been so deeply eroded by lake and river systems that in some parts it appears mountainous, but the absence of sharp edges to the hilltops and the innumerable rounded boulders point conclusively to the fact that at some remote period this immense area was the bed of a vast inland sea.

Of the third area, except in isolated patches, comparatively little is known. The Coast Range of mountains forms a rocky frontier on the west, while the eastern boundary, following the 120th meridian of longitude, cuts the Rocky Mountains at the Peace River Pass and continues north through a rolling prairie region that has never been thoroughly explored. Many large arable areas are found, to which much attention has been turned of recent years, and the extreme northern portions, apart from agricultural possibilities, will be valuable for the precious metals, coal and other minerals which are found in large deposits throughout its length and breadth.

AREA.

The area of British Columbia has been variously set down from 375,000 to 395,000 square miles. From careful surface measurements of the map, the following results have been obtained, according to the present main political divisions:—

	Square Miles.	Acres.
Kootenay	23,500	15,060,000
Yale	24,300	15,850,000
Lillooet	16,100	10,300,000
Westminster	7,660	4,900,000
Cariboo	150,550	96,350,000
Cassiar	150,000	96,000,000
Comox (Mainland)	7,100	4,550,000
Vancouver Island	16,400	10,000,000
	395,610	253,010,000

The foregoing figures are given approximately, to approach round figures as nearly as possible.

RIVERS AND LAKES.

Seven rivers form the natural avenues of transportation for British Columbia, the Fraser, Columbia, Thompson, Kootenay, Skeena, Stickine, Liard and Peace.

The Fraser is the great water-course. Rising in the Rocky Mountains, about mid-way along the eastern boundary, it runs almost due west in two branches for some 200 miles, and these joining it flows southerly through the Cariboo, Lillooet and Yale Districts till, near Chilliwhack, it abruptly turns to the west and finds an outlet to the Pacific through the Gulf of Georgia. Several tributaries of importance add to its volume, among them being the Thompson, draining the Kamloops and Shuswap Lake areas, the Chilcotin, Lillooet, Nicola, Harrison and Pitt. From its last westerly turn it flows through a wide alluvial plain, mainly deposited from its own silt. It is navigable for vessels drawing 20 feet to New Westminster, about 15 miles from its mouth, and light draught boats can travel to the small town of Yale, 95 miles further inland. Another stretch of 60 miles in the far interior is also navigable for small craft, from Quesnel-mouth to Soda Creek, in Cariboo. The waters of the Fraser teem with salmon, and the canneries near its mouth give employment to many thousand men during the fishing season.

The Columbia rises almost in the south-east corner of the Province and runs north about 150 miles to where the Canoe River runs into it, when, turning in an abrupt semi-ellipse, it takes a southerly course and, draining the watershed of the Arrow Lakes, leaves the Province in the vicinity of Rossland. Though interrupted by a number of rapids it is navigable to a very large extent, and steamers ply regularly between Windermere and Golden and both north and south from Revelstoke.

The Peace River lies only partly within the Province, but will in the future be of great importance. Mr. F. W. Valleau followed this river from its source to the eastern boundary of the Province, and found many indications of agricultural possibilities.

The Thompson, so called, is practically two distinct streams flowing at right angles to each other into the eastern end of Kamloops Lake. The South Thompson connects that body of water with the Shuswap Lakes to the east, while the North Thompson, having its source in the Clearwater Lakes, Cariboo, flows due south through a wide valley suitable, with irrigation, for agricultural purposes. For a considerable distance both rivers are navigable.

The Skeena is second in importance of the rivers wholly within the Province, and is navigable nearly 200 miles from its mouth. Hazelton, 150 miles inland, is at present the most easterly point having steamboat connection, which lasts about seven months each year, or during the season of high water. The total length of the Skeena is 300 miles and its general course south-west.

Although for the last few miles of its course the Stikine River runs through Alaska, it forms the main artery of communication at present for that portion of the Province known as Cassiar District. It has been regularly navigated for many years for a distance of 130 miles, the eastern steamboat terminii being Glenora and Telegraph Creek.

OKANAGAN RIVER.

NEAR SUMMERLAND.

Many natural depressions are filled by lakes in British Columbia, the principal ot which are tabulated below, the areas being transcribed from the reports of the last census of Canada :—

Lakes.	Area in Acres.	Lakes.	Area in Acres.
Adams..............	33,280	Owikano............	62,720
Atlin (part)	211,680	Quesnel.....	94,080
Babine..............	196,000	Shuswap............	79,150
Chilko	109,760	Stuart	141,120
Harrison............	78,400	Tacla.....	86,240
Kootenay	141,120	Tagish (part).......	58,180
Lower Arrow........	40,960	Teslin (part)........	78,400
Okanagan	86,240	Upper Arrow........	64,500

CLIMATE.

Varied climatic conditions prevail in British Columbia. The Japanese current and the moisture-laden winds from the Pacific exercise a moderating influence upon the climate of the coast and provide a copious rainfall. The westerly winds are arrested in their passage east by the Coast Range, thus creating what is known as the "dry belt" east of those mountains, but the higher currents of air carry the moisture to the loftier peaks of the Selkirks, causing the heavy snowfall which distinguishes that range from its eastern neighbour, the Rockies. Thus a series of alternate moist and dry belts are formed. The climate of British Columbia, as a whole, presents all the conditions which are met with in European countries lying within the temperate zone, the cradle of the greatest nations of the world, and is, therefore, a climate well adapted to the development of the human race under the most favourable conditions. As a consequence of the purity of its air, its freedom from malaria, and the almost total absence of extremes of heat and cold, British Columbia may be regarded as a vast sanitarium. People coming here from the east invariably improve in health. Insomnia and nervous affections find alleviation, the old and infirm are granted a renewed lease of life, and children thrive as in few other parts of the world.

In his first report, Professor Macoun, of the Geological Survey, stated as follows :— "The cause of the mild and moist climate of the Pacific Coast is precisely the same as that of Western Europe. A stream of warm water a little south of the Island of Formosa, on the eastern coast of China, a current analagous to the Gulf Stream, is observed moving to the north-east. It passes east of Japan, while a part of it enters Behring Sea, the remainder passes south of the Aleutian Islands and ameliorates the climate of Alaska to such a degree that the annual temperature of Sitka in latitude 57 deg. is higher than that of Ottawa, in latitude 45 deg. 25 min., the mean annual temperature of the former being 44.8 deg., while the latter has only 37.4 deg. Esquimalt, within three miles of Victoria, in latitude 48 deg. 25 min., has a mean annual temperature of 47.4 deg., only three degrees higher than that of Sitka, which is nine degrees further north. With these facts, the temperature of Sitka and Esquimalt before us, it is very easy to forecast the future of the region west of the Cascades, between Victoria and the Stikine River. The Queen Charlotte Islands being more insular than Vancouver, must have a climate even milder, and hence they may be set down as of equal value.......It only remains for me to add that as years roll on, and our possessions become developed, the value of this second Britain will come so vividly before our people that men will ask with astonishment why such ignorance prevailed in the past. To-day there are 400 miles of coast-line in our western possessions clothed with a forest growth superior to anything else in the world at present ; its shore indented with multitudes of harbours, bays and inlets, teeming with myriads of fish ; its rocks and sands containing gold, iron, silver, coal and various other minerals. And, besides all this, a climate superior to England in every respect, both as regards heat and moisture, and yet men ask what is it all worth ? I answer, worth more than Quebec and all the Maritime Provinces thrown in, and sceptics may rest assured that the day is not far distant when my words will be accepted as truth."

Apart from the coast region, the climate conditions of which obtain for a considerable distance up the valleys of the Fraser and Skeena Rivers, the Province enjoys, on the whole, a climate that is unexcelled elsewhere. As is natural, the more elevated portions are subject to greater extremes of temperature, but the heat of summer is not of a depressing character, and while, in some localities, a severe degree of cold is met with, the bracing atmosphere largely removes the unpleasantness.

Taking the various districts seriatim, as local topographic differences cause somewhat abrupt changes of climate, the conditions in various parts of the Province may now be described with more particularity.

In the Kootenay District, which embraces the drainage area of the Columbia River, comprising the mountain belt of the Selkirks and the western flank of the Rockies, the high average altitude renders the air rarified and bracing. There is a sufficient precipitation of moisture, from 18 to 20 inches of rainfall per annum, and a snowfall of from 1 to 3 feet. In summer the thermometer rises as high as 80 or 90 degrees in the shade, but the nights are comparatively cool. At times, in the winter, there are cold spells, when the mercury falls considerably below zero, but these are of short duration. It is seldom that any damage is caused by drought, and though in swampy lands there are occasional summer frosts, their effects are removed by drainage and cultivation.

Throughout the great interior plateau a much drier climate is found, the total precipitation being from 7 to 12 inches, according to locality. Luxuriant vegetation is confined to the borders of lakes and water-courses, while the general landscape presents the usual round-topped hills and bunch grass of typical pasture or range land. In the many valleys thriving farms show the effects of careful cultivation, and wherever vegetation has been practised the result is seen in ample crops.

South of Shuswap and Kamloops Lakes the climate presents the mean between the dryness of the bunch-grass country and the humidity of the coast. The land is largely park-like in character, somewhat rolling, and vegetation luxuriant and varied. There is sufficient rainfall for all purposes, and the climate closely resembles that of Central Europe.

The many valleys cutting the Coast Range have distinct climatic peculiarities. Sheltered as they are by the surrounding hills from bleak north winds, the warm breezes from the coast are freely wafted through them. The sun's rays are concentrated on the side hills with almost tropical intensity, and even on the higher benches orchards and vineyards yield enormous crops.

As soon as the mountains are left behind and the Pacific littoral is reached, there is an astonishing change in conditions. Where vegetation has been left in its virgin state there is almost an impenetrable undergrowth, from which rise luxuriantly huge forests of fir, pine and spruce. This is accounted for by the heavy rainfall, which increases towards the north. But the winters are short and temperate, and emphasised more by a heavy rainfall than any other climatic change. The high mountains of Vancouver Island break the force of the heavy storms sweeping eastward over the Pacific, and heavy gales are infrequent in the inland waters.

Probably the driest point on the coast is in the vicinity of Victoria. Harvest time is rarely unsettled, and there is seldom any difficulty incurred in reaping the crops. During many winters there is no perceptible frost, and delicate plants thrive throughout the year in the open air. Any severe weather that may arise usually comes in short spells during January and February.

That part of the northern interior embracing the Cariboo and Chilcotin country has a more severe climate than the Kootenay. This is accounted for by the higher altitude, but considering this, the climate is very moderate as compared with that of Central Russia, lying within approximately the same latitude.

That portion of the Peace River Valley lying within the Province enjoys a much milder climate than the more elevated district farther south. Mr. F. W. Valleau, who traversed this district during the summer of 1900, reported that the country east of Parle Pas Rapids, is of the first class. At Fort St. John, situated some 50 miles from the eastern boundary, he said : "Grain and roots of all kinds do well and ripen." In confirmation of this Professor Macoun writes regarding Mount Selwyn, the date being July 11th :—

"When we left the river in the morning the thermometer stood at 84 deg., and on the top of the mountain in latitude 56 north, 7,500 feet above the sea, it stood at 82 deg."

As to conditions at Fort St. John, he fully bears out Mr. Valleau's statement. Professor Macoun's conclusions are summarised as follows :—

"That from the middle of April until the first week in November the ground is fit for the plough, that winter is actually shorter on Peace River than in Manitoba, and that 1,200 miles north-west of Fort Garry a milder temperature prevails in autumn than at that point."

As to the northern country in general, the remarks of the same authority may well be carefully considered :—

"There can be no doubt that when the forest is cleared, by whatever cause, the soil will become drier and the climate will become considerably milder. Owing to the latitude the sun's rays fall obliquely on the forest, and, as a natural result, there is little evaporation. As Germany was to the Romans, so much of our North-West to us —a land of marsh and swamp and rigorous winter. Germany has been cleaned of her forest, and is now one of the finest and most progressive of European countries. May not the clearing of our North-Western forests produce a similar result in the distant future of British Columbia?"

———

AGRICULTURAL AREAS.

Gold was the lode-stone which first attracted attention to British Columbia ; next the fame of its forests and fisheries spread, and lumbering and salmon fishing assumed the importance of great industries. The agricultural possibilities were overlooked or ignored by the miner, lumberman and fisherman, and for many years the world at large was ignorant of their existence. The opening of the country by the trunk line and branches of the Canadian Pacific Railway, however, disclosed the fact that the agricultural and pastoral lands of British Columbia are not the least valuable of its assets and that they are not confined to a small proportion of the total acreage. Professor Macoun, after careful personal investigation, says :—

"The whole of British Columbia, south of 52 degrees and east of the Coast Range, is a grazing country up to 3,500 feet, and a farming country up to 2,500 feet, where irrigation is possible."

This is a most important statement and its truth is being confirmed by the practical experience of settlers who have established themselves in the country. Within the boundaries thus roughly defined by Professor Macoun the capabilities of the soil are practically unlimited. All of it that is not too elevated to serve only for grazing purposes will produce all the ordinary vegetables and roots, much of it will grow cereals to perfection, while everywhere the hardier varieties of fruits can be successfully cultivated. As far north as the 54th degree it has been practically demonstrated that apples will flourish, while in the southern belt the more delicate fruits, peaches, grapes, apricots, etc., are an assured crop. Roughly estimated the extent of these fertile lands may be set down at one million acres, but this figure will probably be found far below the actual quantity capable of cultivation when the country has been thoroughly explored. The anticipation of such a result is justified from the fact that at several points in the mountains, even in the most unpromising looking localities, where clearing and cultivation has been attempted it has proved successful. In several instances also, bench land, pronounced only fit for pasturage by "old timers," has been broken and cropped with very satisfactory results. The agricultural lands just mentioned are located as follows :—

		Acres.
Okanagan		250,000
North and South Thompson Valleys		75,000
Nicola, Similkameen and Kettle River Valleys		350,000
Lillooet and Cariboo		200,000
East and West Kootenay		125,000

West of the Coast Range are several extensive tracts of arable land of the richest quality, notably the Lower Fraser Valley, Westminster District, Vancouver Island and adjacent islands in the Gulf of Georgia. These sections of the Province are recognised as agricultural districts, and are fairly well settled, but much of the land is still wild and untilled. North of the main line of the Canadian Pacific Railway, on the Pacific Slope, and but partially explored, are vast areas of agricultural and grazing lands, which will be turned to profitable account when the country is a few years older. Much of this northern region is fit for wheat growing, and all of it will produce crops of the coarser cereals, roots and vegetables, except the higher plateaux, which will afford pasturage to countless herds of cattle, horses and sheep. Some of these districts best known and in which settlements have been established, are Chilcotin, Nechaco, Blackwater, Bulkley, Ootsa, Kispyox, Skeena and Peace River Valleys, and they are estimated to include some 6,500,000 acres. That this is a conservative estimate is clear from the fact that the late Dr. Dawson and Professor Macoun credited that portion of Peace River Valley lying within British Columbia with 10,000,000 acres of wheat land.

The agricultural lands of the Province are so widely distributed and so intersected by mountains, that, in the absence of surveys, in many instances even of an exploratory nature, it is impossible to describe them comprehensively or in detail. In the prairie provinces east of the Rocky Mountains, the contour of the country admits of easy and inexpensive subdivision into townships and sections, and the surveyors' field-notes furnish precise information as to the nature of the soil, timber, etc. The prospective settler in those provinces has, therefore, little difficulty in choosing a location, but in British Columbia he is, as a rule, called upon to make a special trip to the district in which he proposes to establish himself and stake out his pre-emption after having satisfied himself of its suitability.

The lands in the Railway Belt (20 miles each side of the main line of the Canadian Pacific Railway), owned and administered by the Dominion Government, are partly surveyed into townships, but, taking the Province as a whole, the rule is that a settler must seek for and stake his land at his own expense. This handicap to the rapid settlement of the vacant lands of the Province will no doubt be removed through the Dominion Government adopting a system of surveys in the Railway Belt in the near future, and the Provincial Government may take similar action with respect to Provincial lands, but up to the present no provincial administration has found it compatible with the revenue to inaugurate such a system.

In the settled portions of the Province, along the established lines of travel and in the neighbourhood of the cities and towns, there is very little good land left for preemption, but there are many desirable tracts of land and farms more or less improved, which may be purchased from the owners at prices which vary according to locality and extent of improvements—running all the way from $5 to $1,000 per acre, the latter being for matured orchards and carrying the good-will of well established business.

AGRICULTURAL PROGRESS.

Less than fifty years ago British Columbia was shown on maps of North America as "New Caledonia," and was held as a fur preserve by the Hudson's Bay Company under lease from the British Government. To the world at large it was a hyperborean wilderness, a home of savage men and wild beasts. One day gold was discovered, thousands of treasure hunters rushed in, and sudden and important changes occurred. The territory was created a Crown Colony with a responsible government, laws were enacted and enforced in accordance with British precedent, roads and trails were made to the "diggings," civic, educational and religious institutions were established, and British Columbia emerged from obscurity and became the Mecca of a vast army of sturdy pilgrims from all parts of the world.

The primary object of the new comers was gold, and the fortunate ones succeeded in winning about $30,000,000 in the period between 1858 and 1868, but the needs of the miners encouraged ventures in other industries, and in due course British Columbia's timber and fisheries came to be regarded as nearly equal in importance with her gold mines. During the halcyon days of placer mining agriculture was ignored—for who would waste energy in planting potatoes in soil that produced crops of nuggets—but when the golden harvests became lighter and the work of mining harder, many miners turned to farming, some from necessity, others for congenial employment. Cultivated fields and cattle ranches slowly began to appear in the beautiful valleys, on the lake fronts and river banks. Few of these early cultivators took their new occupation seriously—to most of them it was a stop gap to permit the prosecution of their real work of prospecting, while to others it was little else than a pastime. The minority, practical farmers who were in earnest, made money and to-day their fine residences, embowered in flowers and shrubberies, surrounded by well tilled fields and fruitful orchards, are the envy as well as the incentive of every new settler. The industry and intelligent efforts of these pioneer farmers demonstrated the capabilities of the soil of British Columbia for producing in perfection every cereal, fruit and vegetable which can be grown in the temperate zone. As these prosperous holdings are well distributed throughout the southern portions of the Province, those who come after have but to follow the example set to attain success.

LARGE HOLDINGS.

The tendency in the early days when land laws were lax or non-existent was to stake large areas of land. In this way many of the most fertile valleys were monopolised by a few individuals, who owned from 1,000 to 30,000 acres, many more than they could possibly cultivate or otherwise utilise. These big estates are now being sub-divided and sold in small parcels, with the result that small farms and orchards are becoming numerous on ground which was held for years as pasture or merely for purposes of speculation. The breaking up of these large ranches is one of the most hopeful signs of the times, as it insures a large increase in the industrial population as well as the bringing under cultivation of very considerable areas of land in different parts of the Province and the establishment of new communities, all contributing to the general prosperity.

A NEW ERA.

Since the opening of the southern portions of the Province by the Canadian Pacific Railway main line and branches, the construction of other short railways and the establishment of steam navigation on some of the principal lakes and rivers, the advance in all branches of agriculture has been steady, but the great opportunities offered by the prairie country for quick profits in wheat checked the westward tide of immigration, and it is but recently that settlers in large numbers have begun to cross the Rocky Mountains and establish homes in British Columbia. The advantages offered by the Province, however, are so manifest that no sooner has a new comer established himself than he becomes an enthusiastic immigration agent and hastens to advise his friends and old neighbours to "pull up stakes and come to the Garden of Canada." The ideal conditions of soil and climate, in the midst of beautiful and inspiring scenery, and the ready sale at good prices for everything produced are fully appreciated by men who have been "grubbing along" in the worn out fields of the older countries and their glowing reports are inducing thousands of farmers in Eastern Canada, the United States and the British Isles to sell out and secure land in British Columbia, which is destined to become, in a measure, the Orchard of the Empire, as the prairie provinces are its granary.

MARKETS.

It is an axiom in trade that "there is no market like the home market," and in this respect British Columbia is singularly blessed, for there is no country in the world which offers such exceptional advantages in the way of markets for farm products. The mining and logging camps, with which the whole country is dotted, employing thousands of men ; the numerous working mines and smelters with their large staffs of employees ; the railways, operating and under construction, and the lake and river steamers, are all liberal patrons of the farmer at prices unaffected by competition, for imported articles do not disturb local trade, and in every case home products are preferred to those from abroad. The established cities and towns and the new ones which are constantly springing up, with the opening of new mines and the establishment of new industries, afford splendid markets to the farmer, who deals directly with the consumer or retailer for cash—the trading system in vogue in older countries being practially unknown. Fruits and early vegetables not disposed of locally find an unlimited market east of the Rocky Mountains and in the Coast cities of the Province. Eggs, butter, milk and cream are always at a premium, the local production falling far short of supplying the demand. In many towns fresh milk is hard to get, and it is unknown in the mining, lumbering and railway camps, where the imported condensed substitute is used. The importations of these articles into British Columbia for an average year throw light on the possibilities for dairying and poultry-raising in Southern British Columbia. They are :—

Butter	$1,179,511
Condensed milk and cream	165,000
Eggs	339,000
Poultry	73,700

If cheese, which is not made in quantity in British Columbia, be added, $333,342, we have a total of over $2,000,000 sent out of the Province annually for articles which can be profitably produced at home.

Again, in the matter of fresh meats, and pork, ham, bacon and lard, the yearly importations aggregate $2,136,366, as well as $800,000 worth of beef cattle, sheep and swine, all of which should be raised by the farmers of the Province.

Although British Columbia has begun to export fruits the home market falls far short of being supplied, for we find that in the same year (1904) the Province imported $800,000 worth of fruits and fruit products, viz. :—Apples, other fruits (not tropical), canned fruits, jams and jellies. The importation of apples may be accounted for by the demand in the early spring and summer months, when no home grown stock is available, which has to be supplied from New Zealand and Australia. The "other fruits" represent berries and early fruits grown in California and brought in before the local fruits have matured. The jams, jellies and canned fruit, however, should and will be produced in the Province as the fruit industry develops, and in good time all the other products of the ranch, farm, dairy and orchard, of which the Province now imports over $7,000,000 worth annually, will be won from the fertile valleys and hillsides of British Columbia. There is no fear of over-production in any branch of agriculture, for in the future as in the past, the farmers will not be able to supply the ever increasing demand, created by the march of industry. Should a day arrive when they find themselves with a surplus the great mining camps of the north will provide a market for more than they can offer.

While on the subject of home markets attention may be called to the fact that of 3,181 tons of fruit shipped by freight over the Canadian Pacific Railway in 1905, 1,669 tons were consigned to points within the Province.

There is, therefore, practically no risk to the farmer in settling in British Columbia. His market is at his door and will be for many years, and he can confidently assure himself of such prices for his produce as will give him a comfortable living and enable him to lay by a "nest egg" every year, in anticipation of his old age. If he is possessed of sufficient capital to start on a comfortable scale he should become independent and well to do in a few years. Even with limited means there are no difficulties in the way which may not be surmounted by industry and perseverance.

MIXED FARMING.

To the man of small capital mixed farming affords the most promising means of making a comfortable livelihood in southern British Columbia. To engage exclusively in fruit growing, one is obliged to provide for the period from the setting out of the trees till they come into bearing, thus requiring an income from other sources, while in mixed farming returns may be counted on from the start. A few acres planted in small fruits, early vegetables, potatoes, carrots, onions, cabbages, etc., with fowls, some cows and pigs, will give a man an assured income the first season, and will not interfere with his planting a variety of fruit trees, which will become profitable later. Another advantage o mixed farming is the fact that a man and his family can attend to the work, which occupies them pleasantly the year round, while the special farmer, with but one crop to depend upon, has to cultivate a larger area and hire help during the short periods of seeding and harvest, and has nothing to occupy his time the remainder of the year. Large farms, and specialties in agriculture, should only be attempted by men of sufficient means to tide over long periods of unproductive idleness.

As an example of what can be done on a 10-acre farm in British Columbia, the following statement of early fruit and vegetables shipped from Gellatly, B. C., by D. E. Gellatly & Sons is submitted :—

EARLY FRUITS AND VEGETABLES SHIPPED BY D. E. GELLATLY & SONS.

Shipments.	By Express. lbs.	By Freight. lbs.	Total. lbs.
Beets	120	......	120
Beans, green	1,028	......	1,028
Corn, green	998	.....	998
Cabbage	815	3,711	4,526
Carrots	985	3,075	4,060
Cucumbers	3,295	.....	3,295
Citron		4,090	4,090
Egg plant	151	......	151
Melons	2,436	......	2,436
Onions	200	1,030	1,230
Parsnips		1,450	1,450
Pumpkins		275	275
Potatoes	1,780	11,065	12,845
Peppers	170	......	170
Rhubarb	760	1,000	1,760
Raspberries		700	700
Strawberries	3,775	6,725	10,500
Turnips	1,060	155	1,215
Tomatoes	44,035	25,228	69,263
Totals	61,608	58,504	120,112

Total....................60 tons 112 lbs.

PLANTS.

Tomato	85,000
Cabbage	10,000
Strawberry	80,000
Raspberry	3,000
Total	178,000

All this was raised on a 10-acre clearing in heavy bush in Okanagan Lake District, the fourth year after Mr. Gellatly located the land.

RED POLLED CATTLE, VANCOUVER ISLAND.

A BUNCH OF BLACKBERRIES

Mr. J. M. Durr has four acres of bottom land on St. Joseph's Creek, south of Cranbrook. He favours small farming for men of limited means, and declares that 10 acres of garden will pay better than a ranch of 300 acres, and is much more easily managed, besides requiring very little capital. He has 3½ acres under cultivation. Off ninetenths of an acre he raised 12 tons of first class potatoes last season. An acre and a half of cabbages yielded the enormous quantity of 15 tons, which sold for 8 cents per pound. He also raised 5 tons of carrots from a quarter acre. Mr. Durr's profit for the season after retaining all the vegetables required for his own use, was $800, equal to over $200 per acre.

Mr. R. Lounsbury, who has four acres of bottom land on St. Joseph Creek, started with five cows and $200 cash. He sells milk in Cranbrook at 8 to 10 cents per quart. His stock now consists of ten cows, and from these he derives an income of $3,000 a year. He supplements his dairy business by raising a few hogs, which sell readily at 10 cents a pound live weight.

In the Fraser River Valley 35 tons of timothy is reported from one acre, and several instances are recorded of small farms clearing $300 to $500 from a single acre of strawberries. These examples are not isolated nor applicable to any particular district; wherever bottom land is properly cultivated or bench land irrigated, the same results are attainable.

SMALL HOLDINGS.

The opportunities for profitable diversified farming are practically unlimited. The demand for every product of the farm is great and ever increasing, the present supply being wholly inadequate for the local market. Under a system of small land holdings, with diversified field culture, every object of cultivation is highly profitable, because produced by labour that might otherwise be unproductive.

The results of a scheme for the encouragement of small holdings, adopted by the Prussian Government a few years ago, furnishes a good example of the benefits to be derived from small farms and mixed farming. In one instance an estate of 1,370 acres was divided into fifty-one holdings, of which thirteen were under 6½ acres, fifteen under 25 acres, twenty from 25 to 62 acres, and three above that area. The majority (32), however, were classed as "one-horse farms," of an average size of about 32 acres. This land when worked as one farm supported 70 persons, whereas 300 were provided with a living after the sub-division. There were 27 horses, 88 cattle, 120 pigs and a few fowls on the big farm, but in two years after the sub-division the live stock consisted of 70 horses, 212 cattle, 340 pigs, and 771 head of poultry. In another case an estate of 8,400 acres was split up into 158 farms, varying in area from 6 to 62 acres and over. The results were, an increase of population from 915 to 1,456 and the doubling of the number of live stock. In the case of another sub-divided estate, the population increased from 232 to 600, and the live stock from 834 to 3,061 head.

IRRIGATION.

The introduction of irrigation has wrought great changes in agricultural methods, but its advantages are not generally understood. Mixed farming is especially profitable on irrigated lands, for it has been proved that under this system seemingly worthless land is made to produce four times as much as the choicest soil cultivated under the old method. There is nothing intricate or difficult to learn in connection with irrigation, and men quickly appreciate its great advantages. It renders them independent of the elements in the conduct of their farm work, so that they have only to study the needs of their locality and adjust their products to the demand, thus deriving a continuous income without fear of failure from drought or excessive rain.

Under the "Water Clauses Consolidation Act, 1897," and amending Acts, unrecorded water may be diverted from any natural source for irrigation or agricultural purposes generally. The scale of fees is the same for industrial purposes, and is calculated on a sliding scale. For a record fee of $10.75 per 100 miner's inches up to $110.75 for 500 inches; $260.75 for 1,000 inches; $560.75 for 2,000 inches; $680.75 for 5,000 inches; $880.75 for 10,000 inches, and so on according to the quantity of water actually required. For industrial purposes there is an annual fee calculated according to the same sliding scale. No annual fee is charged on water recorded and actually used for agricultural purposes. A miner's inch of water represents a flow of about 100 cubic feet per hour, equal to about 623 gallons, or 14,950 gallons per day, 24 hours.

Generally speaking, there is abundant water within reach, but there are sections where the height of the land above the water level or distance from the source of supply stands in the way of individual attempts at irrigation, but the work may be accomplished by co-operation and with the expenditure of capital. The supplying of water to these higher plateaux is, however, a matter for future consideration, as there is sufficient land capable of irrigation at comparatively small cost to meet the requirements for some years to come. In Okanagan, Similkameen and Kamloops districts companies have purchased large tracts of land, formerly used as cattle ranges, which they are subdividing into small holdings of ten acres and upwards, and constructing reservoirs and ditches, which will provide an unfailing supply of water. These companies are already reaping the reward of their enterprise, as the land is being rapidly sold to actual settlers, who are planting orchards and engaging in mixed farming. The example set by the Canadian Pacific Railway in Alberta in wresting over 2,000,000 acres from barren sand and low-producing grain fields, and making them yield millions of bushels of wheat, is one which cannot be overlooked by British Columbians, who, witnessing the transformation which is taking place on their eastern border, cannot fail to profit by the lesson. It is, therefore, safe to predict that the next few years will witness the reclamation of many hundreds of thousands of acres of bench lands from pasturage to flourishing orchards and farms, the homes of thousands of prosperous settlers.

DRY FARMING.

As will be noted in the chapter describing the Province by districts, there are many hundred thousands of acres which may be classed as arid land, and to which it seems impossible to supply water, unless some genius shall arise with a new scheme and methods now unknown to irrigation experts. In Southern Yale alone there are nearly two million acres which are practically valueless except for pasture, and as it takes many of these acres to support a single head of stock, a very extensive region seems doomed to remain indefinitely an almost uninhabited wilderness. Irrigated and sown with fodder crops, these lands would feed ten times the number of cattle; but planted with fruit, grain and vegetables, each forty acres would support at least from three to five people, or, at a conservative estimate, a population of 200,000. It has been demonstrated in Nebraska, Wyoming, Kansas, Colorado, Texas, Utah, Arizona, New Mexico and other

States lying partly or wholly within the boundaries of the American Desert, that, under the system of "dry farming" invented or discovered by Mr. H. W. Campbell, of Lincoln, Nebraska, wherever the annual rainfall averages as high as twelve inches, as good crops can be raised without irrigation as with it. Mr. Campbell has successfully introduced his "dry system" in almost every portion of the arid zone from Western Dakota to Texas, with most gratifying results. "The Campbell System of Dry Farming," as it is called, consists simply in the exercise of intelligence, patience and tireless industry. Its underlying principles are : First, to keep the surface of the land under cultivation loose and finely pulverised. This forms a soil mulch that permits the rains and melting snows to percolate readily through to the compacted soil beneath, and that at the same time prevents the moisture stored in the ground from being brought to the surface by capillary attraction, to be absorbed by the hot, dry air. The second is to keep the sub-soil finely pulverised and firmly compacted, increasing its water-holding capacity and its capillary attraction, and placing it in the best possible physical condition for the germination of the seed and development of plant roots. The "dry farmer" thus stores water not in dams and artificial reservoirs, but right where it can be reached by the roots of growing crops. The results of dry farming in Sedgwick County, Colorado, last year are given as follows :—Wheat, 35 bushels to the acre ; corn, 50 bushels ; potatoes, 200 bushels ; rye, 30 bushels ; oats, 65 bushels ; millet, 2 tons ; cane for forage, 5 tons. This is but one of many instances of the benefits of "dry farming" reported from different sections of the American Desert.

There are at least two cases in British Columbia in which crops are being successfully grown on land which had been considered worthless for agricultural purposes on account of its altitude and the impossibility of getting water to it. One of these examples of "dry farming" is on what is known as the Commonage, near Vernon, and the other on the uplands near Midway. The success which has attended these experiments will doubtless encourage others to take up scientific farming—for that is just what dry farming is—and in time much of the land that is now given up to sage brush, cactus and bunch-grass, will yield plentiful crops of grain, fruit and vegetables.

FRUIT-GROWING.

FRUIT-GROWING is one of the infant industries of British Columbia, but it is growing rapidly and is quite certain ere many years to rival mining, lumbering or fishing. A few years ago the man who would venture to describe the Kootenays as fruit-growing districts would be looked upon as a visionary or an imbecile ; to-day all Southern British Columbia is acknowledged to be the finest fruit country on this continent. Not only will it produce fruit in abundance, but the quality of its fruit is superior to that grown in any other part of America. Certain varieties of fruit attain perfection in certain localities—for instance, the Fameuse apple develops it best qualities on the Island of Montreal—but taking a collection of British Columbia fruit it is larger, better coloured and better flavoured than any similar miscellaneous lot, the product of any other country. Proof of this is not far to seek. In 1903 Messrs. Stirling and Pitcairn, of Kelowna, on Okanagan Lake, shipped a trial carload of apples to Great Britain. The shipment consisted of Spys, Baldwins, Ontarios and Canada Reds. They arrived in Glasgow, Scotland, on November 9th, in splendid condition, and sold at six shillings per box, or about $1 more per barrel than the choicest Eastern Canadian apples—reckoning three and a half boxes to the barrel. The British Columbia apples aroused much interest amongst fruit-dealers as well as consumers, and many letters were received by the consignors from persons eager to secure shipments of the splendid fruit. In the year following, 1904, the British Columbia Department of Agriculture forwarded a collection of British Columbia fruit to London, England, for exhibition purposes. It consisted of apples, pears and plums, including the following varieties : Apples—Fall Pippins, Kings, Vanderveres, Twenty-ounce Pippins, Blue Pearmains and Oranos, from Lytton ; Ribston Pippins, Wolfe Rivers, Wealthies and Snows, from Kelowna and Lytton ; Warners, Kings, Canada Red, King of Tompkins, Ontario, Jonathan, Northern Spy, Belle of Boskoop, Baldwin, St. Lawrence, Greening, Golden Russett, Alexander, Blenheim Orange, Wagoner and McIntosh Red, from Kelowna ; Wealthy, Ribstons and Gravensteins, from Victoria. Pears—Beurre Clairgeau, Easter Beurre, Beurre d'Anjou and Howells, from Kelowna, and plums from Victoria. The exhibit was greatly admired, and evoked the highest enconiums from the newspapers. The London Times, while hesitating to declare the fruit superior to the best English specimens, admitted that they very nearly approached them in colour, shape and flavour, even after having travelled 6,000 miles by railway and steamship. The Royal Horticultural Society's appreciation of the fruit was shown by the award of the Society's gold medal and diploma.

PRIZE FRUIT.

One result of this exhibit was the deluging of the Agent-General of British Columbia (Hon. J. H. Turner, Finsbury Circus, London), with letters from prominent fruit-dealers anxious to do business with British Columbia fruit-growers. To momentarily satisfy the clamor for British Columbia fruit, and to emphasize the fact of its good qualities, the Department of Agriculture shipped in cold storage a full carload of assorted fruits to London in the fall of 1905, in charge of Mr. R. M. Palmer, Provincial Horticulturist. This fine collection was the chief attraction at the Royal Horticultural Fruit Show at London, England, and at several provincial shows, and was awarded many prizes. The Royal Horticultural Society prize winners were :—

PROVINCE OF BRITISH COLUMBIA (for collection), gold medal.

J. C. GARTRELL, Trout Creek, silver-gilt Knightian medal.

J. R. BROWN, Summerland, silver Knightian medal.

THOS. W. STERLING, Kelowna, silver-gilt Knightian medal.

COLDSTREAM RANCH, Vernon, silver-gilt Knightian medal.

THOS. G. EARL, Lytton, silver Knightian medal.

MRS. J. SMITH, Spence's Bridge, silver Knightian medal.

KOOTENAY FRUIT-GROWERS' ASSOCIATION, Nelson, silver Banksian medal.

J. L. PRIDHAM, Kelowna, silver Banksian medal.

After going the rounds of the fruit shows and securing unqualified approval everywhere, this collection was broken up and sold to fruit-dealers at the highest prices. Several of the leading fruit firms of Great Britain have placed orders for this season's fruit, so it may be confidently stated that the fruit trade with the Old Country has been firmly established.

Following up the success of last year the Department of Agriculture will forward a commercial exhibit of fruit to London this autumn, where it will be shown at the Royal Horticultural Fruit Show, and at as many provincial fairs as possible. This exhibit will be in charge of Mr. Palmer, and will be made up of the best fruit available from all parts of the Province. The Canadian Pacific Railway Company generously co-operates with the Government of British Columbia in the collection and transportation of the fruit, furnishing cold storage cars and cold storage space on its Atlantic steamships free of charge.

To show the progress of the Province's fruit business it is only necessary to quote the shipments for the last four years, which are as follows :—

	By Freight. Tons.	By Express. Tons.	Total. Tons.	Increase. Tons.
1902	1,469	487	1,956	.....
1903	1,868	676	2,544	568
1904	2,161	864	3,025	481
1905	3,181	1,176	4,357	1,332
	8,679	3,203	· 11,882	2,401

An increase of over 50 per cent. in four years.

The total shipments for four years, 11,882 tons, are far from representing the whole crop, the greater portion of which is consumed locally.

The increase in fruit acreage has also been great within recent years. In 1891 the total orchard area was 6,431 acres ; in 1901 it had only increased to 7,430 acres, but between that and 1904 the increase was jumped to 13,430, and in 1905 to 29,000 acres. This increase in acreage for 1905 means the planting of about 1,000,000 young trees.

The quality of the peaches and grapes grown in Southern British Columbia can scarcely be excelled, the crisp, dry air and bright sunshine combining to impart a lusciousness and flavour lacking in the fruit of hot countries. The recent discovery of fig trees growing wild on Vancouver Island, near Nanaimo, has suggested the possibility of the successful cultivation of this fruit, especially in the southern districts, and no doubt the experiment will be made in the near future. Almonds, walnuts, chestnuts, nectarines, apricots, olives, and other semi-tropical fruits, have been successfully grown.

MAKING AN ORCHARD.

The setting out and care of an orchard until it becomes a source of profit requires considerable outlay of cash and personal exertion, but the results after a few years furnish ample compensation.   The cost of setting out twenty acres of apple trees in Southern British Columbia is about as follows :—

*Twenty acres, at $100 an acre..................	$2,000 00
Fencing	200 00
Preparing land	100 00
Trees (968) at 12½ cents each...............	121 00
Freight, etc...	20 00
Setting out, at 5 cents each	48 40
	$2,489 40

*Irrigated land.

Root crops and small fruits, planted between the trees for the first year or two, and red clover up to the fifth year, should more than pay for the trees.   The fourth year the trees should produce some fruit—probably $100 worth.   The cost of maintenance for five years, with the original cost and interest, would amount to $4,836.22, or $242 per acre, less the value of roots, clover and fruit.   In the sixth year the orchard should produce $850 worth of fruit, in the seventh $3,200, and in the ninth $5,800, after which it should pay a net annual profit of $125 to $150 per acre—an assured income for life of $2,500 to $3,000 a year.

ACTUAL RESULTS.

This estimate of profits is not based upon paper and pencil but is justified by actual experience.   Mr. T. W. Stirling, Bankhead Ranch, Kelowna, says :—

This orchard of about 16 acres will produce about 160 to 170 tons this present year (1905).

In 1903 it produced 140 tons.

In 1904 it produced 130 tons.

In 1905 it produced 160 to 170 tons.   And probably has not yet reached maximum produtcion.

Apples (Jonathan) planted in 1900 produced this year 100 ℔s. a tree.   (Fruit worth $1.50 per 40-℔. box, f. o. b. packing-house.)

Last year these trees yielded, as four-year olds, 60 ℔s. a tree.   Next year's crop may be estimated at 200 ℔s. per tree.

One and one-third acres of Bartlett pears produced 16 tons of fruit, or about 800 boxes.   Selling price, $1.35 per box, f. o. b. packing-house, $1,080.

One and one-third acres of Beurre d'Anjou pears produced 17 tons, or 850 boxes. Selling price, $1.40 per box, f. o. b. packing-house, $1,190.

Two and one-third acres of Italian prunes produced 32 tons, or 3,200 crates.   Selling price, 60 cents per crate, $1,920.

One acre of plums produced 12 tons, or 1,200 crates.   Selling price, 70 cents a crate, $840.

Over $5,000 from six and one-third acres !

The actual experience of many fruit-growers is highly satisfactory to them, and a temptation to every man who desires to make money pleasantly to set up in the business. In Okanagan there are instances of $500 to $600 gross profit per acre.   At Kelowna 9 tons of pears and 10 tons of prunes per acre are not uncommon.   Near Nelson, 14 acres produced 1,000 cases of strawberries and 94 tons of roots, netting the owner $100 per acre. This land was formerly a cedar swamp. At Lytton to-day grapes, averaging 4 ℔s. to the bunch, were grown in the open. On the Coldstream Ranch, near Vernon, 20 acres produced $10,000 worth of Northern Spy apples. At Peachland one acre and a half gave a return of $700 in peaches. Tomatoes to the value of $1,500 per acre were grown on Okanagan Lake. A cherry tree at Penticton produced 800 pounds of fruit. These cases are by no means exceptional or confined to any single district, similar ones could be cited from almost any part of the Province. Apples and pears produce from 8 to 15 tons of fruit per acre, according to variety, and the average price is $26 and $30 per ton respectively. Plums, prunes, cherries and peaches invariably bear largely, and the prices are always satisfactory, if the fruit is properly picked and packed.

Fruit-packing has been brought to a fine art in British Columbia, the methods used being considered perfect by experts, and other countries are following her lead in this most important matter. Careless or dishonest packing is not tolerated, offenders being severely punished.

PEACHES AND GRAPES.

Peaches are successfully grown in many parts of Southern British Columbia, and in every case the fruit has attained a good size, ripened fully and possessed an exceptionally fine flavour. Peach-growing gives promise of becoming an important industry in Okanagan, where the area of young orchards is increasing rapidly. Many of these are bearing, and peaches, from now on, will become a noticeable item in fast freight and express shipments. So far the shipments have been very small, as nearly all the peaches grown find ready sale on the spot, and there has been no surplus with which to supply even the Provincial markets. The small lots exported have been in the nature of experiments—samples with which to demonstrate the capabilities of the country.

Peaches grow to perfection in all the valleys south of the main line of the C. P. R., and as this fact becomes generally known more attention will be given to their cultivation.

One advantage of peach-growing is the fact that the trees come into bearing earlier than apples, so that, under favourable circumstances, four-year-old peach orchards will yield as high as $300 worth of fruit per acre.

Speaking of the future of peach-growing, Mr. Thomas Cunningham, Fruit Inspector, says:—

"They are splendid peaches (alluding to samples of the first crop grown by Mr. Thompson Elliott, of Peachland, Okanagan), but only an index of what is to come with the second crop. This second variety is an excellent type of fruit, and will equal any of the imported stuff. Of course it does not come yet in sufficient quantities to displace the imported article, but at the present rate of progress, it will be only a few years now before British Columbia will be supplying the Canadian market and exporting huge quantities.

"Since last October over 200,000 trees have been planted, and the coming year will see a great increase over last. The cultivation of peaches has been confined chiefly to Peachland and Summerland and that district, but now large territories are being opened up in other of the mountain valleys. Shatford Bros. have opened a section of 40,000 acres around Penticton, in the Okanagan, and a short time will see every acre of it stocked with peaches, apricots and the finer fruits. Then there is the Similkameen Valley. The new railroad through that district will throw open large areas of land. Already nearly 50,000 trees have been planted."

Grape culture on a commercial basis can scarcely be said to be established in the Province, but wherever their cultivation has been tried in the southern districts it has proved successful. The experience of Mr. Thomas G. Earl, of Lytton, who may be styled the pioneer grape grower, is that nearly every variety of grape will ripen in the "dry belt," and that in most cases they will come to maturity about two weeks earlier than in Ontario.

The fact that grapes of excellent quality and flavour can be grown in quantity sufficient to supply the large and steadily increasing demand having been established, horticulturists in the "dry belt" will be encouraged to set out vineyards, and in time that part of British Columbia will rival Ontario's famed Niagara Peninsula as a producer of grapes and peaches. British Columbia grapes are as yet a novelty on the market, but their superior merits will in time win them a leading position.

OTHER FRUITS.

Nectarines, apricots, figs, almonds and several other of the less hardy fruits and nuts have been tried in a small way with success, and men of experience are not wanting who express the opinion that the sunny slopes of the lake country and the Boundary will produce any fruit or vegetable which is grown for 300 miles south of the International Boundary Line.

FRUIT CROP OF 1905.

W. J. Brandrith, Esq., Secretary-Treasurer of the British Columbia Fruit-Growers Association, reports to the Bureau of Provincial Information, regarding the fruit crop in the Lower Fraser Valley and prices for the season of 1905, as follows :—

" Strawberries were very light, except where mulching was practised ; in those cases about 75 per cent. of an average crop was gathered. Early berries suffered somewhat from rain, but the last of the season showers would have been welcomed. Good, firm berries realised fair prices, while soft, inferior fruits, as usual, demoralised the local market to a certain extent. Prices ruled from 80c. to $3 per crate, the bulk of the crop being sold for about $2

" Raspberries were about 20 per cent. below the average. This and the shortage in the strawberry crop was mainly attributed to a sharp frost in November, 1904, which heaved the ground and broke the roots. Prices for raspberries were satisfactory, ranging from $1.50 to $2.40 per crate.

" Blackberries, too, were below the average, for the same reason, but good prices were realised, selling from $2.40 to as high as $4 per crate.

" Cherries were a splendid crop, but the early varieties suffered considerably from the early June rains and the brown rot ; the average price paid was about 4½c. per tb. The later cherries were marketed in good condition and realised very good prices, ranging from 6 to 12½c. per tb.

" Gooseberries yielded very well, and prices ranged from 6 to 10c. a tb.

" White currants were a full crop, but the demand is very limited, 3½c. per tb. being the highest price I heard offered for them.

" Red currants also were a full crop, with prices ranging from 3½c. to 6c. per tb.

" Black currants on the Delta lands were never known to fail, and those who had established a connection on the prairies received remunerative prices, the figures ranging from 4 to 10c. per tb.

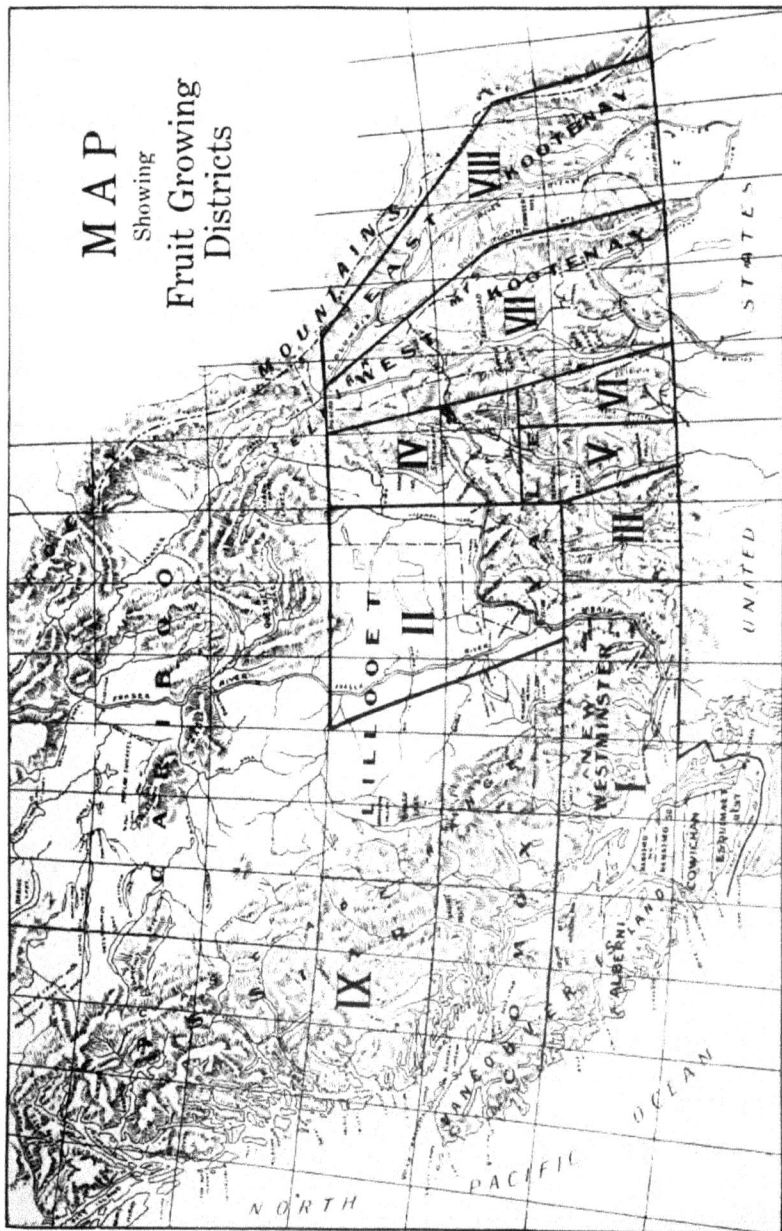

MAP

Showing

Fruit Growing
Districts

"Plums were a full crop and large quantities were marketed in very good condition. The brown rot was very bad in some localities and on some varieties. Prices ranged from 30c. to 80c. per crate, in a few cases $1 being paid for choice samples.

"Crabapples were an average crop and fetched about 2½c. a ℔., on an average.

"Pears were about 90 per cent. of a full crop, but prices were well maintained, ruling from 75c. to $1.50 per box.

"Early apples were a full crop and sold well, No. 1 fetching an average of 80c. per box.

"Fall apples were nearly a full crop, and about the same prices prevailed as for early apples. Winter apples were about 75 per cent. of a crop and sold for the highest prices ever received in British-Columbia, ranging from 80 to $1.75 per box, according to variety and condition.

"Taking the fruit-growers of the Lower Mainland, west of Hope, we find that, with a few exceptions, more interest is being taken than in former years. Better varieties, that is, varieties suited to the environment, are being planted. Spraying is being more systematically and intelligently carried on, and better methods of cultivation are being adopted, resulting in a most marked improvement in the fruit put on the market. There are still some growers who think that anything is good enough to sell, and these flood the market with fruit which should be put to some other use, thus spoiling the market for good fruit."

AVERAGE PRICES.

Average prices throughout the Province were as follows :—

No. 1 apples, from October 1st, 1905, to March 31st, 1906, were $1.27 per 40-℔. box, f. o. b. shipping point. The early varieties started out at $1 net, and during the latter part of February and March as high as $2 per box was being paid for strictly No. 1 in carload lots. The average prices of other fruits for the season of 1905 were : Pears, $1.38 per 40-℔. box ; prunes and plums, 75 cents per 20-℔. box ; peaches, $1.15 per 20-℔. box ; strawberries, $2.30 per 24-basket crate ; raspberries, $2.19 per 24-basket crate ; blackberries, $2.40 per 24-basket crate ; gooseberries, 5½ cents per ℔. ; crab apples, 2½ cents per ℔. ; tomatoes, 5½ cents per ℔. ; currants, 7 cents per ℔. ; cherries, 9 cents per ℔.

FRUIT-GROWING AREAS.

Mr. Maxwell Smith, Dominion Fruit Inspector, describes the fruit-growing areas of the Province as follows :—

"No. 1 might be called the South-Western Coast District, which includes the southern half of Vancouver Island, adjacent islands, and what is usually called the Lower Mainland. Here the production of small fruits may be said to be more successful, and consequently more profitable, than that of the tree fruits. Nevertheless, there are a number of very excellent varieties of apples, pears, plums, prunes and cherries which grow to perfection in this district, besides many different varieties of nuts, and, in especially favoured spots, peaches, grapes, nectarines, apricots and other tender fruits.

"In most parts of this district the mild character of the climate and the excessive moisture during the winter season are very favourable to the development of fungous diseases, and it is therefore necessary to practise persistent and systematic spraying of the orchards, clean cultivation of the soil, and a thorough system of under-drainage in order to get the most profitable results.

"District No, 2 includes the valleys of the Upper Fraser, the main Thompson, and North Thompson, the Nicola and Bonaparte Rivers. Here there are practically none of the above-named difficulties to contend with, but the question of water to irrigate the lands is one requiring serious consideration, as without an abundant supply of water in the dry belt it is impossible to be sure of a crop every year. The prospective fruit-grower, however, does not have to contend with heavy forests along the Thompson River that have to be encountered on the Coast. The fruits grown are of the very highest quality and include all the varieties mentioned in connection with District No. 1.

"The largest quantity of grapes shipped annually from any one point in the Province are produced near the junction of the Fraser and Thompson Rivers.

"District No. 3 may be briefly described as the valleys of the Similkameen and its tributaries, portions of which are perhaps the most tropical in climatic conditions of any part of British Columbia and most favourable locations for the cultivation of grapes, peaches and other delicate fruits, wherever sufficient water for irrigation purposes is available.

"No. 4 includes the districts surrounding Adams, Shuswap and Maple Lakes, and the valley of the Spallumcheen River. Here the natural rainfall is sufficient, and splendid apples, pears, plums and cherries are successfully grown. The climatic conditions in this district resemble very much those of Southern Ontario, and a fruit-grower with fixed ideas from the latter Province might be more successful in this district than he would on irrigated lands. The timber is, generally speaking, light, and the land rich.

"No. 5 is the great Okanagan Valley, stretching from Larkin southward to the International Boundary. The vicinity of Kelowna, in this valley, contains the largest area of fruit lands of any one place in the Province. Peaches are now being shipped in large quantities from the Okanagan, and all Northern fruits are successfully grown by the irrigation system. Improved modern methods are in general use by the growers in this district, and the industry is perhaps more advanced than in any other part of British Columbia.

"No. 6 is generally called the Boundary or Kettle River country, and although the smallest of all the districts named, the quality of the land is excellent, and the climatic conditions all that could be desired. Where a sufficient water supply is obtainable, there is no trouble in producing fruit of the highest quality.

"No. 7 is West Kootenay, an enormous fruit-growing district where only a little progress has been made on the southern portion, but sufficient to indicate the possibilities and the superior quality of the fruit which may be raised along those lakes and streams. The neighbourhood of Nelson and Kaslo has accomplished wonders in the past few years, but the shores of the Arrow Lakes are practically untouched by the hand of the fruit-grower, and the valley of the Columbia, from the Big Bend south to Arrowhead, affords opportunities little dreamed of by many of those in search of fruit lands. In the greater part of this district irrigation is only necessary in the very dry seasons.

"District No. 8 is the country known as East Kootenay, and is separated from No. 7 by the Dogtooth range of mountains. It is traversed by the Upper Kootenay River from Thunder Hill southward to the Phillips Ranch, on the International boundary, and from Thunder Hill northward by the Upper Columbia River, to the Big Bend. In the southern portion of this district there are immense stretches of thinly-wooded lands suitable for fruit-growing purposes, and the valley of the Upper Columbia River has many choice locations for the enterprising fruit-grower. The lack of transportation facilities is a great hindrance to the development of the fruit lands of the Upper Columbia.

"District No. 9 comprises the Coast Region from Jervis Inlet to Skeena River. There is little known of its capabilities, but, undoubtedly, it has a few surprises in store for the future. Though in small quantities as yet, apples, peaches and grapes have been successfully grown on the Skeena. The first apple trees were planted at Hazelton in he spring of 1901, and fruited in the fall of 1904."

"A BEAUTIFUL ART."

His Excellency Earl Grey, Governor-General of Canada, who recently visited British Columbia, is greatly impressed with the future possibilities of the fruit industry. In his reply to the address of the Royal Agricultural Society, on the occasion of the opening of the New Westminster Exhibition, His Excellency said :—

"Fruit-growing in your Province has acquired the distinction of being a beautiful art as well as a most profitable industry. After a maximum wait of five years, I understand the settler may look forward with reasonable certainty to a net income of from $100 to $150 per acre, after all expenses of cultivation have been paid.

"Gentlemen, here is a state of things which appear to offer the opportunity of living under such ideal conditions as struggling humanity has only succeeded in reaching in one or two of the most favoured spots upon the earth. There are thousands of families living in England to-day, families of refinement, culture and distinction, families such as you would welcome among you with both arms, who would be only too glad to come out and occupy a log hut on five acres of a pear or apple orchard in full bearing, if they could do so at a reasonable cost."

DAIRYING AND LIVE STOCK.

——o——

DAIRYING pays handsomely, especially in cases where the farmer is not obliged to employ skilled labour to do the milking and butter-making. The local demand for butter is constantly increasing with the population and the prices secured are far higher than in the East. In 1904 the creameries of the province produced 1,119,276 pounds of butter, which was sold at an average of 26½ cents per pound, or $296,608, little more than twenty-five per cent. of the value of butter imported. The production for 1905 was 1,456,343 pounds from the creameries, while the dairy product was about 400,000 pounds, 1,850,000 pounds in all, valued at $490,000. Quite a large proportion of the imported article was forwarded to Yukon, but that only serves to show the great possibilities for dairying in British Columbia. The province possesses many elements necessary to constitute it a great dairying country, the products of which should include cheese and condensed milk. There are extensive areas of pastoral land in the interior, while increased cultivation in the lower country will form the necessary feeding ground. With a plentiful supply of good water and luxuriant and nutritious grasses, there is every required facility added. The coast climate is most favourable to the dairying industry. Clover, one of the most valuable plants in cultivation, is practically a weed in British Columbia, west of the Cascade Range. Once it gets established in the soil it is almost impossible to get it out. Lucerne, or alfalfa, is succeeding admirably. In Okanagan Valley, Thompson River Valley and many other points, three heavy crops of this nutritious fodder are produced annually.

There are sixteen co-operative and private creameries established in the province, all doing well and earning satisfactory dividends. The Provincial Government aids the establishment of co-operative creameries and cheese factories by loaning the promotors one half the cost of the buildings, plant and fixtures, repayable in eight instalments with interest at five per cent., the first of such instalments to be paid at the expiration of three years, and the other seven annually thereafter.

Returns furnished to the Department of Agriculture give the following details of the output of the creameries for 1905 :—

	Pounds of Butter.	Average Price per lb.
Abbotsford	14,251	·239
Alberni	13,358	·27
Chilliwack	222,704	·27
Comox	70,241	·26
Courtenay (Alex. Urquhart)	20,000	·26
Cowichan	146,431	·28¼
Delta	109,939	·27⁷¹
Eden Bank	181,083	·27½
Nanaimo	57,565	·26⁴
New Westminster	240,000	·26
Okanagan	23,963	·27
Salt Spring Island	30,254	·27
Sumas (O. Bowman)	38,230	·25
Surrey	26,389	·26¼
Victoria	204,810	·28
White Valley	57,125	·26½
Total Production	1,456,343	·266
Value		$387,387

A company known as the British Columbia Cheese Co., with Mr. Ernest G. Sherwood as manager, and Mr. F. S. Rolph as factory superintendent, began the manufacture of cheese about April 1, 1906, at Langley. The factory has a capacity for handling five tons of milk per day, which would make about 1,000 pounds of cheese. The cheese produced is of good quality and is finding ready sale. This is the first cheese factory built in the province, and its success has already encouraged the establishment of another, which is being constructed at Murray's Corner.

LIVE STOCK.

Cattle-raising on a large scale was once one of the chief industries of the Province, and many of the large ranches are still making money, but the tendency of late has been for smaller herds and the improvement of the stock. The efforts of the British Columbia Stockbreeders' Association have proved successful in this direction. The Association imports and sells to its members every year a certain number of young pure-bred stock, purchased in Eastern Canada by a special agent, who visits the principal stock-markets in the interests of the farmers. In 1904 the Association imported and distributed 43 cattle, principally Shorthorns ; 10 mares, Clydes; 13 sheep, Hampshire Downs ; 14 pigs, Yorkshire ; 33 fowls, White Leghorn. The bulls sold from $100 to $150 and up to $500 ; the mares averaged about $300. At a sale held by the Association at New Westminster, in March, 1906, the following prices were realised :—Shorthorns, $65 to $152 ; Holsteins, $50 to $100 ; other breeds, $50 to $100 ; Suffolk stallion, $300 ; Clydesdale stallion, $595 ; Shropshire ram, $30 ; ewes, $15.

While the Province is capable of raising all the beef, mutton and pork required for home consumption, a very large quantity is imported, the money sent abroad annually amounting to about $3,000,000. The parts of the Province particularly adapted to cattle-raising are the interior plateaux and the Fraser River Valley, though there is scarcely a district in which the keeping of a few head will not pay well, for the high prices prevailing justify stall feeding. The development of irrigation should stimulate the cattle industry and make the Province self-supporting in respect of beef.

Sheep-raising is another branch of agriculture capable of great expansion. In the past the ranchers of the interior objected to sheep, as they are such close feeders, and sheep-raising was confined chiefly to southern Vancouver Island and the Gulf Islands, where considerable numbers were produced. These are the most favourable parts of the Province for sheep-raising, though they do well in many localities in the interior.

Hogs, in small farming, are probably the most profitable of live stock, owing to the general demand for pork, bacon, ham and lard, and much attention is now being given to raising them. Over $1,000,000 of hog products are imported annually, and prices are always high, so that the farmer can never make a mistake in keeping a small drove of pigs. The breeds which mature earliest are the Berkshire and Poland China. The increased production of hogs has encouraged the establishment of some small packing houses, but there is room for very extensive expansion. Hogs thrive in every part of the Province, and are in demand at all seasons, especially animals weighing from 125 to 150 pounds, suitable for fresh pork.

The demand for good horses, especially heavy draft and working animals, is always increasing, and prices are consequently high. Formerly horses were raised in great numbers in the interior without much attention to their quality, and in consequence great bands of wild horses became a nuisance and a menace to the farmers and ranchers to such an extent that the Legislature had to adopt measures for their destruction. The quality of horses has been much improved of late, and although the "cayuse," the native pony, will always be prized for its hardihood and endurance, the tendency everywhere is for a better class of animal. The horses exhibited at the recent Dominion Exhibition at New Westminster compared favourably with those of any country in the world.

A substantial increase has been made in the live stock industry in the past year, as the following figures will show :—

STATEMENT OF LIVE STOCK SHIPMENTS FROM POINTS ON THE PACIFIC DIVISION OF THE C. P. R. TO COAST POINTS AND POINTS EAST OF LAGGAN DURING 1904 AND 1905. (In Tons of 2,000 ℔s.)

To Coast Points.		To Pts. East of Laggan.		Total, 1904.	Total, 1905.
1904.	1905.	1904.	1905.		
Horses..... 142 tons.	1,151 tons.	720 tons.	627 tons.	862 tons.	1,778 ton·.
Cattle .....5,996 ″	7,037 ″	170 ″	395 ″	6,166 ″	7,432 ″
Hogs ...... 114 ″	361 ″	....	14 ″	114 ″	375 ″
Sheep ..... 11 ″	84 ″	....	....	11 ″	84 ″
Total. .6,263 tons.	8,633 ″	890 tons.	1,036 tons.	7,153 tons.	9,669 tons.

Total increase for 1905, 2,516 tons.

PURE BRED STOCK.

As already noted, the Dairymen's and Live Stock Associations are doing splendid work in securing to the farmers of British Columbia a better class of live stock. The efforts of the Association in this direction are materially assisted by the Canadian Pacific Railway, which grants a freight rate of one-half the regular rates on all importations of pure bred stock, the only condition to granting such rate being the production of uniform record certificates in every case. The Company insists that "all Record Certificates accepted by the railway must be of uniform size and appearance, and bear the seal of some central body recognised by the Department of Agriculture." While this rule protects the railway company against fraud, it acts as a double safeguard to the importer and purchaser of high-bred animals.

The Live Stock and Dairy Industry of British Columbia.

BY F. M. LOGAN, B. S. A., LIVE STOCK COMMISSIONER AND DAIRY INSPECTOR.

The live stock business in this Province is varied, as well as profitable. There are the great cattle ranches of the Upper Country, the pure bred herds of the Lower Mainland, the dairy herds of the creamery districts, and the grades all over. Many years ago, even before British Columbia could boast of a railroad, ranchers were utilising the fertile valleys among the mountains for the production of beef. Under favourable conditions and good management, some of these herds have increased until their owners now reckon them by the thousand, rather than by the individual animal.

There are many ranches worthy of mention, but perhaps the most noted is that of the Douglas Lake Cattle Company, situated at the head of the Nicola Valley, about 50 miles from Kamloops, on the main line of the C. P. R. The stock on this ranch consists of about 15,000 head of cattle and 2,000 horses, most of which are well bred and many of them registered animals. One can scarcely imagine the land area necessary for the maintanance of such herds.

A story is told of a man who was given permission to let a horse run in one of these pastures for a few days, and it took him a week to find his horse again. Some of the fences enclosing these fields extend for 40 or 50 miles; this will give some idea of the size of these ranches. There are many other ranches of vast proportions, well worthy of mention if space permitted. Among these would be the Gang Ranch, owned by the Western Canadian Ranching Co., which carries about 10,000 head; The Bostock Ranch, owned by Senator Bostock, situated at Ducks, B. C.; then, down the Okanagan Valley, situated near Vernon, is a large ranch owned by Price Ellison, M. P. P., as well as the Greenhow, and O'Keefe ranches, either of which is worth nearly a half million dollars, and a score of others. The Ellis Ranch, at Penticton, was also a very large estate, but has recently been cut up into 10-acre lots and is being sold for fruit-growing purposes. This, by the way, should be done with several others. The possession by one man of 10,000 or 20,000 acres of tillable land, when he is not in a position to cultivate more than a few hundred, is bound to retard the progress of the country. It is only a matter of time when all these ranches will be divided into smaller farms, supporting ten times their present population; the big ranch will then be a thing of the past. Beef can be produced on these ranches for less than $2 per cwt., and sells at the Coast for $8 or $10. This explains how some ranch owners have become millionaires. They have, of course, done pioneer work, and aided materially in developing the country, so no one will envy them a well-earned retirement, with all the luxury money can buy.

PURE BRED HERDS.

Among the progressive, well-to-do farmers of the Coast, especially along the Fraser River, may be found some very good herds of pure bred cattle, the foundation stock being largely imported from Ontario. The Shorthorns are the most numerous of the beef breeds, while the Jerseys seem to be the most popular in the dairy districts. There are, however, some excellent Holstein herds on Lulu Island, and near Duncan, on Vancouver Island.

The pure bred horses are confined almost entirely to stallions, the Clydesdales being decidedly more numerous than any of the other breeds. The various breeds of pure bred hogs are represented, as well as several breeds of sheep. Beef production on the Coast will never be carried on extensively, as the land is too expensive for that purpose, but probably considerable will be done in raising pure bred animals to supply the ranches of the upper country.

DAIRY STOCK.

The dairy herds along the Coast are on the increase and will continue to be so for some time. Only a few years ago there were no creameries in the Province, and now there are about twenty in operation. Most of these are doing a large, successful business, which will make an increased demand for dairy stock, so, all things considered, the outlook for the live stock industry of the Province is certainly good.

LIVE STOCK ASSOCIATIONS.

Two associations have been recently formed, one known as the British Columbia Stock Breeders' Association, which has for its object the improvement, importation and sale of live stock, as well as holding stallion shows, winter fairs and auction sales. Winter fairs have proven of great educational value wherever conducted ; with the absence of side-shows, lacrosse games, horse races, etc., it is much easier to keep the attention of those present upon topics relating to agriculture than it is during the fall fairs, when these attractions are in progress ; therefore, much information is gained by those who attend. An auction sale was held during March of this year which proved a success. It brought together the buyers of the ranching sections and the farmers of the Coast districts, who have pure bred stock for sale, under the most favourable circumstances, and a large number of animals changed ownership, with the least possible expense to both buyer and seller. In the future this auction sale and winter fair will no doubt be an annual event.

DAIRYMAN'S ASSOCIATION.

A new association was formed during February of this year, for the purpose of promoting the dairying interests of the Province. A Dairy Inspector has since been appointed and the Dairy Act amended, empowering the Inspector to prohibit the sale of any milk, cream, butter or cheese which he considers unfit for human consumption ; and if such prohibition is unheeded, a penalty of $50 may be imposed for each offence. This authority, accompanied by a careful inspection of the stables, stock, dairies, creameries, cans, utensils, etc., should do much to improve the dairy products of the Province.

British Columbia as yet produces only about half as much butter as she consumes, so there is ample room for the business to increase, and with a short winter, excellent grazing, and abundance of winter feed, and high prices for butter, few occupations will pay better.

Live Stock Conditions, 1905-6.

BY G. H. HADWEN, ESQ.

Cattle went into winter quarters in very fair condition, pasture having been good on most of the ranges of the Upper Country. Owing to large tracts of country in the C. P. R. belt having been recently leased, a good deal of fencing has been done in the last two or three years, which is necessarily causing a certain amount of readjustment in the herds of this section, and the open range is getting to be a thing of the past. Ultimately, owing to the possibility of allowing the bunch-grass to recover, the country will carry more head of cattle than it does at present, but the altered conditions will bring about different methods in stock-raising and are not in favour of the larger outfits.

At the present time the pastures are reserved for winter feed, and the cattle are turned out in the spring in the timber and the still open range. This method allows the bunch-grass seed to ripen and fall, and the recovery of this grass, especially in a showery season, is very marked. As the open range becomes more limited the summer beef will be thin, and this difficulty is bound to become more acute as the country is fenced up, so that three propositions will present themselves to stockmen, viz., to reduce the herd, to turn the beef on the fenced land, or to sell store cattle. I will refer briefly to the possibilities of marketing of store cattle in British Columbia.

LORD SUFFIELD APPLES.

PURE BRED JERSEY BULL

Although hay is cheap this year on the Lower Fraser, as a general rule it is worth too much money at the Coast to feed cattle profitably, owing to the demand in the towns and to the more remunerative dairying on the farm. In the North-West, at the present time, the cattle interests are depressed, owing to the low prices, and very few stockers have been purchased from outside points, except by one or two of the larger firms, and the average man has bought very little for the past two or three years. But, I think, we may look for better times in cattle and a probable market for calves, year-lings and, perhaps, 2-year-olds in this direction, under certain conditions.

It must be remembered that Alberta is fencing up very quickly and range conditions are changing equally rapidly. The districts now raising fall wheat will, under the new conditions, raise more cattle than under the old range system, only in small lots, but will only fatten a limited number, and the calves and yearlings will be for sale just as they are offered in Manitoba and Assiniboia. Here the cows are herded on some vacant land or pasture and milked during the summer, and get often little more than straw to eat during the winter. Hence the British Columbia calf is generally a better grown animal than the average Ontario or Manitoba skim milk calf, and is nearer by three or four days' travel, which are two important points in his favour. There are, moreover, still districts to be opened up in the West and in northern Alberta, which is more of an oat country; the farmers are winter-feeding to a certain extent and have bought stockers in former years, and if prices improve would purchase again.

In connection with this may be mentioned that Mr. Roper, of Cherry Creek, has already shipped 1,000 yearlings to be pastured on the Sarcee Reserve, and the Aberdeen ranch has also shipped in a similar way. If this method were carried out entirely, it would mean adopting a system in vogue in parts of Texas, where there is not feed enough to fatten steers. There only breeding stock is kept, the yearlings being shipped to the North-Western States, where feed is more plentiful and breeding conditions less favourable. From what I can find out, I should say that the average calf crop in British Columbia is better by 20 per cent. than in the Middle-West.

Should the embargo be raised in the Old Country by the new Government (the subject is once more being warmly discussed), the effect on British Columbia prices would be considerable, possibly one half a cent. It would mean that a very considerable number of North-West cattle would be shipped as stores, and of those sold as butchers' cattle (which is the class that comes to British Columbia during the winter months). This would relieve the situation considerably and prices would improve all round.

The dairying industry has been growing steadily; all the creameries have shown increased outputs, and prices have been good. Milch cows have been selling at high prices. One herd of grade cows sold at auction on the Lower Fraser from $90 to $135; these were specially good animals. Horses have been and are still high in price, with no prospect of a drop. Farm chunks are worth in the neighbourhood of $200 and draughters at $300. In consequence of the general neglect to breed light horses in favour of the draught breeds, drivers and saddlers are scarce and likely to be.

There is little change in regard to sheep in the Province, very few being kept and no large flocks. Prices are high in accordance with the general scarcity of sheep throughout the continent and the improved price of wool. Ewes are quoted from $7 to $9. The Coast market has been in part supplied from the Middle-West this winter. Swine have suffered in some districts with hog cholera, but, on the whole, do not show any marked change.

POULTRY RAISING.

POULTRY RAISING is an important branch of general farming which is gradually developing in British Columbia, but not to the extent which its importance warrants. The home market is nowhere nearly supplied either with eggs or poultry, large quantities being imported from Manitoba, Ontario, California, Washington, Oregon. In 1904 the value of eggs and poultry imported amounted to over $400,000, and good prices prevail at all seasons, the average wholesale prices for eggs on the Coast being : Fresh eggs, 30 cents per dozen ; case eggs, 22 cents per dozen ; while the retail price for fresh eggs averaged 37½ cents per dozen, ranging from 25 cents to 70 cents. Fowls bring from $5 to $8 per dozen ; chickens, $4 to $7 ; ducks, $5 to $11 ; geese, $1 to $1.50 each, and turkeys, from 22 to 30 cents per pound.

A practical poultry raiser, who has made a success of the business on Vancouver Island, says :—" I have no hesitation in saying that there are good profits in the business, conducted on a strictly commercial basis. In fact, I know of no other branch of agriculture which is so profitable, having in view the amount of capital to be invested and the expense of conducting it. . . . . . Properly managed, in any number, poultry ought to reap a profit of at least $1 per head per annum."

Another poultryman, writing from Cranbrook, Kootenay, says :—" Eggs in this locality have been selling all winter at from 60 to 65 cents, and in several cases at 75 cents per dozen. In March they were worth 40 cents per dozen. I do a little in the egg and poultry line myself. We have 400 hens, and they have laid fairly well; but the eggs are sold fresh laid, people coming to the house for them. During the season I have sold in the mining town of Moyie many dozens of birds at 22 cents per pound. My birds are dressed. I have a steady market for 100 pounds of poultry flesh per week at the price named. It is not easy to buy birds fit for market about here. The farmers all along the foot-hills of the Rockies and through the line up to Strathcona give no attention to poultry, and miss a good many dollars by their neglect of what would pay them well. They seem quite content to allow a Calgary dealer, who gets his supplies from the United States, to bring in all the turkeys, geese and ducks consumed in the whole district during the Christmas holidays. The Chinese population of this town— Cranbrook—alone are extensive buyers of chickens and ducks. An experienced man with incubators and brooders would make lots of money here."

Mr. J. Arnould, of Sardis, B. C., has been very successful with his poultry. Speaking from a farmer's standpoint, he says :—

" I see no reason why a larger number of poultry should not be kept. I am keeping a large flock of White Leghorns, and by the use of a 360-egg Cypher incubator and brooders, I have had no difficulty in getting most of the hatching and brooding of the chicks finished up before the middle of April, when one has, of course, to attend to seeding operations and the usual spring work. I find the Leghorns lay well all winter if properly fed and attended to, though I have never used any artificial heat, even in the coldest weather, and by closing them up carefully at night and having well-built houses, they seldom get frozen combs or are checked in laying, even when the thermometer is down to zero, as the severe cold seldom lasts more than a few days, and they do not appear to suffer from a few days' confinement, if given plenty of room and plenty of litter to search in. Of course, with Plymouth Rocks and Wyandottes there is still less danger of frost, and they appear to do equally as well. I let the poultry have free range in summer, and keep them in separate pens in winter, and so far have never regretted giving them as much attention as any other branch on the farm ; and after deducting the price of their food, etc., I have always had a good balance to show for my labour and money invested."

A farmer who lives near Colquitz, Vancouver Island, gives the following results from 150 hens for the year 1905 :—

RECEIPTS.

From sale of eggs	$375 00	
From sale of chicks	50 00	
From increase of flock	25 00	
		$450 00

EXPENSES.

100 bushels wheat @ $1.05 per bushel	$105 00	
50 bushels barley @ 60 cents per bushel	30 00	
Sundries	10 00	
		145 00

Net profit $305 00

This shows a net profit of $2 for each hen, not including labour, which yields a handsome return for the money invested.

Every portion of British Columbia is suitable for poultry-raising. In the Coast districts hens, ducks and geese can be bred to great advantage, and the dry belts and uplands are particularly well adapted to turkeys.

With such facts before them, it is a matter for surprise that many farmers in British Columbia send to the nearest store for their eggs and fowls. Eggs and chickens are by-products on every well-conducted Eastern farm, and they add considerably to the annual income, as well as providing agreeable and healthful variety to the family's bill of fare.

GENERAL FARMING.

WHEAT is grown principally in the Fraser Valley, Okanagan, Spallumcheen, and in the country around Kamloops in the Thompson River Valley, and is manufactured at Enderby, Armstrong and Vernon. Until the northern interior of the Province is brought under cultivation through the construction of railways the wheat area will not be increased. Wheat is only grown on the Mainland Coast and Vancouver Island for fodder and poultry feeding.

Barley of excellent quality is grown in many parts of the Province.

Oats are the principal grain crop, the quality and yield being good, and the demand beyond the quantity grown. Rye is grown to a limited extent, and is used for fodder.

The average yields of grain and prices are as follows :—

```
Wheat, bushels per acre.....  25.62 ; Price per ton..............$33 15
Oats      "      "     .....  39.05 ;    "       "    .............. 27 00
Barley    "      "     .....  33.33 ;    "       "    .............. 28 00
```

These averages are very much exceeded in many cases, and according to nature of soil and local conditions. In the matter of oats as high as 100 bushels to the acre is not an uncommon yield.

Potatoes, turnips, beets, mangolds and all other roots grow in profusion wherever their cultivation has been attempted. Sixty-eight tons of roots to a measured acre is recorded at Chilliwhack, and near Kelowna, on Okanagan Lake, 20 acres produced 403 tons of potatoes, which sold at $14 per ton. The Dominion census places the average yield of potatoes at 162.78 bushels to the acre. The average price of potatoes is $14 to $16 per ton, while carrots, turnips, parsnips and beets sell at an average of about 60 cents per bushel.

The Okanagan, Agassiz and Chilliwhack Districts are well suited to hop-growing and produce large quantities, unexcelled in quality. British Columbia hops command good prices in the British market, and most of the crop is sent there, though recently Eastern Canada and Australia are buying increasing quantities. The yield of hops averages 1,500 pounds to the acre and the average price is 25 cents per pound.

Besides the nutritious bunch-grass which affords good grazing to cattle, horses and sheep on the benches and hillsides, all the cultivated grasses grow in profusion wherever sown. Red clover, alfalfa, sainfoin, alsike, timothy and brome grass, yield large returns—three crops in the season in some districts and under favourable circumstances. Hay averages about 1½ tons to the acre and the average price was $17.25 in 1904.

Tobacco-growing has proved successful in several districts, notably in Okanagan, where a leaf of superior quality is produced. Tobacco of commercial value will grow in almost in any part of Southern British Columbia, and there is no reason why the farmers of the Province should not cultivate it in a small way for their own use, as is the custom n many parts of Quebec and Ontario.

The importance of apiculture is beginning to be recognised and a considerable quantity of delicious honey of home production is found in the local markets. As the area of cultivation extends, bee-keeping should become a profitable adjunct of general farming.

The Coast Districts and many of the lowlands of the Interior are well suited to cranberry culture, which is being tried in a small way, but with success, by settlers on the West Coast of Vancouver Island.

Celery, another vegetable luxury, is grown in limited quantities, but the soil and climate warrant its cultivation on a more general scale. Celery properly grown and packed would command good prices, and an unlimited market.

Sugar beets grow to perfection in several localities, but their cultivation on a large scale has not been attempted.

Indian corn, melons and tomatoes are profitable items in the output of the small farmer, and are successfully grown in all of the settled districts.

SHIPMENTS OF AGRICULTURAL PRODUCTS.

The following table shows the quantities of agricultural products shipped over the Canadian Pacific Railway from all points on the Pacific Division for 1903, 1904 and 1905 :

	Grain and Hay, tons.			Vegetables, tons.			Fruit, tons.		
	1903.	1904.	1905.	1903.	1904.	1905.	1903.	1904.	1905.
January	948	822	789	70	130	89	8	7	56
February	671	522	486	87	203	121	37	8	19
March	766	444	636	374	804	342	49	37	63
April	1,012	336	261	979	1,219	202	28	34	60
May	443	408	650	800	618	156	3	9	21
June	771	336	357	242	122	35	....	....	2
July	337	249	202	142	166	235	11	13	10
August	285	1,099	265	356	331	210	75	152	265
September	769	895	1,448	537	715	779	531	736	996
October	448	986	1,343	2,396	1,672	2,647	857	839	1,184
November	872	380	949	608	759	498	230	229	308
December	943	676	624	286	199	154	41	61	48
	8,265	7,653	8,010	6,877	6,938	5,468	1,870	2,116	3,032

The above figures do not represent the total shipments of produce and fruits, being only what was carried on the main line of the C. P. R. A great deal of agricultural produce is carried by the river and coasting steamers, figures for which are not obtainable.

DESCRIPTION BY DISTRICTS.

———0———

TO adequately describe the various districts of British Columbia would require much more space than is permitted in this Bulletin. The following notes are therefore confined to an attempt to convey an idea of their agricultural capabilities, the branches of the industry to which they are severally best adapted, and the quantity of available land which each contains. As previously remarked, the absence of a general system of surveys renders it difficult to define with accuracy the areas of agricultural land, so that the figures given are in every instance approximate. It must also be borne in mind that by far the greater portion of the land bordering the rivers, creeks and lakes, and adjacent to the railways, is settled upon or in the hands of individuals, so that, with few exceptions, the settler seeking a pre-emption will have to make his selection at some distance from present lines of traffic. This fact, however, should not discourage the newcomer, as the rapid progress of the country will insure him roads and transportation facilities within a reasonable time. Meantime, however isolated he may be, he can count upon making a comfortable living off his land, supplemented by game and fish, which are plentiful everywhere.

———·—·

THE KOOTENAYS.

East Kootenay forms the south-eastern portion of British Columbia and is famed for the immense coal measures through which the Crow's Nest Railway runs for many miles both in Alberta and British Columbia. These mines are estimated to be capable of yielding 10,000,000 tons of coal a year for 7,000 years. Several mining companies are developing properties, and those that have reached the producing stage are turning out close to 1,000,000 tons annually. There are also extensive deposits of petroleum in this section of the district, but these have not as yet reached the commercial stage. Entering the province by the Crow's Nest Pass Railway, which crosses the Rocky Mountains through the Crow's Nest Pass, one descends into the magnificent Kootenay Valley, watered by the Kootenay and Elk Rivers and several smaller streams. The scenery along this route and in the valley is indescribably picturesque. Sheltered to the north and east by the Rocky Mountains and open to the south and west, the climate is exceptionally mild and healthful. The bottom lands will produce all kinds of crops in perfection, but the uplands require irrigation, which may be obtained from the Elk River and other streams. Fruit trees planted late in the fall stand the winter and thrive (a severe test for nursery stock) and wherever orchards have been established they are doing well. Conditions vary somewhat with locality everywhere, more especially in a mountainous country, but speaking generally of this district, there is no fairer valley in British Columbia and none better fitted for general farming, fruit-growing, dairying and cattle-raising. From Kootenay Landing, on the west, to the Alberta boundary on the east, the country is more or less all suited to agriculture ; portions of it are exceedingly fertile, while the rest can be made equally rich by irrigation. Much of the land is open and rolling, a beautiful park-like region, ideal for stock raising, a profitable industry, as there is a good home market for horses, cattle, sheep and hogs.

COLUMBIA VALLEY.

Going northward from the Crow's Nest branch of the Canadian Pacific Railway, the Columbia Valley is entered at Canal Flats. The scenic beauties and fertility of this magnificent valley baffle description. Dominated on the east by the Rocky and on the west by the Selkirk Mountains, the diversity and grandeur of scene from every point of view is fairly bewildering. The only present means of transportation is by stage from Cranbrook and Fort Steele to Wilmer and Windermere and by steamer from Windermere and Golden, on the Canadian Pacific Railway main line, or vice versa, but a railway will shortly traverse the valley from north to south. The advent of such a road would be followed by an inrush of settlers and the development of the minerals and timber, for the country is well endowed in both respects. Quite a few settlers have taken up land in the valley, and despite the lack of quick communication with the outside world, they are all doing well. Grains, vegetables and fruits flourish and cattle and sheep thrive on the nutritious bunch-grass which covers the benches and hillsides. The snowfall is so light that the live stock winter out and winter feeding is the exception.

The lands in the Kootenay and Columbia Valleys from Golden, south to Tobacco Plains, on the border of the United States, are mostly in the hands of the Government and the Canadian Pacific Railway Company. Mr. J. F. Armstrong, Government Agent at Cranbrook, reports to the Department of Lands and Works :—" There is now no land in this district open for pre-emption which is fit for agriculture. . . . . The want of water for irrigation prevents most of the vacant lands being available for agriculture. I have sent many intending settlers to see these, but they have proved unsatisfactory and I would not recommend them to anybody."

With regard to North-East Kootenay, Mr. J. E. Griffith, Government Agent at Golden, reports : " There is no agricultural land left," excepting about 1,000 acres on No. 3 Creek, north-west of Wilmer. This land is lightly timbered and requires irrigation. All the other vacant land, says Mr. Griffith, is practically useless except for pasture, as water cannot be got to it.

KOOTENAY AND ARROW LAKES.

The Columbia River after flowing northward for about 100 miles from its source, the Columbia Lakes, turns southward, and, with the Arrow Lakes, drains a large area, mostly mountainous and well timbered, till it crosses the International Boundary near Trail, This country, together with the large valley formed by the Lardeau River and Kootenay Lakes, comprises West Kootenay, which was until recently regarded as a purely mining and lumbering district. There are, however, many patches of good land in the district, and much that can be reclaimed by dyking. The bottom lands are a rich black alluvial and very fertile, the benches being a gravelly loam well adapted to fruit-growing. At the southern end of Kootenay Lake there is a tract of 47,000 acres of meadow land which has been partially reclaimed, and is very productive. Wherever fruit trees have been planted in this district they have given great satisfaction, and fruit-growing on a commercial basis has now been established in the Kootenay Lake section. The shores of Kootenay Lake are proved to be well suited to fruit-growing, and many orchards have been planted. On the West Arm, near Nelson, this is especially noticeable, the number of plantations indicating that in time the whole available lake front will be one immense orchard. Peaches of large size and exquisite flavour are grown in the Nelson District.

The country bordering and south of the Arrow Lakes includes some large areas of good land, which will all be settled upon and cultivated before many years. Summer frosts may nip the tenderest shoots on the lands in the creek bottoms, but wherever the land is ten or more feet above the water level there is no danger of damage. Professor Thomas A. Sharpe, who made a trip of observation through Southern British Columbia in the summer of 1905, says of this part of Kootenay : " If I owned land anywhere along this section I would have no hesitation in planting an orchard."

Potatoes yield 300 to 400 bushels to the acre, and other roots give equally heavy returns. Garden vegetables grow to perfection, and even the more delicate varieties are seldom injured by frost or drought. Although the acreage and aggregate yield of vegetables, grain and root crops are increasing yearly, the quantity produced is far from satisfying the local demand. Mining, lumbering and railway construction employ large numbers of men whose wants the farmers cannot supply, so that very much which the country can produce has to be imported. Figures to illustrate this important point are not available in the case of agricultural products in general, but in the matter of fruit the return of freight shipments over the Canadian Pacific Railway for 1905, shows that of 3,061 tons carried, 1,669 tons were delivered within the province and were absorbed by the home market.

In the Columbia Valley and throughout the Kootenay Districts, the high altitude rarifies the air and makes it bracing. The rainfall averages from 18 to 20 inches per annum, the lesser precipitation being in East Kootenay, and the snowfall varies from one to three feet. The winters extend from December to March, snow not falling to lie, as a rule, earlier than the end of December, Navigation on the Upper Columbia closes early in November, and on the Arrow Lakes and Lower Columbia about the end of that month. The Kootenay Lakes do not freeze over. During the winter the mercury occasionally drops below zero, but these cold " snaps " are of short duration, and, owing to the dryness of the air, the cold is not severely felt. The highest summer temperature varies from 80 to 90 degrees in the shade, with compensatingly cool nights.

The Upper Columbia Valley, forming the northern portion of Kootenay, varies in width from one to five miles, being bounded on the east by the Rocky and on the west by the Selkirk Mountains. Much of the land is low and swampy, but when reclaimed by dyking is very fertile. Most of the land is at present only valuable for the wild hay which it produces in abundance. There is also considerable land lightly timbered and easily cleared, which brought under cultivation and irrigated would produce splendid crops of all kinds, including small fruits, apples, pears, plums, etc. The foothills are sparsely timbered and afford excellent pasture. The Government land available for pre-emption throughout the Kootenays is limited in extent and is confined to bench lands, which cannot be irrigated. The C. P. R. has some good land, which may be bought at reasonable prices. Many large holdings and improved farms are also on the market, at prices ranging from $10 to $50 per acre.

YALE.

Yale occupies a large area to the west of Kootenay, extending to the 22nd degree of west longitude, and from about 49 degrees to 52 degrees north latitude. The whole occupies an area of about 15,500,000 square acres, and lies almost wholly within the dry belt of the Province, although, from its extent, it has a variety of soil and climate. It includes the rich valleys of the Okanagan, the Nicola, the Similkameen, the Kettle

KOOTENAY STRAWBERRIES.

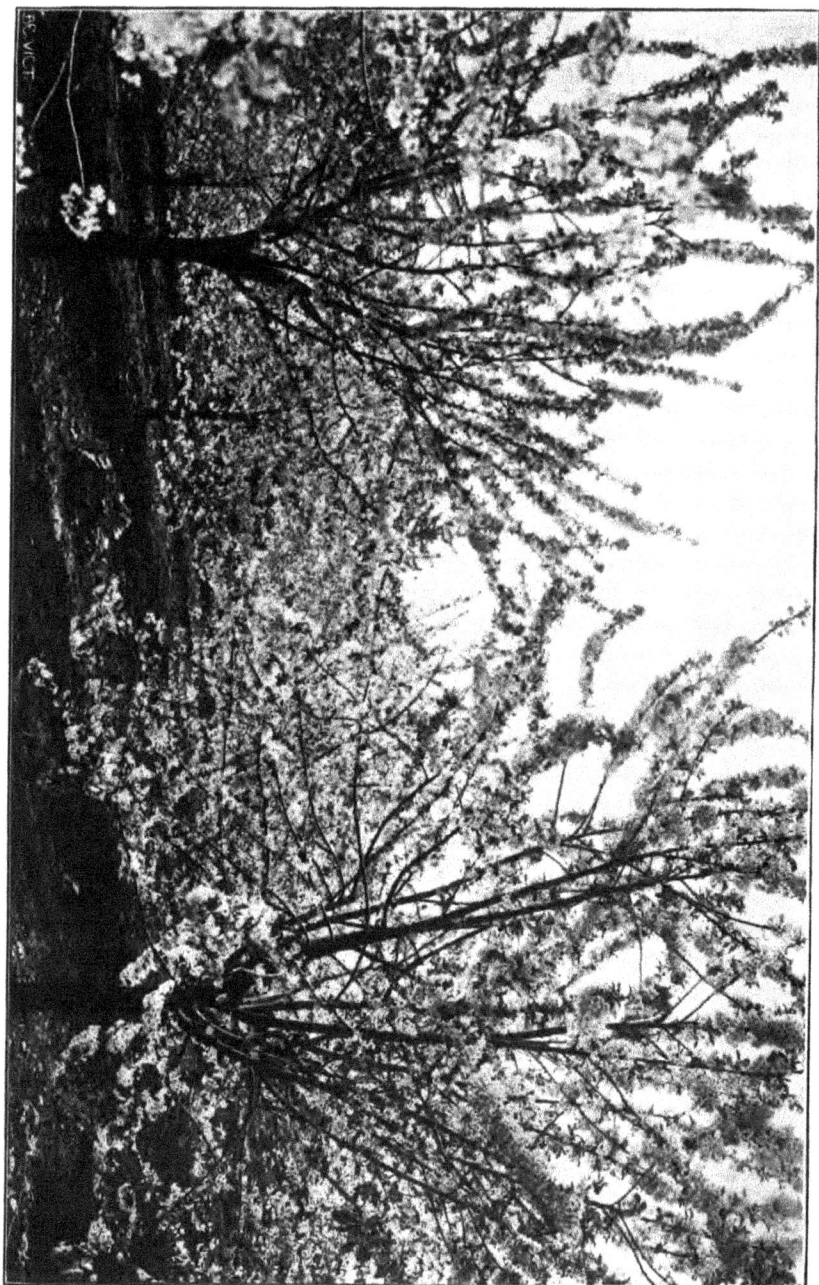

CHERRY BLOSSOMS.

River country, and the valleys of the North and South Thompson, in the vicinity of Kamloops. It possesses, perhaps, the largest area of purely agricultural and pastoral lands of any district in the Province. The valleys of the Okanagan District raise excellent wheat, which is milled at three local grist mills.

Mr. G. C. Tunstall, Government Agent at Kamloops, furnishes the following report on vacant lands in his district :—

" The North Thompson, beyond the Railway Belt, and a few of the nearer and more accessible valleys in that direction, have hitherto attracted the great majority of home-seekers. The land situated in the more remote valleys running at right angles to the river, including the benches at a considerable distance from the road, are comparatively unknown, as settlers are averse to taking up pre-emptions which are remote from means of communication, and are reached by steep ascents. The areas available for settlement, under present conditions, exist on the flats and benches bordering on the river. They are isolated and not continuous, and this feature, in addition to the localities of which but little or nothing is known, renders an estimate of any value impossible."

Yale contains large cattle ranges and many large herds of cattle, and in addition gives excellent promise as a fruit-growing district, the range of products including tomatoes, watermelons, grapes, peaches, almonds, etc. Fruit-growing has made rapid strides, and though yet only in its incipiency, promises to become a very important industry. The Canadian Pacific Railway passes very nearly through the centre of the district, a little to the north ; while the Shuswap & Okanagan branch, from Sicamous to Vernon, affords communication southward, which is continued to the International Boundary by means of the Okanagan and other lakes, forming a system of water stretches parallel to those referred to in the Kootenays. Railways are projected from Spence's Bridge and Midway into the Similkameen, and from Midway to Vernon.

SIMILKAMEEN.

According to a report made by Mr. C. A. R. Lambly, Government Agent at Fairview, there are, approximately, 1,800,000 acres of Government land in Similkameen, 1,100,000 of which are lightly timbered, 426,000 acres of pasture land, and 284,500 acres of arable land, all of which requires irrigation. Mr. Lambly says :—" Owing to meagre transportation facilities, markets are poor and prices are high in consequence. Wages are, on an average, about $2.50 per day for unskilled labour. Agricultural prospects are excellent. All the arable land requires irrigation. Timbered lands show light and medium wooded country, not heavy timber. A large percentage of the timbered land, especially the thinly wooded bench lands, is suitable for pasture."

Roots and vegetables reach perfection on the rich, black lands of the valley ; carrots, onions and parsnips do well on the sandy benches. Red clover and timothy are the chief hay grasses. Indian corn grows well, but is not used for forage. The valley is, above everything, a fruit country, and will in time be the first in British Columbia. The benches are entirely free from summer frosts. Apples, pears, plums, apricots, cherries, peaches, strawberries and all small fruits—with irrigation—are a grand crop and of good flavour. Grapes are not planted as they should be ; in fact, very little attention has been paid to fruit. At present the valley is entirely devoted to cattle-raising in a rough and ready way ; but in the future, when better methods are adopted, the valley should produce a large amount of butter. Portions of the hillsides and mountains are very steep, but the tops are covered with bunch-grass and winter grass ; the bottom is rye-grass, which stands up through the snow and makes good winter forage.

Stock-raising is at the present time the prevailing industry, but on account of the lack of irrigation and railway communication the agricultural capabilities of the country are restricted to the narrowest limits. The whole country from Spence's Bridge and from Kamloops to Princeton and the Valleys of the Similkameen and Tulameen Rivers are splendidly adapted for fruit-growing, especially in the way of apples, and, to a greater or less extent, to the growing of peaches, grapes, melons and tomatoes.

All the land suitable for agriculture or pastoral purposes in the Nicola and Similkameen countries has been taken up and is being held for future development. Including the Kettle River country, Midway and Rock Creek, Keremeos and Osoyoos, the Nicola Valley, etc., the districts immediately tributary to a railway route amount in the aggregate to 350,000 acres. Of this amount about 250,000 is prairie and meadow land, but of the whole amount only about 15,000 acres is under cultivation.

THE BOUNDARY.

West of the Kootenays lies the Boundary country, which forms the extreme southern part of Yale District. It is about 40 miles from east to west and extends for 50 miles north from the International Boundary. The character of the district, while varied, is not very different from that of other parts of the great interior plateau of British Columbia, save that the highest elevations seldom exceed 5,000 feet. Most of the hills are wooded to their summits, with open slopes, facing the south, east and west, plentifully carpeted with bunch-grass, a natural beef producer, while the valleys offer excellent openings for farming and fruit-growing, the higher benches requiring irrigation. The climate is mild and healthful, presenting no extremes of heat or cold. The snowfall in the valleys is light and spring opens early. The winter is confined to eight to ten weeks of frosty weather, the mercury occasionally falling below zero, but the cold is not extreme nor protracted. The summers, like those of the Kootenays, are warm without being oppressive and the nights are always cool. The atmosphere is clear, the prevailing condition being bright sunshine both winter and summer, and the air is crisp, dry and bracing. The average rainfall is 10.8 inches, and snowfall 27 inches, which would represent 7 to 12 inches on the level.

Between Lower Arrow Lake, its eastern boundary, and the divide between the Kettle River Valley and Okanagan Lake, the Boundary possesses many fertile valleys and wide stretches of rolling prairie, all more or less wooded. The beautiful Kettle River Valley includes from 40,000 to 50,000 acres of farming lands, a rich black loam averaging 18 inches, with a sandy clay subsoil, while lesser areas are situated on Boundary Creek, Anarchist Mountain, or Sidley, Rock Creek, and on the North and West Forks of the Kettle River. All the soil of these valleys and their benches is alike fertile and capable of producing grain, fruits and vegetables, even in the higher altitudes, as at Anarchist Mountain (3,500 feet altitude) where hardy grains and vegetables do excellently and yield heavy crops. This should be proof positive of the fertility of the soil. The abundance of water and the variety of the native grasses makes the Sidley section an ideal dairy country. Hogs and poultry-raising have proved very profitable to those who have tried them. Most of the land not already taken up is in the railway belt. One of the unique conditions found in this section is, that despite its high altitude, neither drainage nor irrigation is required.

There is a fine plat of land near Midway, which the settlers testify is free from summer frosts and yields splendid crops of potatoes, barley, oats and vegetables. Fruit is successfully grown although the number of trees planted as yet is not large. Those that have come into bearing show a healthy, clean growth, and produce good crops. A peculiar fact was noted near Midway by Professor Sharpe, *i. e.*, "potatoes and fruit trees on the uplands north of the town were doing very well without irrigation." Other observers have made similar, remarks with regard to other parts of the southern districts, and there is little doubt that some of the bench lands, of exceptional depth and fertility, will yield good returns for many years without irrigation if extra care be given in the way of tillage and rotation of crops.

Between Cascade and Carson (12 miles) there is a fine tract of land, about 20,000 acres, in a beautiful valley about two miles wide. Most of the cultivated land here is devoted to mixed farming and dairying. Fruit-growing is rapidly becoming popular in this section, about 20,000 trees representing the progress made so far. The land is admirably adapted to apples, pears, plums and berries ; cherries, grapes and peaches are also grown to a small extent. Five carloads of fruit were shipped last fall (1905) to eastern points.

Three lines of railway are now projected to traverse the Boundary, so that its future as a desirable field for agriculture and other industries is assured. Even now, with transportation facilities confined to a comparatively small portion of the district, the farmers are thriving and steadily growing rich. The numerous mining camps scat-tered over the country-side, the logging-camps, lumber mills, and smelters, provide markets at the very farm gate, indeed it is almost a rule for the buyer to seek the market, thus reversing the order prevailing in older countries, and the prices are, to say the least, satisfactory, for instance, oats are $30 per ton ; potatoes, $20 ; hay, $25 ; beef cattle, 3½ to 4 cents per pound, live weight ; hogs, live weight, 7 cents ; eggs average 30 cents ; butter, 25 to 30 cents.

OKANAGAN.

Lying west of the Kettle River Valley, and divided from it by a comparatively low watershed, is the Okanagan District, which forms an irregular strip of country stretch-ing from Sicamous, on the main line of the Canadian Pacific Railway, southward to the International Boundary. The district includes Spallumcheen, White and Creighton Valleys, Mable and Sugar Lakes, Priest's Valley, and the Commonage in the north, and Okanagan Lake, Okanagan Mission Valley, Penticton and Trout Lake in the south. Okanagan District has been appropriately named the garden of the Province, for in no portion of British Columbia is cultivation more general and successful. The district is traversed from Sicamous to Vernon by the Shuswap and Okanagan branch of the Canadian Pacific Railway, which connects at Okanagan Landing, at the head of the lake, with Canadian Pacific Railway steamers running to Penticton at its southern end. The railway runs for almost its whole length (51 miles) through a magnificent farming country, a large part of which is open, some lightly wooded, and the rest more heavily, but all very fertile when brought under cultivation. Many large farms in this section are devoted to wheat, which yields well and is a sure crop. The wheat ground locally, at Armstrong, Enderby and Vernon, makes an excellent flour. This part of the district is especially adapted to mixed farming, dairying and fruit-growing. The soil produces large crops of vegetables of all kinds and fruit of excellent quality, while native and cultivated grasses grow luxuriantly. The rainfall in this section of Okanagan is sufficient for all purposes, and irrigation is not necessary. The climate is bracing and pleasant, fairly hot in summer with cool nights, cold in winter, averaging 44.7, with occasional dips to zero and below. Snow lies from three to five months, the average fall being about 37 inches.

As Okanagan Lake is approached the climate is much milder and drier, and -from Vernon southward irrigation is necessary on all the bench lands. Here luxuriant vegetation is wholly confined to the borders of the lakes and water-courses, while the higher benches and round-topped hills present the characteristic semi-barren appearance of this class of pasture land. Appearances are deceptive in this case, however, for those bare hillsides and benches are transformed into fruitful fields and orchards by the application of irrigation. The country on the west side of Okanagan Lake is generally hilly and broken by ravines formed by water-courses from the higher elevations in the background. These water-courses will furnish sufficient water for irrigation if a system for storing it is provided. Many individual settlers and land companies are putting in the necessary embankments and ditches. A peculiarity of the Commonage, a large tract of high land near Vernon (regarded as only fit for pasture and not considered worth taking up by early settlers, as no water was available to irrigate it), is the fact that good crops of grain and vegetables are being raised on portions which have been cultivated. There are probably other areas of high land in this district and other parts of Southern British Columbia which will turn out as well with anyone bold enough to make the experiment.

Southern Okanagan, with the Similkameen country farther west, is destined to become the great peach and grape-producing section of British Columbia. At Peachland, Summerland, Penticton and other points, a great many peach trees have been planted, and the fruit is of fine quality and exquisite flavour, commanding the highest prices wherever offered for sale. Grapes are successfully cultivated at various points, but their culture is not general, and the quantity exported is inconsiderable. Mr. T. G. Earl, of Lytton, has gone extensively into peach and grape growing, and his success is encouraging many others to plant vineyards. Tobacco of excellent quality is grown in Okanagan Valley to a limited extent, the existing excise laws discouraging large plantations.

There is a brisk demand for lands in Okanagan, the prices ranging from $10 to $300 per acre, according to location and irrigation facilities.

THOMPSON RIVER VALLEY.

This section, including Shuswap, Ducks, Grande Prairie, Kamloops, Campbell Creek, Cherry Creek, North Thompson and Tranquille, lies to the westward of Salmon Arm, and although the station of Shuswap is but 15 miles from Notch Hill, the climate changes very materially, so much so that the whole of this region is so devoid of rain that irrigation is necessary in every part. The valley of the Thompson, on both sides of the river, is fertile in the extreme, and there are many places where water is obtainable from the tributary streams which flow from the adjacent hills. There are also extensive tracts which, on account of the absence of available water, are arid wastes during the summer. An intelligent system of irrigation, there being ample supplies of water in the mountains, would be the means of adding an immense area to the agricultural capabilities of this part. Such a system has been undertaken successfully near Kamloops by the Canadian Real Properties, Ltd., with the effect of stimulating fruit-growing to a large extent, all this part being exceptionally well adapted for the purpose, the apples exhibited at the Kamloops shows bearing abundant evidence to this fact. At Grande Prairie the supply of water is ample, so much so that it is not always used with discretion.

The valley of the North Thompson, which includes Adams Lake, Louis Creek and the Barrier River, is well adapted to mixed farming and fruit-growing, the soil along the rivers being mostly a rich, sandy loam. There are in this section about 50,000 acres of excellent arable land available for pre-emption. This land is outside of the Railway Belt and is all good bottom land on both sides of the North Thompson, lightly timbered with fir, poplar, cottonwood and willow, and extends for 80 miles up the river.

The Lower Thompson Valley includes Ashcroft, Spence's Bridge, Lytton and Savona, and is admirably adapted to fruit-growing; peaches, grapes, apples, tomatoes, melons, etc., are grown in profusion where irrigation is possible. The valleys in this section are from 700 to 2,500 feet above sea level. The general contour of the country is a succession of round-topped or sloping mountains, intersected by numerous narrow valleys containing more or less agricultural land. The upper benches and mountain slopes afford good pasturage, few of them being heavily timbered, and cattle-raising is the main industry. All the unoccupied land in this section lies within the Railway Belt and is under the jurisdiction of the Dominion Government.

SHUSWAP LAKE.

In Shuswap Lake is included Craigellachie, Sicamous, Salmon Arm, Notch Hill and Tappen Siding, comprising all that portion of the country between Craigellachie and Shuswap, on the main line of the Canadian Pacific Railroad. This section is peculiar, in that it is all wooded more or less heavily, and that the precipitation is ample for agricultural purposes.

The land generally is of a good quality, and grows fine crops of clover, grass and vegetables. It is an ideal dairying section. The area under cultivation, owing to the heavy nature of the timber and the comparatively new settlements, is necessarily circumscribed, but when more land is cleared there is no doubt that the exceptional advantages enjoyed in the way of transportation facilities will bring this part to the front as a fruit-growing and dairying section. The timber consists principally of Douglas fir, cedar and spruce, intermixed with birch and poplar. Trout fishing in the river is very good when the water is low. Willow grouse abound in the bottoms and blue grouse on the hills, with larger game further afield. Apples, cherries, plums, pears, etc., grow to perfection on the bench lands, while strawberries and other small fruits do well on the bottom lands.

NICOLA.

Nicola, including Upper, Central and Lower Nicola, is a fine pastoral country, and where irrigation has been established, grain, fodder crops and vegetables are raised successfully. The principal industry is cattle and horse-raising. Fruit does well, but is not grown to any great extent, owing to lack of transportation facilities. The C. P. R. is now building a branch from Spence's Bridge through Nicola and Similkameen which will furnish an outlet for agricultural produce and fruit and doubtless give an impetus to fruit-growing. The scenery in this section is particularly beautiful and the shooting and angling are unsurpassed, the lakes and ponds abounding in aquatic birds and the streams teeming with trout. There are no Crown lands available for pre-emption, all the agricultural and pasture lands having been granted or sold to farmers and stock-raisers.

LILLOOET.

Lillooet District comprises the valley of the Upper Fraser, and includes Lillooet, Pavilion, Big Bar Creek, Empire Valley, Dog Creek, Gang Ranch, and Chilcotin. The district is handicapped by lack of transportation facilities, and by the fact that a great deal of the land is so elevated that irrigation is impracticable. On the bottoms and lower benches, where irrigation is possible, excellent crops are produced, including the cereals, roots, vegetables, hay, alfalfa, apples, pears, peaches, cherries, plums, melons, tomatoes, etc. The production of fruit and agricultural products is practically limited to the local demand, as the cost of freighting is too high to admit of profitable exportation. Large ranches are the rule, and cattle-raising the chief occupation.

An exception to the arid conditions prevailing in Lillooet District is found at Pemberton Meadows, where the land is low-lying and requires drainage. All the land suitable to farming and fruit-growing is occupied or held by individuals, many of whom are offering their holdings for sale at prices ranging from $10 per acre upwards.

Mr. F. Soues, Government Agent at Clinton, reports as follows :—

"The Fraser River runs through the district from its northern boundary, latitude 52 degrees north, to a few miles below the town of Lillooet, a distance of over 120 miles. Anywhere in the valley of the Fraser in that distance the cereals, vegetables and fruits of a temperate climate grow to perfection. Wheat and potatoes may be especially named. If the water in the river could be used on the benches on both sides, enough could be grown to supply the present population of the Province in flour, vegetables and fruit, second to none. Science in the future may devise ways and means of lifting the water of the Fraser at a minimum cost ; at present it is absolutely valueless, and an exceedingly dangerous stream at all times. The only other source of note for purposes of irrigation is the Bonaparte River and its tributaries. This river is wholly on a much higher level than the Fraser, and both it and the various tributaries can be diverted. The waters in them are long since covered by record. Bounded on the north by the 52nd parallel, the extreme south-west is an imaginary line drawn through the middle of Lillooet Lake.

"In this corner of the district are situated the Pemberton Meadows, concerning which there appears to be a very great amount of public misconception. A glance at any of the Provincial maps will show that they are situated in a narrow gorge in the centre of the Coast Range of mountains, and the meadows so called are being formed annually by the silt and débris brought down by the Upper Lillooet River from the live glaciers and eternal snow-peaks of the Coast Range. This gorge, valley or canyon—all three terms are applicable—has its origin on the divide between the head waters of Bridge River (falling into the Fraser four miles north of the town of Lillooet), and thence by Lillooet Lake, Tenass Lake, Lower Lillooet River, and Harrison Lake, finally discharging into the main artery of the Province, the Fraser River, at Harrison mouth. Both sides of this waterway are bounded by mountain peaks covered by eternal snow, with numerous live glaciers.

"The effect of a June and July sun may be imagined, but it would require to be seen to thoroughly understand the amount of denudation and erosion that is an annual occurrence. In 1890 the Government decided to survey the meadows, and above on both sides of the Upper Lillooet, wherever there was available land. The survey of the meadow portion was under 5,000 acres, and of this 4,526 acres were bought up by land speculators—3,246 acres classed as second-class land, and 1,280 as third-class land.

Nearly the whole of the above was shortly disposed of to various parties. In December, 1903, the District Assessor offered for sale upwards of 2,400 acres for unpaid taxes, and as there were no bidders the whole reverted to the Crown. There have been no improvements of any note made on the original surveys.

"Areas of land fit for production of general crops is a question that cannot be answered. The whole valley of the Fraser River, up to an elevation of 2,000 feet on both sides, is suitable for general crops, provided always that water can be had for irrigation. So far as the east side of the river is concerned, every stream is long since covered by record. Up to 3,000 feet in favoured localities with a south-western aspect, mixed farming can be done with profit. Above that point, however, it is not a success in these northern latitudes; the season is short at best, and even that is liable to monthly frosts. The lower part of the Bonaparte Valley, say for a distance of 20 miles, is good for mixed farming, although on a much higher level above the sea than that portion of the Fraser Valley named. Any part of the Fraser Valley in the district, for a distance of say 60 miles above Lillooet, is just as suitable for fruit, i. e., apples, pears, plums, and peaches, as any part of the Province, and there is not the slightest hesitation in making this assertion. As a matter of fact the industry is not required, for the simple reason there is no market. Wherever this branch of horticulture has been tried—and it should be borne in mind by men with, as a rule, very rudimentary knowledge—anywhere at an elevation of from 500 to 1,000 feet above the sea the fruits named are grown to perfection. There need be no doubt entertained about this statement; a visit in the fruit season to the town of Lillooet and neighbourhood will satisfy anyone. The various farmers along the valley of the river, for a distance of 60 or 70 miles, grow a few trees for their own use. It should be here remembered that while these men can grow a good hay or grain crop, their knowledge of how to manage and care for their fruit trees is exceedingly limited. Irrigation is in evidence everywhere, with the exception of the south-west corner already referred to. Without it, agriculture of every kind would come to a standstill. The topography of the district is widely different. The settled portions from 600 to 3,500 feet above the sea level; the mountainous portions, by far the greater, rise in the Coast line to perpetual snow 8,000 to 9,000 feet, and it is very doubtful if there are not several peaks over 10,000 feet. Climatic conditions are equally varied. The same may be said of both rain and snowfall; parts deluged and laden with snow, and in a distance of a few miles both absent. No two summers or winters alike. Extremes of temperature—90 degrees to 100 degrees in the shade in parts of the Fraser Valley, and 40 degrees to 50 degrees below zero as the extreme in winter. The latter figures, however, have not been reached for several years past."

WESTMINSTER.

This rich agricultural district is bounded on the north by Lillooet District, on the west by the Gulf of Georgia, on the east by Yale District, and on the south by the State of Washington. It includes some 4,900,000 acres, of which 350,000 are suited to agriculture, most of it being exceeding fertile. This applies especially to the Fraser River Valley, the soil of which is a rich black loam capable of producing bountiful crops of all kinds. The bottom lands of the Fraser Valley are subject to spring freshets, but that drawback has been minimised by the construction of dykes which check the overflow and render many thousand acres available to the farmer, stockman and fruit-grower. The ection of the valley south of the river includes the municipalities of Delta, Surrey,

48

Langley, Matsqui, Sumas and Chilliwack, and the unorganised district eastward to Hope. On the north side of the river are the municipalities of Richmond, South Vancouver, Burnaby, Coquitlam, Maple Ridge, Mission, Dewdney, Nicomen and Kent. Vancouver and New Westminster, two of the most important cities of British Columbia. "All the lands in the Fraser River Valley and in the vicinity of the Cities of New Westminster and Vancouver have been taken up years ago," reports Mr. C. C. Fisher, Government Agent for New Westminster District, "and the only land that can be pre-empted is situate north and west of Howe Sound, and that land is, I understand, very mountainous and also heavily timbered." There are many improved and partially improved farms, and considerable wild land, for sale at reasonable prices, ranging from $5 to $75 and $100 per acre according to locality and the extent of the improvements.

The Valley of the Fraser is the Centre of the farming industry and, speaking generally, there are few, if any, agricultural districts in the world of similar extent of equal fertility. All the grains and grasses yield heavy crops, roots and vegetables of all kinds are exceedingly prolific and commonly of abnormal size, fruits, small and large, attain perfection, and cattle, horses, sheep and swine thrive wonderfully. The district presents opportunities in every branch of the farming industry to men possessed of sufficient capital to purchase land and experience to work it to the best advantage. The municipal system is more general in Westminster than in any other part of the Province and the district is well provided with roads, bridges and transportation facilities, both rail and water Some of the largest saw-mills in the Province are situated on the Lower Fraser and at and near New Westminster are many salmon canneries which give employment during the fishing season to many thousand men and add materially to the wealth and prosperity of the district.

SOUTH SIDE OF FRASER RIVER.

From the mouth of the Fraser River to Hope is one of the most fertile sections of the Province ; the land being mostly composed of silt, is an alluvial deposit of great depth. In this section are included the municipalities of Delta, Surrey, Langley, Matsqui, Sumas and Chilliwack, and the unorganised district between the last named and Hope.

Delta is low-lying, with little or no timber ; soil very rich in all elements of plant food. The greater portion of Surrey lies much higher. The timber consists of deciduous trees, principally of alder, maple (three kinds), cherry and birch ; and coniferous trees of Douglas fir, spruce, cedar and hemlock. The Great Northern Railway runs through its centre to South Westminster, and to Ladner and Port Guichon, on the Delta ; thence a ferry conveys freight cars to Sidney, where they are landed and conveyed to Victoria by rail. There are co-operative creameries at Delta and Chilliwhack, and proprietory creameries at Sumas and Chilliwack, which together manufacture a large quantity of butter and are of great benefit to the dairymen throughout this section. Besides these creameries, there is one situated at New Westminster, on the other side of the river, and many of the farmers of Surrey find it more advantageous to take their cream to that point.

Langley, the district next above Surrey, has some low-lying land subject to overflow, and, therefore, requiring dyking ; the major portion, however, is out of the reach of floods, and much of it is covered with a second growth of the deciduous trees aforementioned, the original forest of cedar and fir having been destroyed by fire at some bygone period, leaving only the huge trunks, some still standing, but mostly fallen, to tell the

GATHERING PLUMS.

"POULTRY-RAISING PAYS WELL."

tale. The cedar logs, although dead for many years, are still sound, and are utilised by the settlers for many purposes. Matsqui and Sumas have a large area of low-lying land in the vicinity of the Fraser and Sumas Lake, which requires dyking. That at Matsqui has, in point of fact, been reclaimed, and the dykes have so far stood well. The land about Sumas Lake is not so favourably situated, as the Chilliwack River flows through it, and it is therefore subject to flood from that source, as well as from the Fraser. Chilliwhack is the banner district of this part; it is well situated, and much of the land is beyond the reach of floods; those parts which lie low are dyked and so reclaimed. The soil, as it is all through the previously-mentioned districts, is fertile beyond description, and therefore crops of all kinds are produced in the greatest perfection. Dairying is carried on extensively and is increasing in importance daily. Grain-growing is, of course, not prosecuted to any great extent, as the land can be put to much more profitable uses. Root crops are largely grown, as well as fruit, for which this district is celebrated. It is also a fine district for the production of honey, as white clover grows everywhere and remains in bloom through the summer. The valley above the district of Chilliwhack becomes more contracted, owing to the mountains approaching nearer the river. All the land is good, but covered with a thick undergrowth, principally of vine maple and some large timber.—*Report of Department of Agriculture.*

NORTH SIDE OF FRASER RIVER.

From the mouth of the river to Yale, including the Municipalities of Richmond, South Vancouver, Burnaby, Coquitlam, Maple Ridge, Mission, Dewdney, Nicomen and Kent, and the unorganised districts adjacent to and to the eastward of the last-mentioned municipality. In this area are situated two of the principal cities of the Province, viz., Vancouver and New Westminster, and the Canadian Pacific Railway runs through its entire length. There is more high land in this section than on the south side, but also a large area of low lands, liable, where not protected by dykes, to floods when the snows melt in the mountains, during the months of May and June. These lands are equally fertile with those on the other side of the river, and much the same conditions prevail.

The Municipality of Richmond includes Lulu and Sea Islands, which are formed by the north and south channels of the Fraser, access to which is had by steamers, waggon and railroad; the two latter connect it with Vancouver and New Westminster and the former with Victoria. Most of the milk supply for the City of Vancouver comes from this district, which, on account of the plentiful supply of green, succulent food, is admirably adapted for such purposes. Little or no butter is manufactured, it being more profitable to sell the milk. These islands, lying low, require dyking, and the soil, as it is on the other side of the river, is of great depth and fertility, producing enormous crops of all kinds. Part of the middle portion of Lulu Island is peaty, and, lying lower, as it does, than that nearer the river, is difficult to reclaim. This class of land is, however, of comparatively small extent. The Municipality of South Vancouver is bounded on the south by Richmond, on the west by a Government reserve, on the north by Vancouver City, and on the east by Burnaby. The land in the municipality is high and out of the reach of flood, and is, as a rule, very heavily wooded. Dairying, for supplying milk to Vancouver, is also carried on in this section; also fruit-growing and some market gardening. The Municipality of Burnaby is bounded by the last-named municipality on the one side, and on the other side by Coquitlam Municipality, the other two boundaries being Burrard Inlet and the Fraser. New Westminster is situated in this municipality, on the Fraser. The land is high and similar in its characteristics to South Vancouver, for the most part a rich loam, and intersected by gravelly ridges. There are many small holdings in this section devoted to fruit and market gardening, for which it is well adapted. One of the best district exhibits made at the New Westminster Agricultural Show last year came from Burnaby. An electric tram line runs through this municipality, connecting the towns of Vancouver and New Westminster ; good roads also intersect it in several directions. Coquitlam is a large, irregular, municipality lying to the east of Burnaby, and extending north to the Coquitlam and Pitt Lakes. It consequently embraces part of the Pitt Meadows, which are low-lying and are dyked. The Coquitlam and Pitt Rivers run through this municipality and empty into the Fraser above New Westminster ; the main line of the Canadian Pacific Railway also runs through it and is the chief means of communication. The land is partly wooded and partly low meadows ; the former is, as a rule, high, and does not require dyking. It is all fertile, with the exception of the dry, gravelly, fir-clad ridges. Maple Ridge Municipality follows, bounded on its southern side by the Fraser, and running back some six miles it embraces the greater part of the Pitt Meadows, which were partly dyked by a syndicate some years ago, and are now for sale in lots to suit purchasers, the price being about $40 per acre. The land in these meadows is exceedingly fertile and well adapted for dairying. The principal places are Port Hammond, Port Haney and Wharnock, all stations on the Canadian Pacific Railway. With the exception of the Pitt Meadows, the land in this municipality is high and out of the reach of floods. It is wooded, but not heavily, with a second growth of fir and cedar, and the usual deciduous trees. Lillooet River, a fine trout stream, flows through almost the entire length. This is a fine fruit section, and a great deal of attention is being paid to the industry. It is also good for dairying, and were a creamery established hereabouts it would prove to be one of the most successful in the Lower Fraser districts. Some sheep are produced, the country being fairly well adapted for the purpose ; pigs are also produced in limited numbers.

Mission Municipality adjoins Maple Ridge to the eastward. Stave River flows through Mission, and affords splendid trout fishing at certain seasons. A milk-condensing factory and creamery is situated on the Fraser at a point where the Canadian Pacific Railway crosses the Fraser. This establishment has had the effect of promoting dairying greatly, not only in this municipality but in Matsqui, on the opposite side of the river, and the vicinity generally. Dewdney is a small municipality to the eastward of Mission. Except in the back part, it is subject to overflow when the river is in flood. The higher portions are rather heavily wooded. Much the same conditions exist here as in Mission. The Municipality of Nicomen comes next, and embraces the island of Nicomen. This is a fine, fertile island, partly open, but mostly wooded with cottonwood or poplar, which is easily cleared ; but, unfortunately, it is not dyked, and, consequently, it is liable to be flooded at high water. The mountains at this point approach the Fraser very close, greatly narrowing the valley and contracting the area of agricultural land. To the eastward of Nicomen is Seabird Island, an Indian Reserve and an unorganised district, extending to the Municipality of Kent. Agassiz, where the Dominion Experimental Farm is situated, is the chief place of this district. Harrison, at the mouth of the river of that name, is the point where a ferry for Chilliwhack connects with the trains. At Harrison Lake, five miles from Agassiz, where the river takes its rise, are the celebrated hot springs. At Agassiz there are two good hotels and many fine farms, where a good deal of fruit is produced, as well as other crops. It is also a most favourable locality for the production of hops, of which a considerable quantity was cultivated, but in consequence of some adverse seasons and the low price of hops in England, most of the yards were abandoned. A few are still in existence, however, and are successfully carried on. The land is mostly lightly timbered, principally with birch, alder, etc. Some of the land near the Fraser requires dyking, and all requires draining. This is also a good district for dairying. Above Kent there is not much cultivation, the mountains gradually drawing in until Yale is reached, where they finally close in on the Fraser altogether, forming the canyons of the Fraser.—*Report of the Department of Agriculture.*

CARIBOO AND CASSIAR.

Although Cariboo and Cassiar are the two largest districts in British Columbia, estimated to contain 96,000,000 and 100,000,000 acres respectively, they are the most sparsely inhabited and the knowledge of their resources is very limited. In the early sixties Cariboo and Cassiar were invaded by a great army of placer miners, who recovered about $50,000,000 in gold from the creeks and benches. Hydraulic mining on a large scale is being carried on by several wealthy companies at different points in the district with fair success, and individual miners and dredging companies are doing well in Atlin, Recently large deposits of gold and silver quartz were found in Windy Arm, east of Atlin, and give promise of rich returns. Large coal measures have been located on the Telqua River and at other points, and copper ore is found in many localities. The country is lightly timbered and promises in time to become an important cattle-raising and agricultural district, as there are many fertile valleys, which, even now, despite the absence of railways, are attracting settlers. In the southern part of Cariboo, along the main waggon road, are several flourishing ranches which produce cattle, grain and vegetables, finding a ready market in the mining camps. At some points, where the altitude is not too great, apples of good quality are grown, but, as a general rule, the crops consist of oats, barley, clover, sainfoin, red top, beets, potatoes, carrots and the general run of hardy vegetables. Irrigation is necessary all along the Cariboo Waggon Road, but much of the land is too elevated to permit the application of water. There is a good supply of timber everywhere for all purposes, and cattle-raising and dairying are profitable occupations.

After passing the height of land at the headwaters of the Fraser River and its branches, the country slopes gradually to the north and west, forming a series of valleys of more or less extent, which contain many thousand acres of good agricultural and pasture land. The part which lies east of the Rocky Mountains is an almost level plateau, with a slight dip to the valleys of the Peace and Smoky Rivers. Owing to the depth of these valleys and the absence of rock, the conditions for drainage are perfect. According to Professor John Macoun, the finest tract of land lies between the Smoky and the Peace River. In what is known as the Peace River country, both north and south of the river, there are from 5,000 to 10,000 square miles of fine agricultural land lying within British Columbia. This country is largely prairie and poplar copse, and everywhere the soil is good. North of Fort St. John, on the plateau back from the river, common grasses attain a height of from four to six feet, and vetches are found eight feet in July.

Professor John Macoun, in summing up his impressions of Northern Cariboo, says :

"After having seen the growth of vegetables and cereals at Dawson, in the Yukon District, and remembering what I had seen on Peace River, the Nechaco, Lake Babine and the reports from the Skeena and Stikine, I am led to believe that the day of a general awakening has come, and we can now say that Northern British Columbia will, in the future, support a very large population on its own productions. Throughout the whole region, including the Yukon District, fodder for horses and cattle in any quantity can be grown. At Dawson, clover and timothy were found last season to do remarkably well. Oats, barley and wheat were found in the same field. The two former were ripe on August 23rd, and the wheat so far matured that, after drying, the ears looked ripe. Last month I sent three ears of wheat grown at Dawson in latitude 64° 15', to the Experimental Farm in this city, to have it tested. The report received the other day was '100 grains planted, 100 grains sprouted, and 100 grains were vigorous, and no weak plants were produced.' Such a report as the above shows that all lands suitable to grow wheat in the Peace River region, Northern British Columbia and the Yukon District, have climatic conditions suitable for the growth of all necessaries in a civilized community."

Cassiar, which lies west of Cariboo, is better known as a gold mining district and the home of big game than for its agricultural possibilities. Mr. James Porter, Government Agent at Telegraph Creek, reports that "owing to the ever likely summer frosts and other disadvantages to be met with under present conditions, the district as a whole cannot be considered as offering inducements to settlers from an agricultural point of view. However, when the country is opened up through its mineral resources, and large numbers of people are attracted to it on that account, there will be a ready and easy got at market for farming products, and I feel sure that farming, under these conditions, will follow and be made profitable too, as, on the whole, there are many places throughout the country where the soil is of excellent quality."

There is a great deal of very good land, fit for agricultural purposes on portions of Dease and Liard Rivers, that will some day be put to profitable use. In the Klappan River Valley there is abundance of good land and much open grass country. The old Ashcroft-Telegraph Creek trail passes through this country. In the valley of the Tanzilla River (which is a branch of the Stikine) there are about 10,000 acres of farming and pasture land, about 6,000 acres of which is more or less heavily timbered. There are also some small areas of arable and pasture land in the Stikine River Valley. On Turnagain River, about 75 miles east and south-east of McDame Creek, there are reported to be large tracts of open grassy country, suitable for stock-raising, if not for general farming. The trail from Peace River to Yukon, now being constructed by the North-West Mounted Police, should traverse some of this land.

The difficulties which lie in the way of settling the Cassiar District are the lack of roads and transportation facilities and the want of a local market, as the demands of the mines operating there at present are limited.

VANCOUVER ISLAND.

Vancouver Island, which is about 285 miles long, with an average width of about 60 miles, is separated from the British Columbia mainland by the Gulf of Georgia and the Straits of Haro and Juan de Fuca, and bears a close resemblance to Great Britain in its geographical position as well as in climate and certain natural characteristics. The climate, mild and moist as in England, is warmer and brighter, with less average rainfall, the summers being invariably dry with continuous sunshine, while the winters are much less foggy with frequent spells of crisp, bright weather. Holly, ivy, broom, gorse, box, heather, privet and other shrubs grow in perfection, and all the favourite English flowers are seen in the fields and gardens. Wall flowers, primroses and violets bloom the year round and in the early summer the whole country is transformed into a vast rose garden, wild and cultivated varieties flourishing everywhere. The climate and the flowers are, however, far from being the most important natural assets of this favoured region. Its timber is the finest in the world and of great extent; its coal measures are practically inexhaustible; the deposits of other minerals—iron, copper, gold and silver— are vast and but slightly developed; its fisheries rival those of the Atlantic, and its soil is of wonderful fertility, capable of producing every grain, fruit, root and vegetable grown in the temperate zone.

The coast of Vancouver Island is deeply indented with bays and arms of the sea, forming numerous deep-water harbours, providing good shipping facilities for the mines, lumber mills and other industries, and numerous streams and lakes afford access to the interior. The country on the southern and eastern coast is comparatively level, while the interior is broken by mountains and heavily wooded valleys. The greater part of the agricultural land is covered with big trees and thick underbrush, but the quality of the soil will well repay clearing, as wherever the timber has been removed and the soil cultivated the results are highly satisfactory. Along the eastern coast are several areas of open land occupied by successful farmers, fruit-growers, dairymen and poultry-raisers. Wheat is not generally grown as mixed farming is found to be more profitable. Apples, pears, plums, cherries, prunes, and all kinds of small fruits grow luxuriantly, and peaches, nectarines, apricots, grapes, almonds, filberts and other nuts, are produced with a little extra care and attention. Fig trees, found growing wild near Nanaimo, encourage the belief that this fruit can be successfully cultivated. Tomatoes, melons, and other tender vegetables, ripen well and give big returns. Such is the fertility of the soil that a small patch of from 10 to 20 acres well cultivated will produce a handsome profit after supplying a comfortable living for an average sized family.

The agricultural settlements on Vancouver Island, near Victoria, along the line of the Esquimalt & Nanaimo Railway, and at Comox, are the oldest in British Columbia, and the excellence of their products has more than a local reputation. Island poultry, Island mutton and pork, Island strawberries, cherries, plums and apples, and Island butter, command the highest prices and such is the demand that little ever crosses to the Mainland—the local markets absorb all and ask for more. Cattle, sheep, swine and poultry do well on the Island, the climate being so mild as to permit their roaming at large and picking up an abundance of green food the year round. Dairying is a profitable and growing industry, although the local market is still far from being supplied, while the progress of mining, lumbering and fishing is constantly creating new demands and the Oriental trade, as yet in its infancy, assures a continuance of good prices in the future. The average price of butter at first hand is 25 cents per pound.

In the Esquimalt, Metchosin, Sooke, Lake, Victoria, North and South Saanich, Goldstream and Highland Districts, which adjoin the City of Victoria, there is considerable good land suitable for poultry-raising, dairying, fruit growing and market gardening. Malahat District also contains areas of arable land, some of which is heavily timbered, which might be profitably utilised for poultry, dairying, fruit-growing and sheep-raising.

Further north lies the famous Cowichan Valley, noted for its beauty of scenery and fertility of soil. Cowichan, including the districts of Comiaken, Quamichan, Chemainus,

Somenos, Sahtlam, Seymour and Shawnigan, is one of the most flourishing settlements on the Island. The soil of the Cowichan Valley is of peculiar richness, being strongly impregnated with carbonate of lime, with usually a depth of two to three feet and a subsoil of blue clay and gravel. The soil is suited to all kinds of crops, but is particularly adapted to fruit, which grows in great abundance and of excellent quality and flavour. The roads throughout the district are the best on Vancouver Island—where bad roads are almost unknown—thanks to the efforts of the local Municipal Council. Ve·y little wheat is grown, the area under cultivation being too limited, but oats are a principal crop, yielding 60 bushels to the acre. Peas produce between 30 and 40 bushels per acre, potatoes from 400 to 600 bushels, hay from two to three tons. Apples, pears, plums, cherries and small fruits give big returns. Sheep-raising is carried on to a considerable extent, a ready market for sheep and lambs being found at Victoria, Ladysmith and Nanaimo. Hogs pay well and thrive, and poultry give good returns, the prices of eggs and fowls being always high.

West of Duncan, in the Cowichan Valley, there is a large area of good land, that portion on the north shore of Cowichan Lake being an almost level country admirably adapted to farming. From the lake to the Nitnat River and Barkley Sound the country is more rugged and heavily timbered and is reported to be one of the richest mineral sections of British Columbia. The same remarks apply to the land in the vicinity of Ladysmith, and in the Nanaimo, Mountain, Cedar, Oyster, Bright, Cranberry, Douglas, Wellington, Nanoose and Cameron Districts. Mountain is broken, with considerable good land in the Millstream Valley, and the uplands furnish excellent grazing, with large and small timber of good quality. Cedar and Cranberry Districts very much resemble Cowichan and possess large areas of good farming land. North of these districts the character of the soil changes, inclining to be sandy and gravelly, in patches, but around Qualicum it again reverts to a rich loam of the best quality. A good deal of land is under cultivation in the country lying between Nanaimo and Comox, but much of the best of it is still unreclaimed, and many thousand acres will be available when cleared of timber.

Wellington, the present terminus of the Esquimalt & Nanaimo Railway, was formerly a town of considerable importance, but, since the closing of the coal mines in that district, has lost much of its trade. There is considerable good land in the neighbourhood, the area of which will be increased by clearing.

Extending from the northern boundary of Nanaimo lies Comox District, considered one of the best agricultural and dairying sections of Vancouver Island. Sixty miles long, with an average width of seven miles, between the sea and the mountains, is a bench of undulating land admirably adapted to cultivation. Parts of it are heavily timbered, and there are many marshes and beaver meadows easy of reclamation. Several valleys cut through from the mountains to the sea, and these are specially fertile. All of this bench land will produce crops. Where it is too light for growing cereals or roots, it will give large returns in hay and alfalfa. The growth is marvellous, a bit of burnt land sowed with grass seed will become a tangled mass of vegetation within a year. Cattle fatten on the native grasses and vetches in a wonderfully short time. Butter-making and poultry-raising are carried on as adjuncts to the regular farm work, but so far have not been engaged in systematically. A good local market for everything produced is afforded by the Union Coal Mines, with headquarters at Cumberland. These mines employ between 800 and 1,000 men, who, with their families, are good customers of the neighbouring farmers, paying liberal prices for everything they consume. Grain is raised in considerable quantity, but only for feeding stock. Oats yield well, and sell for 1½ cents per pound. Butter averages 25 cents per pound; beef, 7½ cents to 10 cents by the carcass. Cows are worth $50 to $70 each. Lambs, $5 to $6; ewes usually breed twins. Hogs bring 8 to 9 cents live weight. Eggs sell from 25 to 60 cents per dozen. Apples, pears, plums, prunes, cherries, and all small fruits, are cultivated to a limited extent, and all produced is of excellent quality. Much of the wooded land in Comox District is easily cleared, being chiefly alder, and the swamps are not difficult to drain. The swamp bottoms are remarkably fertile, producing splendid crops of grain and vegetables.

Communication is had with Comox by waggon road from Wellington (the present terminus of the Esquimalt & Nanaimo Railway), and by steamers running to Nanaimo, Victoria and way ports. A short line of railway connects Cumberland with Union Wharf, the shipping point of the Union Coal Mines.

Alberni Valley, at the head of Alberni Canal, about 20 miles long and from six to eight miles wide, is destined to become an important district from an agricultural standpoint, as it is the centre and natural distributing point for a large and rich mineral district. It is 110 miles distant from Victoria and 55 miles from Nanaimo, being connected with the latter by waggon road. A very large area of good agricultural land can easily be brought under cultivation by clearing and drainage. The soil generally is a clayey loam and very productive, being well adapted for fruit-growing and dairying. A very considerable part of the fertile Alberni Valley lies within the Esquimalt & Nanaimo Railway Grant, and is included in the areas which the Company will render fit for cultivation and offer for sale to settlers. At present Alberni has a tri-monthly steamboat service with Victoria, and stage coach communication with Nanaimo. It is quite possible that the Esquimalt & Nanaimo Railway will build a branch line to Alberni as soon as settlement of the lands is assured and business conditions warrant the outlay.

In Alberni District there are about 4,600 acres of Government land open to preemption. These are located as follows:—Sproat Lake, 480 acres; Central Lake, 880 acres; Taylor River, 300 acres; Barkley, west of Nitnat Lake, near Pachena Bay, 800 acres. Some of the islands in Barkley Sound, near the Pacific Cable Station and the whaling station, are fairly well adapted to sheep-raising. In the inlets of Clayoquot Sound there are small patches of heavily timbered land aggregating about 600 acres, and in Rupert District, between Cape Scott and the West Arm of Quatsino Sound, there are 1,000 acres, much of which is low and swampy. Apart from these areas, the bulk of the land is covered by timber leases and licences, and is not open to settlement. South of Barkley Sound, in and about Port San Juan, there is considerable arable land, but it is all heavily timbered and difficult to get at, owing to the absence of roads.

Besides the districts which have been briefly described, there are several valleys and benches of prairie lands in the interior of Vancouver Island fit for agriculture when proper transportation facilities have been provided. Much of this portion of the Island is unexplored, but the Esquimalt & Nanaimo Railway Company has parties in the field who will examine and report upon the agricultural possibilities of the land lying within the company's grant, it being the intention to attract settlers to these interior valleys as soon as practicable. The numerous lakes and streams and the mountains, which divide the valleys, afford a most attractive diversity of scenery, which combined with the salubrity of the climate, which is drier and warmer than on the coast, will make these interior districts very desirable for residence when they have been thrown open to settlement by the building of branch lines of railway. Existing reports on the interior are to the effect that there are considerable areas of grazing lands on the high plateaux and in the foothills of the Island Range.

The only land open to pre-emption in the Nanaimo Land District is situated on Lasqueti Island, and consists of 6,863 acres. Mr. Marshal Bray, Government Agent at Nanaimo, reports that all the land in his district is owned by the Esquimalt & Nanaimo Railway Company, with the exception of Gabriola, Denman and Hornby and other islands which are all taken up. The land on Lasqueti Island is unfit for anything but sheep pasture, as it is rocky and covered with scrubby timber and brush.

Besides the lands above mentioned there are considerable areas in the interior valleys and along the north-east, north-west and northern coasts. The lands bordering the sea are, as a rule, low-lying and wet, though portions could be reclaimed by inexpensive dyking and drainage. These lands are well suited for cattle-raising and dairying, while farther inland a better quality of land is found, capable of producing good crops of vegetables and small fruits, but on account of the prevailing humidity, little of it is fit for the cultivation of cereals. The area of land adaptable to agricultural purposes has been variously estimated. Taking the reports of surveyors of 14 townships, out of 44, which extended northward from Reef Point (just south of the entrance to Quatsino Sound), on the West Coast, and westward to Alert Bay, there are 143,000 acres of grazing lands and 12,740 acres of agricultural lands included in those 14 townships. Taking a line drawn from Campbell River to the headwaters of Salmon River, and thence to Kamuntzen Lake, a total distance of 166 miles, there lies adjacent to such line an area of 140,000 acres, much of it of excellent quality. The character of the country is diversified, but the greater portion is well timbered and watered, with many patches of open prairie and grazing lands. These estimates do not include the extensive valleys of the interior which remain unexplored, but which, according to the reports of prospectors, timber-cruisers and trappers, will be found equally available for settlement.

There are numerous islands in the straits which separate Vancouver Island from the Continent, many of which support prosperous communities of farmers and fruit-growers. The climate is mild and equable, and, being sheltered from the north winds, the more delicate varieties of fruits and vegetables do well. Apples, plums, pears, peaches, apricots, nectarines, grapes, figs, melons, tomatoes, corn, etc., grow to perfection. Sheep-raising is one of the chief industries and is highly profitable. The more northerly islands are, as a rule, densely timbered, but many clearings have been made which demonstrate the soil to be first class, capable of producing general crops of all kinds, including apples and small fruits. Farmers living on the islands and all along the coast have an endless preserve of fish and game right at their doors, so that they are assured of an abundance of food at all seasons.

A successful colony, known as the Danish settlement, was established about nine years ago at Cape Scott and along the San Josef River, the north-western end of Vancouver Island. There are several townships in that portion of the Island which are more or less suited to agriculture and dairying, but much of the land is low and wet, requiring drainage. The Cape Scott settlers report that they are doing well and are satisfied with their surroundings.

The meteorological record for 1902 at Cape Scott (latitude 50.48° N., long. 128.27° W.) is as follows :—Highest temperature, June, 81.5° ; lowest, January, 16° ; average, 46.3° ; rainfall, 137.76 inches ; snowfall, nil.

FRUIT PACKED READY FOR SHIPMENT.

It will be observed that the rainfall is excessive and the snowfall *nil.*. This is characteristic of the West and North-West Coast, and is accounted for by a warm ocean current setting in to that shore, the evaporation from which is condensed by the snowy mountains of the coast line and falls in the shape of rain. The greater part of the above precipitation, viz., 92.72 inches, occurred during the months of January, February, March, November and December.

WEST COAST OF THE MAINLAND.

This includes the various settlements at Howe Sound, Squamish, Froek, Bute Inlet, Bella Coola, etc. Communication with these settlements is maintained altogether by steamers, there being no roads, and the character of the country being of such a nature that their construction is all but impossible, and must of necessity remain in abeyance until the poplation is sufficient to justify it. The country is, without exception, thickly wooded, principally with Douglas fir, spruce, hemlock, red cedar, balsam fir, maple, alder, birch, and other woods, and a great variety of underbrush. The coast line is indented along its whole length with deep fiords, which run many miles into the interior, and at the heads of which are large streams. The shores of these fiords are, except where rivers debouch, almost invariably precipitous. At the mouth of the streams and along their valleys are generally flats, fit for agricultural purposes after they are cleared of timber. As may be imagined from the description given, the precipitation is excessive, consisting almost entirely of rain, the snowfall, owing to the influence of the sea, being comparatively small, and the temperature, from the same cause, never very low. The markets of this part consist mostly of loggers' camps and Indians.

Fishing is carried on at various points along this coast, and logging for the mills at Burrard Inlet, Chemainus, etc. The Island of Vancouver, which lies parallel with this coast for a distance of some 300 miles, protects it for that distance from the force of the Pacific Ocean, and the waters of the straits are therefore comparatively calm. The whole of the region is naturally very sparsely populated, farmers or ranchers being in the minority ; in fact, most of the inhabitants are engaged in other pursuits, fishing, lumbering, etc., and even those who are engaged in agriculture for part of the year take advantage of the fishing season to supplement their incomes. Railroads now projected, and which it is believed will in the course of a few years be constructed, will have the effect of very materially altering the state of affairs all along this coast.—*Report of the Department of Agriculture.*

THE NORTH-WEST COAST.

The country lying along the coast and the numerous adjacent islands between Wakeman Sound and the Kitimat River were explained in 1895 by George D. Corrigan, P. L. S. His report deals in detail with the country examined, which includes the valley of Wakeman Sound, valley of Kingcome River, Ah-ta Valley, valley of Thompson Sound, Quascilla and Bay River valleys, Mereworth Sound, Cape Caution, Branham Island, Shelter Bay, Bradley Lagoon, Banks Island, Goschen and Porcher Islands, Henry, Stephen, William and Arthur Islands, the valley of the Kitimat, Kildala Valley, Gardner Canal and Swindle Island. The country, with some exceptions, is not suitable for farming, but the mainland coast and nearly all the islands would support fishing communities who would increase their tribute from the sea by keeping a few cattle, hogs and sheep, and growing vegetables.   On Wakeman Sound there is an area of about 15,000 acres of first-class land and at the mouth of Kingcome River 10,000 acres or more which will grow hops, oats and roots.   The land is easily tilled and there is plenty of good timber ; in fact, every requisite for a prosperous settlement.   In the Ah-ta Valley there are about 1,000 acres of good land, which is easily accessible from Kingcome Inlet. On many of the islands there are small patches of arable land and considerable tracts of coarse grass lands which would support flocks of sheep.   On Porcher Island there is a tract of some 25 miles of fairly good land along the coast which might be utilised by drainage.   The valley of the Kitimat is by far the best country visited.   The soil appears very productive, the ordinary garden vegetables, together with cultivated small fruits, giving good returns.   The wild crab-apple bears pleantifully, and there is an abundance of high-bush cranberries, saskatoon and other berries.   The run of oolachans is said to be enormous at the mouth of the Kitimat River, the Indians coming from long distances to fish for them.   The mouth of the river is very shallow and difficult to enter except at full tide.   In the lower valley there are several tracts of fairly good land which could easily be brought under cultivation, but a drawback exists in the fact that much of it is subject to flooding during the early summer freshets.

OUTLYING AREAS.

The great central valleys lying to the north of the settled portions of the Province offer an attractive field to the adventurous pioneer, willing to endure present inconvenience for the certainty of a future competence.   These include the Chilcotin District, the valleys of the Skeena, Bulkley, Kispiox, Nechaco and Blackwater, the country surrounding Ootsa, Francois, Stuart and Fraser Lakes, and the great Peace River Valley, containing between 6,000,000 and 10,000,000 acres, practically all of which is available for agricultural purposes.   Existing conditions in those districts are similar to those of the early days of Western Ontario and the Middle Western States, but unlike the latter they are free from " bad " Indians and life and property is as safe as in the centres of population.   Many settlers have established themselves in those promising districts during the past few summers and all are satisfied with their surroundings and enthusiastic as to the future.   The lack of transportation facilities—now confined to pack trails—is the great present drawback, but that will be remedied within the next few years, as the Grand Trunk Pacific, Canada Northern and Canadian Pacific Railways are preparing to extend their systems through those fertile valleys and there is talk of the Great Northern Railway also building a line which will serve some portion of them.

The great undeveloped areas have attracted a great deal of attention since the announcement that the Grand Trunk Pacific, Canadian Northern and Canadian Pacific Railways were contemplating building through them, and thousands of letters of inquiry have been received from prospective settlers and speculators eager to secure land in these new districts.   To meet the demand for information, this Bureau issued Bulletin No. 9, "The Undeveloped Areas of the Great Interior of British Columbia," (2 editions), and Bulletin No. 22, " New British Columbia," which contains all available information regarding the Bulkley, Kispyox, Ootsa, Nechaco and the other fertile valleys lying within practicable distance of the Coast.

A copy of Bulletin No. 22, with map, can be had on application to the Bureau of Provincial Information.

LAWS AFFECTING AGRICULTURE.

——:o:——

THAT the Government and Legislature of British Columbia have not failed to carefully guard and assist the interests of agriculture is evident from the fact that there are on the Statute Books of British Columbia forty-two Acts, including amending Acts, dealing with subjects either directly or indirectly connected with farming. These have been passed from time to time, as the development of the agricultural industry has seemed to warrant, and the conditions affecting the agriculturist demanded attention.

AGRICULTURAL DEPARTMENT.

First in the order naturally comes the "Department of Agriculture Act," containing provisions for the regulation of the Department of Agriculture, and for defining the powers and duties of the Minister and other officers of the Department. Principal among its provisions is one for the appointment of a statistician for the collecting, abstracting and tabulating of statistics and information of public interest, a general report to be presented to the House at the end of each year. In this connection it may be stated that all persons engaged in agricultural pursuits of any kind are required to furnish replies to official inquiries, and to give such statistics and information as are in their power in regard to the industry under their particular control. Provision is also made for the interchange between the Federal and Provincial authorities of statistics and information bearing on the subject of agriculture.

AGRICULTURAL ASSOCIATIONS.

There are several statutes making provision for organisation and establishment of agricultural and horticultural and other societies, having for their object the mutual benefit and co-operation of agricultural communities. Associations and societies in respect to the following classes of subjects may be formed, namely : (a) Agriculture and horticulture ; (b) benevolent and friendly ; (c) co-operative ; (d) industrial and provident ; and (e) investment and loan. Only the first and third of these associations directly refer to agriculture as an industry, the other being of a general character. In respect to the incorporation of agricultural and horticultural societies, under Chapter 12, Consolidated Statutes, any number of persons may unite themselves into a society for the protection of agricultural, horticultural or fruit-growing industries. The provisions are of such a character as not to be briefly summarised here, but indicate the procedure for the formation of such societies, as to their officers, regulations, property, liabilities and general powers. Part II. of this Act relates to the Horticultural Society and Fruit Growers' Association of British Columbia, which was organised in 1890, and which has been in existence ever since. It has particularly for its object the advancement of the fruit-growing industry. The Act respecting the Co-operative Associations, passed in 1896, although it does not specially refer to agriculture, was really intended to afford a cheap and easy method of forming co-operative societies for the purpose of carrying on any branch of agriculture, or business connected with agriculture. Its provisions are largely technical, simply dealing with the methods of procedure and functions of societies formed under its authority.

FARMERS' INSTITUTES.

One of the most important Acts passed by the Legislative Assembly affecting the interests of the farmer was the "Farmers' Institutes and Co-operation Act," enacted in 1897. This was a decided step in advance, and brought out legislation in line with that existing in Ontario, Manitoba and other Provinces ; and while it contained many of the provisions common to the Acts of Ontario and Manitoba, the measure went still farther in the line of progress, and in several respects may be said to be an improvement on its prototypes. Farmers' Institutes, under this Act, may be organised by a petition to the Minister of Agriculture, signed by fifteen persons resident in the district in which it is proposed to organise, and have for their objects the encouragement and improvement of agriculture, horticulture, arboriculture, manufactures, the useful arts and displays of agricultural products ; holding meetings for discussion and hearing lectures ; for importing and distributing seeds, plants and animals ; for offering prizes for essays and scientific inquiry ; for dissemination of information regarding bee-keeping ; and for carrying on, by co-operation, any industry relating to agriculture. The annual fee of members is 50 cents, which is supplemented by the Government by an amount based on membership, the grant being made conditional upon all the provisions of the Act being complied with. Provision is also made for the organisation of Divisional Institutes, and of a Central Farmers' Institute for the whole Province, and also for the amalgamation with the Central Farmers' Institute of the Fruit Growers' Association or any other existing agricultural association. Authority is taken under the Act by the Lieutenant-Governor in Council to frame rules and regulations defining in greater detail the work of the Institutes and the system under which they may operate. The officers are a Superintendent, who is appointed and paid by the Provincial Government, and Secretary-Treasurers of the local Institutes, who are paid $25 a year by the Government. It is the duty of the Secretary-Treasury of each district to prepare a full report of each year's work and forward it to the Provincial Government.

CO-OPERATION.

Under this Act, too, the principle of co-operation is recognised. Upon application to the Minister, ten or more residents and bonâ fide farmers may engage in and carry on, on a co-operative basis, any of the following, viz.: (a) A Farmers' Exchange, for buying and selling farm produce ; (b) a cheese factory ; (c) a creamery ; (d) a fruit-canning, preserving or evaporating factory ; (e) a mutual credit association for the purpose of receiving deposits and loaning money to its members ; (f) or in any other enterprise that may be approved by the Lieutenant-Governor in Council as coming among the objects and within the meaning of the Act, and such applicants are constituted provisional directors under the Act for managing the affairs of the Association until the first annual election of officers, and possess all the powers of an incorporated company under the "Companies Act," Part I., the "Companies Act, 1862" (Imperial), to hold property, to sue and be sued, make by-laws, and to do all things necessary and pertinent to the carrying on of any business for the mutual benefit and profit of the members subscribing and holding stock : Provided, among other things : (a) That a notice of incorporation containing the names of such applicants be published in the British Columbia Gazette, for which a fee of $10 shall be charged ; (b) that no subscriber may hold or hereafter acquire more than one-tenth of the stock allotted by the Association ; (c) that 25 per cent. of the capital stock be subscribed at the time of making application.

BOARD OF HORTICULTURE.

One more important feature of legislation is the Act providing for the creation of a Board of Horticulture, which has very comprehensive powers with respect to the inspection of orchards, imported nursery stock and fruits. The Board is composed of three members—one representing the Island of Vancouver, one the Lower Mainland, one the Interior or Upper Country, with the Secretary, who is also the Deputy Minister of Agriculture, and the Minister of Agriculture, acting *ex officio*. This Board, which is purely of an official character and under the direct control of the Government, has been in existence since 1894, and has not only a strongly preventive influence in the matter of the spread of pestilential forms of disease and pests, but has exerted an educational influence which is very manifest to-day, and in connection with the Farmers' Institutes has done very effective work in the Province.

Another important Act, known as the "Agricultural Credit Societies Act," was passed in 1898, whereby a certain number of farmers, by joining together and pooling their credit, may obtain advances from the Government, with the object of loaning to other members for various purposes of improvement. So far no societies have been formed under this Act, and the rules and regulations authorised thereunder have never been drafted.

RESPECTING DAIRYING.

Provision is made by the "Dairymen's Association Act" for the formation : (a) Of a Provincial Dairymen's Association, for the general advancement of dairying throughout the Province ; (b) the local Dairymen's Associations, known as Cheese and Butter Associations, for the purpose of carrying on the manufacture of cheese and butter ; and (c) the establishment of creameries on the co-operative system, which, when so established, may, on compliance with the Act, obtain Government aid by way of loans to the extent of one-half the cost of creamery buildings, plants and fixtures, such loans to bear interest at the rate of 5 per cent. and be repayable in eight instalments, the first of such instalments to become due at the expiration of three years from the date of the loan, and the other seven instalments annually thereafter. The Provincial Dairymen's Association is at present doing good work in the importation of thoroughbred stock from the East. By the co-operation of the Government the purchase price of the stock is guaranteed, and the cattle sold locally to purchasers.

By the "Milk Fraud Act" of 1895, provision is made for the "prevention of" adulterated milk and the furnishing of adulterated milk to dairies and creameries.

ANIMALS AND CATTLE ACTS.

The "Animals Act" contains provisions restricting the running at large of certain animals, to prevent injury by dogs, and for the arrest and sale of animals unlawfully at large. It is also enacted that in any action brought to recover damages for injury done by animals of a domestic nature, it shall not be necessary to prove that the owner of the animal knew, or had means of knowing, that the animal causing the injury was vicious or mischievous, or accustomed to do acts causing injury.

There are a series of Acts dealing with cattle in various ways. The "Cattle Farming Act" makes a provision whereby the owners of cattle may entrust them to a farmer for the purpose of securing their care and increase. Under a registered agreement the registered cattle are protected from all claims against and liabilities of the farmer to whom they are entrusted. The "Cattle Lien Act" confers upon agisters of cattle and animals and keepers of livery stables a lien upon cattle and effects left with them, for the value and price of any food, care, attendance, etc. The "Cattle Act" contains elaborate provisions for the protection and marking of cattle, under which registries are established and a mode provided for the registration of brands and marks upon cattle. Penalties are provided for the contravention of the Act; also provision for the mode of transfer of brands and marks, for the inspection of hides, and for a record of cattle and sheep from east of the Cascades into other portions of the Province, so as to guard against stealing. Under the "Breeding Stock Act," the "Cattle Ranges Act," and the "Island Pasturage Act" provision is made for the protection and preservation of cattle ranges, and for their being rendered available on an equitable basis for the use of Provincial settlers.

PREVENTIVE LEGISLATION.

Under another series of Acts for the regulation of various matters there is: (a) the "Contagious Diseases Act," which provides for the appointment of inspectors for the inspection of cattle and for quarantine, and, whenever necessary, the destruction of cattle infected with disease. The provisions of the Act are especially directed towards the prevention and the eradication of tuberculosis and pleuro-pneumonia, and against the transmission of disease by milk; (b) the "Line Fences and Water-courses Act," which provides for the appointment of fence-viewers, the construction and maintenance of boundary fences and ditches, and the settlement of disputes between owners in regard to such matters; (c) the "Fence Act," in which is contained a definition of a lawful fence, and the trespass of cattle in case of land protected and unprotected by lawful fences; (d) the "Bush Fire Act," establishing fire districts and the regulation of fires in fire districts, by which, except under certain conditions, it shall not be lawful for any person to set fires between the first day of May and the first day of October, and also the equipment of locomotive engines with the most improved and effectual means to prevent the escape of fire from furnaces, ash heaps and smoke stacks; (e) an Act for the better regulation of traffic on highways, providing the the passing of vehicles to the left, preservation of roadways west of the Cascade Range, by wide tires for loads over 2,000 pounds, and the prevention of certain unlawful practices which impede traffic or render it dangerous; (f) the "Thistle Prevention Act," and the "Noxious Weeds Prevention Act," the objects of which are sufficiently indicated by their titles; and (g) "An Act for the Extermination of Wild Horses," by which it is made lawful for any person to kill unbranded stallions running at large upon the public lands, provided that such person has first obtained a licence for killing, and has made unsuccessful efforts and reasonable endeavour to capture such stallion, and reports the facts of killing to the nearest Government Agent.

LAND ACTS.

The Acts relating to settlement and taking up of land are dealt with under the general title of land laws. These include the "Homestead Act," "Land Registry and Amendments," the "Land Clauses Consolidation Act," and the amended "Land Act," and the "Water Clauses Consolidation Act," as amended in 1899 and 1900. The "Land Clauses Consolidation Act" contains complete procedure relative to the acquisition of lands required for undertakings or works of a public nature, and in this respect is allied with the "Water Clauses Consolidation Act," which makes provision for the acquirement and regulation of water rights for a number of purposes, including ordinary domestic and agricultural purposes. The provisions of these Acts are too elaborate to be even briefly summarised here.

DYKING AND DRAINAGE.

In view of the fact that there exists throughout the Province large tracts of land which could be rendered available for cultivation by dyking and drainage, careful and extended provision is made for such work in the "Drainage, Dyking and Irrigation Act" for the appointment of commissioners to undertake and carry out such works. By the "Dyking Debenture Loan Act" of 1897 the Government of British Columbia is given authority to authorise the redemption of certain debentures for the construction of dyking works, and to authorise expenditure of additional moneys in strengthening, extending and repairing certain dykes. By a further Act in 1898 the powers of the Government in respect to dyking matters were still further extended, whereby they assumed a liability exceeding $500,000 in respect to the dyking works in Maple Ridge, Sumas, Coquitlam, Pitt Meadows and Matsqui, and of borrowing a further sum of $225,000 with respect to further dyking work in Chilliwhack, Agassiz, Hatzic, Surrey, and New Westminster District generally. These works are of an important character, and have made fit for cultivation various tracts of land, aggregating about 100,000 acres, otherwise subject to overflow.

Altogether, the Province expended $981,219 on the work of reclamation up to November, 1904, and the owners of lands benefited had only paid a small proportion of the assessments which the Government levied to recoup the Province, and these owners represented that the amount payable yearly for interest and sinking fund was more than they could pay, and asked for relief from their burden. Consequently, the Government introduced a Bill at the Session of 1905, which greatly reduced the capital charges against the various municipalities and fixed the rate of interest payable thereon at 3½ %, payable yearly. Repayment of the reduced principal was spread over forty years in the case of Maple Ridge, Coquitlam, Pitt Meadows and Chilliwhack, and forty-three years in the case of Matsqui, with 3½ % interest, payable yearly, after January, 1908. The reductions made by the Act of 1905 are as follows :—

	Total Expenditure by Province.	Amounts to be collected under Act of 1905.	Reductions.
Maple Ridge	$221,981 96	$127,396 00	$ 94,585 96
Coquitlam	151,280 35	57,988 00	93,292 35
Pitt Meadows	88,873 11	34,868 00	54,005 11
Matsqui	209,915 60	125,000 00	84,915 60
Sumas	19,268 29	....	19,268 29
Chilliwhack	289,899 79	200,000 00	89,899 79
Total	$981,219 10	$545,252 00	$435,967 10

Copies of any of the above-mentioned Acts may be had on application to the King's Printer, Victoria, B. C. Price, 25 cents per copy.

LAND.

CROWN LANDS, where such a system is practicable, are laid off and surveyed into quadrilateral townships, containing thirty-six sections of one mile square in each.

Any person, being the head of a family, a widow, or single man over the age of eighteen years, and being a British subject, or any alien, upon his making a declaration of his intention to become a British subject, may, for agricultural purposes, record any tract of unoccupied and unreserved Crown lands (not being an Indian settlement) not exceeding one hundred and sixty acres in extent.

No person can hold more than one pre-emption claim at a time. Prior record of pre-emption of one claim and all rights under it are forfeited by subsequent record or pre-emption of another claim.

Land recorded or pre-empted cannot be transferred or conveyed until after a Crown grant has been issued.

Such land, until the Crown grant is issued, is held by occupation. Such occupation must be a bonà fide personal residence of the settler or his family.

The settler must enter into occupation of the land within thirty days after recording, and must continue to occupy it.

Continuous absence for a period longer than two months consecutively of the settler or family is deemed cessation of occupation ; but leave of absence may be granted not exceeding six months in any one year, inclusive of two months' absence.

Land is considered abandoned if unoccupied for more than two months consecutively.

If so abandoned, the land becomes waste lands to the Crown.

The fee on recording is two dollars (8s.).

The settler shall have the land surveyed at his own instance (subject to the rectification of the boundaries) within five years from the date of record.

After survey has been made, upon proof in declaration in writing of himself and two other persons of occupation for two years from date of pre-emption, and of having made permanent improvement on the land to the value of two dollars and fifty cents per acre, the settler on producing the pre-emption certificate obtains a certificate of improvement upon payment of a fee of $2.

After obtaining the certificate of improvement and paying for the land, the settler is entitled to a Crown grant in fee simple. He pays $10 therefor.

The price of Crown lands pre-empted is $1 (4s.) per acre, which must be paid in four equal instalments, as follows : First instalment two years from date of record or pre-emption, and yearly thereafter, but the last instalment is not payable till after the survey, if the land is unsurveyed.

Two, three or four settlers may enter into partnership with pre-emptions of 160 acres each, and reside on one homestead. Improvements amounting to $2.50 per acre made on some portion thereof will secure Crown grant for the whole, conditions of payment being same as above.

STERLING'S CHERRY ORCHARD, KELOWNA.

PEACH ORCHARDS, PEACHLAND.

The Crown grant reserves to the Crown a royalty of five cents per ton on every ton of merchantable coal raised or gotten from the land, not including dross or fine slack, and 50 cents per M. on timber. Coal and petroleum lands do not pass under grant of lands acquired since passage of Land Act Amendment of 1899.

No Crown grant can be issued to an alien who may have recorded or pre-empted by virtue of his declaring his intention to become a British subject, unless he has become naturalised.

The heirs of devisees of the settler are entitled to the Crown grant on his decease.

Crown lands may be purchased to the extent of 640 acres, and for this purpose are classified as first, second and third class, according to the report of the surveyor.

Lands which are suitable for agricultural purposes, or which are capable of being brought under cultivation profitably, or which are wild hay meadow lands, rank as and are considered to be first class lands. Lands which are suitable for agricultural purposes only when artificially irrigated, and which do not contain timber valuable for lumbering purposes, as defined below, rank as and are considered to be second class lands. Mountainous and rocky tracts of land which are wholly unfit for agricultural purposes, and which cannot under any reasonable conditions, be brought under cultivation, and which do not contain timber suitable for lumbering purposes, as defined below, or hay meadows, rank as and are considered to be third class or pastoral lands. Timber lands (that is, lands which contain milling timber to the average extent of eight thousand feet per acre west of the Cascades, and five thousand feet per acre east of the Cascades, to each one hundred and sixty acres) are not open for sale.

The minimum price of first-class land, $5 per acre; second class, $2.50 per acre; third class, $1 per acre. No settlement duties are required on such lands unless a second purchase is contemplated. In such a case, the first purchase must be improved to the extent of $5 per acre for first class ; $2.50 second class ; and $1 third class.

Leases of Crown lands which have been subdivided by survey in lots not exceeding twenty acres may be obtained ; and if requisite improvements are made and conditions of the lease fulfilled at the expiration of lease, Crown grants are issued.

Leases (containing such covenants and conditions as may be thought advisable) of Crown lands may be granted by the Lieutenant-Governor in Council for the following purposes :—

(a.) For the purpose of cutting hay thereon, for a term not exceeding ten years :

(b.) For any purpose whatsoever, except cutting hay as aforesaid, for a term not exceeding twenty-one years.

The farm and buildings, when registered, cannot be taken for debt incurred after registration ; and it is free from seizure up to a value not greater than $500 (£100 English). Cattle "farmed on shares" are also protected by an Exemption Act.

The fact of a person having a homestead in other Provinces, or on Dominion Government lands in this Province, is no bar to pre-empting Crown lands in British Columbia.

Twenty-one year timber leases are now subject to public competition, and the highest cash bonus is accepted, subject to the 50 cents per M. royalty above mentioned, and an annual rental in advance of 15 cents per acre. The holder must put up a saw-mill capable of cutting not less than 1,000 feet of lumber per day of twelve hours for every 400 acres of land in such lease ; and such mill shall be kept running for at least six months in every year.

How to Secure a Pre-emption.

Any person desiring to pre-empt unsurveyed Crown lands must observe the following rules :—

1. Place a stake or post four or more inches square and four or more feet high—a tree stump squared and of the proper height will do—at each corner of the claim, and mark upon each of the posts his name and a description of the post, for example :—

"John Smith's land, N. E. post (meaning north-east post) ; John Smith's land, N. W. post," and so on.

2. After staking the land, the applicant must make an application in writing to the Land Commissioner of the district in which the land lies, giving a full description of the land, and a sketch plan of it ; this description and plan to be in duplicate. The fee for recording is $2.

3. He shall also make a declaration, in duplicate, before a Justice of the Peace, Notary Public, or Commissioner, in Form 2 of the Land Act, and deposit same with his application. In the declaration he must declare that the land staked by him is unoccupied and unreserved Crown land, and not in an Indian settlement ; that the application is made on his own behalf and for his own use for settlement and occupation, for agricultural purposes, and that he is duly qualified to take up and record the land.

4. If the land is surveyed the pre-emptor must make application to the Commissioner exactly as in the case of unsurveyed lands, but it will not be necessary to plant posts.

5. Every pre-emption shall be of a rectangular or square shape, and 160 acres shall measure either 40 chains by 40 chains—880 yards by 880 yards, or 20 chains by 80 chains—440 yards by 1,760 yards ; 80 acres shall measure 20 chains by 40 chains ; and 40 acres, 20 chains by 20 chains. All lines shall be run true north and south and true east and west.

6. When a pre-emption is bounded by a lake or river, or by another pre-emption or by surveyed land, such boundary may be adopted and used in describing the boundaries of the land.

7. Thirty days after recording the pre-emptor must enter into occupation of the land and proceed with improving same. Occupation means continuous bonâ fide personal residence of the pre-emptor or his family, but he and his family may be absent for any one period not exceeding two months in any year. If the pre-emptor can show good reason for being absent from his claim for more than two months, the Land Commissioner may grant him six months' leave. Absence without leave for more than two months will be looked upon as an abandonment of all rights and the record may be cancelled.

8. No person can take up or hold more than one pre-emption.

9. The pre-emptor must have his claim surveyed, at his own expense, within five years from the date of record.

10. The price of pre-empted land is $1 per acre, to be paid for in four equal annual instalments of 25 cents per acre, the first instalment to be paid two years after record.

11. After full payment has been made the pre-emptor shall be entitled to a Crown grant of the land, on payment of a fee of $10.

12. A pre-emption cannot be sold or transferred until after it is Crown-granted.

67

GOVERNMENT LAND AGENCIES.

The following is a list of Government Agents with whom pre-emptions may be filed. Lands in outlying districts, in which there is no resident agent, are dealt with in the Lands and Works Department, Victoria, Neil F. Mackay, Esq., Assistant Commissioner:

DISTRICT.	GOVERNMENT AGENT.	ADDRESS.
ALBERNI	A. L. Smith	Alberni.
NANAIMO	M. Bray	Nanaimo.
NEW WESTMINSTER	C. C. Fisher	New Westminster.
EAST KOOTENAY (Northern)	J. E. Griffith	Golden.
" (Southern)	J. F. Armstrong	Cranbrook.
WEST KOOTENAY :		
Slocan	E. E. Chipman	Kaslo.
Nelson	R. A. Renwick	Nelson.
Revelstoke	Fred Fraser	Revelstoke.
CARIBOO :		
Barkerville	James McKen	Barkerville.
CASSIAR :		
Telegraph Creek	James Porter	Telegraph Creek.
Atlin	J. A. Fraser	Atlin.
Port Simpson	J. Flewin	Port Simpson.
KAMLOOPS	G. C. Tunstall	Kamloops.
Nicola	George Murray	Nicola.
YALE :		
Vernon	L. Norris	Vernon.
Similkameen, Kettle River	C. A. R. Lambly	Fairview.
Clinton	F. Soues	Clinton.
Yale Division	H. P. Christie	Ashcroft.

USE OF WATER.

Under the provisions of the "Water Clauses Consolidation Act, 1897," and amending·Acts, unrecorded water may be diverted from any natural sources for irrigation or agricultural purposes generally. The scale of fees is the same as for water for industrial purposes, and is calculated on a sliding scale. For a record fee of $10.75 per 100 miner's inches up to $110.75 for 500 inches; $260.75 for 1,000 inches; $560.75 for 2,000 inches; $680.75 for 5,000 inches; $880.75 for 10,000 inches, and so on. For industrial purposes there is an annual fee calculated according to the same sliding scale; but no annual fee s charged on water recorded and actually used for agricultural purposes.

LAND TRANSACTIONS.

:0:

Return Showing Land Transactions for the Year ending 31st December, 1905.

Records issued for the District of

	Victoria	Cowichan	Nanaimo	Coast	New West-minster	Yale	Kamloops	Osoyoos	Lillooet	Kootenay	Cariboo	Cassiar	Total
Pre-emption Records	48	11	3	171	65	7	97	323	53	155	93		955
Certificates of Improvement	18	2	2	1	16	3	10	42	6	85	4	8	189
" Purchase	38	5	4	70	183	14	61	835	59	334	98	13	1,637
Crown Grants	49	33	25	154	94	25	42	249	23	347	10	30	1,064
Timber-Cutting Licences	112		3	280	253	7	38	73	905	1008	7		2,166
" " Hand Loggers'													420
Coal Prospecting Licences	42			38		4	28	19	965	968	9	7	508
Sundry Leases													103

SUMMARY.

	1894.	1895.	1896.	1897.	1898.	1899.	1900.	1901.	1902.	1903.	1904.	1905.
Pre-emption Records	709	630	486	462	467	616	728	646	655	758	885	955
Certificates of improvement	117	113	144	904	144	113	107	168	143	123	909	189
Certificates of purchase	153	334	694	977	765	418	399	551	565	510	969	1,637
Crown grants	159	215	411	766	951	868	1,101	912	1,059	860	953	1,064
Total acreage deeded	47,166.86	95,456	36,821	609,597	371,394	672,148.12	104,723.91	4,632,831.87	87,906.79	83,699.49	107,384.83	133,184.46
Letters received	4,018	5,079	6,532	8,034	9,126	10,993	12,943	13,806	13,546	14,001	15,141	16,699
Timber-cutting licences			68	78	87	87	143	129	525	1,307	1,461	2,166
" hand loggers'		193	189	93	621	309	158	163	190	259	188	420
Coal prospecting licences								223	296	434	960	508
Sundry leases		23	6	10	22	12	39	43	71	106	62	103
Cash received	$73,487.08	$88,315.10	$116,205.83	$107,353.87	$188,256.69	$131,280.90	$179,599.28	$213,329.08	$313,375.01	$473,141.12	$625,070.32	$728,119.07

DOMINION GOVERNMENT LANDS.

All the lands in British Columbia within twenty miles on each side of the Canadian Pacific Railway main line are the property of Canada, with all the timber and minerals they contain (except precious metals). This tract of land, with its timber, hay, water-powers, coal and stone, is now administered by the Department of the Interior of Canada, practically according to the same laws and regulations as are the public lands in Manitoba and the Territories. Dominion Government Agencies are established at Kamloops and New Westminster.

Any British subject who is the sole head of a family, or any male of the age of 18 years, may secure a homestead of 160 acres on any unoccupied land, on application to the local land agent and on payment of a fee of $10. The homesteader must reside on the land for six months in every year, and cultivate at least 15 acres for three years, when he will be entitled to a free grant or patent.

During the year ending June 30th, 1905, Dominion lands in the Railway Belt were disposed of as follows : —

```
Kamloops Agency—
    Homestead entries .................149=  23,840 acres.
    Lands sold ...........................   4,160   "
New Westminster Agency—
    Homestead entries ............ 40=   6,400   "
Grazing lands—
    Leases .....................................391,580   "
Coal lands (Rocky Mountain Park)—
    Twenty-three licences covering.................   8,436   "
```

CANADIAN PACIFIC LANDS.

The Canadian Pacific Railway Company controls large areas of farming, fruit, ranching and timber lands in the Kootenay and Boundary Districts. Generally speaking the prices for agricultural lands are as follows : —

First Class Lands.—Lands suitable for agricultural purposes in their present condition, or which are capable of being brought under cultivation profitably by the clearing of the timber thereon, or which are wild hay meadow lands. Price, $5 per acre.

Second Class Lands.—Lands which are suitable for agricultural purposes only when irrigated. Price, $2.50 per acre.

Third Class Lands.—Mountainous and rocky tracts of land, unfit for agricultural purposes, and which cannot under any reasonable condition be brought under cultivation. Price, $1 per acre.

Any land in the Columbia and Western Land Grant (Boundary District) which contains timber fit for manufacture into lumber to the extent of 3,000 feet board measure, to the acre, does not come under the heading of agricultural land, but will only be disposed of under the provisions of the Company's regulations for the sale or lease of timber lands. In the remaining grants the limit for agricultural lands is fixed at 5,000 feet, board measure, to the acre.

The minimum area sold is 160 acres, and all lands must be purchased in square or rectangular parcels, viz., 160 acres must measure 40 chains by 40 chains ; 320 acres must measure 80 chains by 40 chains ; and 640 acres must measure 80 chains by 80 chains.

Land sold at $1 per acre must be paid for one-fourth cash, and the balance in there equal annual instalments.

Land sold at $2.50 per acre must be paid for one-fifth cash, and the balance in four equal annual instalments.

Land sold at $5 per acre must be paid for one-eighth cash, and the balance in seven equal annual instalments.

Interest at six per cent. is payable on all outstanding amounts of principal, and also on overdue instalments. If land is paid for in full at the time of purchase, a discount of ten per cent. will be allowed on the amount so paid in excess of the usual cash instalment, but no reduction will be allowed on subsequent payment of instalments in advance of maturity. All payments on account of the purchase of lands from this Company, must be remitted direct to the office of the British Columbia Land Commissioner for the Canadian Pacific Railway at Calgary, Alberta ; no agent for the Company being allowed to receive or receipt for money, or to bind the Company by any act whatsoever.

The purchaser of agricultural land will be permitted to use what timber is actually required on the land purchased by him for buildings, fences and fuel ; but any timber cut for sale will be subject to the payment of dues as per the following schedule :—

Lumber, per M feet, B. M. $1 00
Shingle bolts, per cord.................................. .... 1 00
Firewood, per cord 25
Fence posts, per cord 50
Mining props (10 ft. x 10 in. or less), per cord.................... 50
Mining props (larger), each.. 05
Ties, each 02
House logs (20 ft. or less), each 10
Piles, cribbing, timber, telegraph posts, per running foot.......... .... . ½

Such dues are exclusive of all Government royalties, which must be paid by the purchaser. In the case of unsurveyed lands, the purchaser must arrange his own surveys.

The full amount of such of the above dues as are paid to the Company after the payment of the first instalment, on any contract for the purchase of their agricultural lands which may have been in force subsequent to the date of the 1st of October, 1905, will be applied to the credit of the second and succeeding instalments until the full purchase price of the land is made up.

One half of the amount paid by new settlers for fare on the railway lines of the Canadian Pacific Railway in travelling to British Columbia will be applied on account of the first instalment if land is purchased from the Company in this Province.

Timber leases may be secured from the Canadian Pacific Railway Company for timbered areas within the land grant of the Columbia and Kootenay Railway, and timber lands which form part of the British Columbia Southern Land Grant are sold, the rates payable to the Company in each case being as follows :—

Lumber, per M feet, B. M. $1 00
Shingle bolts, per cord... 1 00
Firewood, per cord 25
Fence posts, per cord 50
Mining props (10 ft. x 10 in. or less), per cord..... 50
Mining props (larger), each............. 05
Ties, each 02
House logs (20 ft. or less), each 10
Piles, cribbing, telegraph poles, per running foot .... ½

In addition to these dues, the lessee must pay all the Government royalties and taxes, and arrange and bear the expense of any surveys which may prove necessary to define his limits.

More detailed particulars regarding the Company's agricultural and timber lands can be obtained from J. S. Dennis, Assistant to the Second Vice-President, Calgary, Alberta, and from any of the following local land agents of the Company :—

East Kootenay (Central)—R. R. Bruce, Wilmer.

East Kootenay (Southern)—E. Mallandaine, Creston ; V. Hyde Baker, Cranbrook ; I. H. Wilson, Wardner ; J. Austin, Elko.

West Kootenay—H. M. Bird, Nelson ; W. J. Devitt, Trail ; Thos. Abriel, Nakusp.

Yale District—J. A. McCallum, Grand Forks ; F. W. McLaine, Greenwood ; J. R. Mitchell, Penticton.

Kamloops District—Sibbald and Field, Revelstoke ; F. J. Fulton, Kamloops.

The Company is also interested in the following townsites, where local agents may be consulted as to price of lots :—Elko, Cranbrook, Kimberly, Proctor, Nelson, Lemonton, Nakusp, Arrowhead, Revelstoke, Kamloops, Donald, Gerrard, Castlegar, Cascade, Eholt, Grand Forks, Greenwood, and Midway.

E. & N. LANDS.

The Esquimalt & Nanaimo Railway Company owns 1,500,000 acres of agricultural, timber and mineral lands on Vancouver Island, extending from Otter Point on the south-west coast to Crown Mountain in the Comox District, which include within their boundaries all the flourishing farming, mining, lumbering and fishing communities along the East Coast and the line of the Esquimalt & Nanaimo Railway, a tract recognised to be the choicest portion of Vancouver Island. This magnificent estate is being systematically explored by the company, whose intention it is to clear the land of timber and divide it into convenient sized lots, when it will be offered for sale to fruit-growers, farmers, poultry and dairymen, at reasonable prices and on favourable terms. As the interior is explored it is the intention of the company to extend the railway and build branches into the most desirable valleys, to afford easy access to the agricultural, timber and mineral lands.

Fuller information regarding these lands may be had by application to the Land Department, Esquimalt & Nanaimo Railway Company, Victoria, B. C.

LAND COMPANIES.

In addition to Government and Railway lands, there are several incorporated companies owning large tracts of land which are being sold in small blocks, from five acres upwards, for fruit-growing, poultry-raising and mixed farming. The prices asked by these companies for irrigated land varies from $75 to $250 per acre. Detailed information may be obtained from the managers of these companies. Some of the principal land companies are :—The Coldstream Estate, Ltd., Vernon, B. C. ; White Valley Irrigation & Power Co., Vernon, B. C. ; Kelowna Land & Orchard Co., Kelowna, B. C ; Southern Okanagan Land Co., Penticton, B. C. ; Kettle Valley Irrigated Fruit Lands Co., Vancouver, B. C. ; Fruitlands, Ltd., Kamloops, B. C. ; Peachland Townsite &

Irrigation Co., Peachland, B. C. ; Summerland Development Co., Summerland, B. C. ; Settlers' Association of British Columbia, Vancouver, B. C. ; Canadian Real Properties Ltd., Kamloops, B. C.

PRICES OF LAND.

Apart from the Government and railway company's lands, there is a great deal of desirable land owned by companies and individuals, the price of which varies with locality, quality of soil and cost of clearing or irrigation.

For purposes of comparison the topography and climatic conditions seem to lend themselves to a natural division of the province into the following districts —

1. The Upper Mainland.—All the country to the eastward of the Coast Range, and including the large cattle ranges and what is known as the Dry Belt.

2. The Lower Mainland.—All that portion of the sea coast to the westward of the Coast Range, and including the rich delta lards of the Fraser River. This part of the country is generally heavily wooded with big timber and is the wettest part in the Province.

3. The Islands.—All that portion including Vancouver Island and the adjacent islands. This division partakes somewhat of the characteristic of the two others, and resembles the first in the distribution of the flora and the less precipitation.

Division No. 1 includes the Boundary Country, Similkameen, Okanagan Lake, Okanagan, Shuswap Lake, Thompson River Valley (upper and lower), Nicola, Upper Fraser Valley, Chilcotin and Cariboo Waggon Road. Improved or partly cleared land in the Boundary District is held at about $50 per acre. Similkameen, $25 to $150, the latter being irrigated. Okanagan Lake, $60 to $250 for water fronts, irrigated and improved land, and from $1 to $25 for non-irrigated. Okanagan bush land, $5 to $20 ; partly cleared and improved, $10 to $50, and up to $100 per acre. Shuswap and Upper Thompson Valley, prices about the same as Okanagan. Land may be bought at lower rates than those quoted in Nicola, Upper Fraser Valley, Chilcotin and Cariboo. It is hard to give definite figures as the country is so extensive and conditions are so varied.

Division No. 2 includes Delta, Surrey, Langley, Matsqui, Sumas, Chilliwack, South Vancouver, Burnaby, Coquitlam, Maple Ridge, Mission, Dewdney, Nicomen and Kent, and prices of land vary very much. The improved land is held at $5 to $20, while reclaimed (dyked) land sells from $40 up to $100.

Division No. 3 embraces Victoria, Esquimalt, Metchosin, Sooke, Highland, Lake, Saanich, Cowichan, Nanaimo, Comox, Alberni, San Juan and Fort Rupert Districts and the numerous islands of the Gulf of Georgia. As in other parts of the Province, there are no fixed prices for land. They vary with locality and the estimates of the owners. Wild land, mostly heavily timbered, can be bought from $3.50 to $10 per acre, while improved land ranges all the way from $20 to $200 according to extent and value of improvement.

While some of these prices may be thought high, the cost of clearing the land of timber must be considered, also, that a small farm well located and well tilled in British Columbia will produce more and return bigger profits than a much larger area of land in most other countries.

ROCK CUT, VASEAUX LAKE.

ORCHARD NEAR QUAMICHAN LAKE, VANCOUVER ISLAND.

GENERAL INFORMATION.

TAXATION.

Outside of incorporated cities, towns and municipalities, the taxation is imposed and collected directly by the Provincial Government and expended in public improvements, roads, trails, wharves, bridges, etc., in assisting and maintaining the schools, in the administration of justice.

The rates of taxation imposed by the latest Assessment Act are as follows :—

```
On Real Estate ............. 3-5 per cent. of assessed value of $2,000
    "       "        ........... 1 per cent. of assessed value over $2,000
    " Wild Land . ........  ....................  ........ 4 per cent.
    " *On Coal Land, Class A........................... 1    "
    " **Coal Land, Class B........  .................... 2    "
    " Timber Land .............  ................  ..... 2    "
    " Income of $2,000 or under ....................... 1½   "
    "    "    over $2,000 and not exceeding $3,000....... 1¾  "
    "    "    "   3,000      "         4,000........ 2   "
    "    "    "   4,000      "         7,000........ 3   "
    "    "    "   7,000 ............................ 4   "
```

Discounts of 10 per cent. upwards are allowed for prompt payment of taxes, and the following exemptions from taxation are granted :—

On Personal Property up to $500 (to farmers only).
" Income up to $1,000.
" Pre-empted land for two years from date of record and an exemption of $500 for four years after record.

In addition to above taxes royalty is charged on coal, timber and minerals.

---

*Working Mines.
**Unworked Mines.

EDUCATION.

The Province affords excellent educational opportunities. The School System is free and non-sectarian, and is equally as efficient as that of any other Province in the Dominion. The expenditure for educational purposes amounts to $400,000 annually. The Government builds a school house, makes a grant for incidental expenses, and pays a teacher in every district where twenty children between the ages of six and sixteen can be brought together. For outlying farming districts and mining camps the arrangement is very advantageous. High Schools are also established in cities, where classics and higher mathematics are taught. Several British Columbia cities also now have charge of their own Public and High Schools, and these receive a very liberal per capita grant in aid from the Provincial Government. The minimum salary paid to teachers is $50 per month in Rural Districts, up to $150 in City and High Schools. Attendance in Public Schools is compulsory. The Education Department is presided over by a Minister of the Crown. There are also a Superintendent and four Inspectors in the Province, also Boards of Trustees in each District. According to the last Educational Report, there are 361 schools in operation, of which 13 are High, 65 Graded and 283

Common. The number of pupils enrolled in 1905 was 27,335, and of teachers, 663. The Public School System was established in 1872, with 28 schools, 28 teachers, and 1,028 pupils. Its growth proves that education has not been neglected in British Columbia.

The High Schools are distributed as follows :—Victoria (Victoria College), Vancouver (Vancouver College), New Westminster, Nanaimo, Nelson, Rossland, Cumberland, Vernon, Kaslo, Chilliwack, Grand Forks, Kamloops and Revelstoke. There is a Provincial Normal School at Vancouver, and many excellent private colleges and boarding schools. Victoria and Vancouver Colleges are affiliated to McGill University, Montreal, and have High School and University departments.

SOCIAL CONDITIONS.

The population of British Columbia, widely scattered and composed of many nationalities, is singularly peaceful and law-abiding. Life and property are better protected and individual rights more respected in the isolated mining communities than in some of the great centres of civilization. The province, though new as compared with older countries, enjoys all the necessaries and many of the luxuries and conveniences of modern life. There are few towns which are not provided with waterworks, electric lights and telephones. The hotels are usually clean and comfortable, and the stores well stocked with every possible requirement. There is little individual poverty. A general prosperity is the prevailing condition throughout the country, for none need be idle or penniless who is able and willing to work. The larger towns are well supplied with libraries and reading rooms, and the Provincial Government has a system of travelling libraries, by which the rural districts are furnished free with literature of the best description.

The spiritual welfare of the people is promoted by representatives of all the Christian denominations, and there are few communities, however small, which have not one or more churches with resident clergymen.

All the cities and larger towns have well equipped hospitals, supported by Government grants and private subscriptions, and few of the smaller towns are without cottage hospitals. Daily newspapers are published in the larger places, and every mining camp has its semi-weekly or weekly paper.

ADVICE TO IMMIGRANTS.

There is no country within the British Empire which offers more inducements to men of energy and industry than British Columbia. To the practical farmer, miner, lumberman, fisherman, horticulturist and dairyman it offers a comfortable living and ultimate independence, if he begins right, perseveres and takes advantage of his opportunities. The skilled mechanic has also a good chance to establish himself and the labourer will scarcely fail to find employment. The man without a trade, the clerk, the accountant and the semi-professional, is warned, however, that his chances for employment are by no means good. Much depends upon the individual, for where many fail one may secure a position and win success, but men in search of employment in offices or warehouses, and who are unable or unwilling to turn their hands to any kind of manual labour in an emergency, would do well to stay away from British Columbia unless they have sufficient means to support themselves for six months or a year while seeking a situation.

The class of immigrants whose chances of success are greatest is the man of small or moderate means, possessing energy, good health and self-reliance, with the faculty of adaptability to his new surroundings. He should have at least £300 ($1,500) to £500 ($2,500) on arrival in the Province, sufficient to "look around" before locating permanently, make his first payment on his land and support himself and family while awaiting returns from his first crop. This applies to a man taking up mixed farming. It is sometimes advisable for the new comer to work for wages for a time until he learns the "ways of the country."

To avoid the risk of loss the immigrant from Great Britain should pay the money not wanted on the passage to the Dominion Express Company's office in London, Liverpool or Glasgow, and get a money order payable at any point in British Columbia ; or he may pay his money to any bank in London having an agency in British Columbia, such as the Bank of Montreal, Canadian Bank of Commerce, Bank of British North America, Imperial Bank, etc. This suggestion applies with equal force to persons coming from Eastern Canada or the United States.

United States currency is taken at par in business circles.

The Provincial Government Agent at point of arrival will furnish information as to lands open for settlement, farms for sale, rates of wages, etc.

SETTLERS' EFFECTS FREE.

Settlers' effects, viz. :—Wearing apparel, books, usual and reasonable household furniture and other household effects ; instruments and tools of trade, occupation or employment ; guns, musical instruments, domestic sewing machines, typewriters, bicycles, carts, waggons, and other highway vehicles ; agricultural implements, and live stock for the farm, not to include live stock or articles for sale, or for use as a contractor's outfit, nor vehicles nor implements moved by a mechanical power, nor machinery for use in any manufacturing establishment ; all the foregoing, if actually owned abroad by the settler for at least six months before his removal to Canada, and subject to regulations by the Minister of Customs : Provided that any dutiable articles entered as settlers' effects may not be so entered unless brought by the settler on his first arrival, and shall not be sold or otherwise disposed of without payment of duty until after twelve months' actual use in Canada.

A settler may bring into Canada free of duty live stock for the farm on the following basis, if he has actually owned such live stock abroad for at least six months before his removal to Canada and has brought them into Canada within one year after his first arrival, viz. :—If horses only are brought in, 16 allowed ; if cattle only are brought in, 16 allowed ; if sheep only are brought in, 60 allowed ; if swine only are brought in, 60 allowed. If horses, cattle, sheep and swine are brought in together, or part of each, the same proportions as above are to be observed. Duty is to be paid on the live stock in excess of the number above provided for. For customs entry purposes, a mare with a colt under six months old is to be reckoned as one animal ; a cow with a calf under six months old is also to be reckoned as one animal.

HOW TO REACH BRITISH COLUMBIA.

From the United Kingdom.—Several lines of steamships ply between British and Canadian ports, and full and reliable information regarding routes, rates of passage, etc., can be obtained at the office of the Agent-General of British Columbia, Salisbury House, Finsbury Circus, London ; the office of the High Commissioner for Canada, 17, Victoria Street, London, S. W. ; the office of the Canadian Commissioner of Emigration, 11-12. Charing Cross, London, W. C. ; or to the Dominion Government Agents at Birmingham, Cardiff, Liverpool, Dublin, Belfast or Glasgow.

From the United States through tickets may be bought to any point in British Columbia over any of the transcontinental railways and their branches and connections.

The Government of British Columbia does not aid immigration in the way of assisted passages or special immigrant tickets on steamship and railway lines. Applications for special immigrant rates should be made directly to the agents of the steamship or railway companies.

METEOROLOGICAL.

The following table shows the annual rainfall and snowfall, and the highest, lowest and average temperature at forty stations :—

	Annual Rainfall. Inches.	Annual Snowfall. Inches.	Highest Temperature. Degrees.	Lowest Temperature. Degrees.	Average Temperature.
Midway	6.72	30	96	– 21	41.2
Princeton	9.25	75.2	92	– 26	41.2
Okanagan (Vernon)	11	37	93	– 13	44.7
Griffin Lake	52.30	133	110	– 18	....
Kamloops	8.25	37.2	96	– 10.7	47.5
Nicola Lake	8.73	·46.5	87.5	– 15.5	42.2
Spence's Bridge	6.87	82.08	104	– 13	....
Lillooet	5 to 8	35 to 60	85 to 95	– 10 to – 20	....
Barkerville	20	120	82	– 28	34.2
Stuart Lake	8.51	74.8	88	– 39	33.2
Golden	13	100	88.5	– 16.5	...
Tobacco Plains	14.54	41.4	91	– 25	42.6
Cranbrook	17.38	....	....	....	....
West Kootenay	18.73	91.9	85	– 8	44.50
Pilot Bay	....	....	90	– 3	46.8
Ladners	33.47	17.5	82	2	....
Chilliwack	59.20	29.3	92	10	49
Matsqui	58.25	20.8	92	8.5	48.9
New Westminster	59.73	35.1	90.7	2.0	48.9
Agassiz	51.88	28	95	1.0	47.5
Nicomen	70.94	13	94	9.0	49.5
Vancouver	64.39	30	86	6.0	48.9
Point Garry	37.72	17	78.8	7.0	47.8
Goldstream Lake	62.60	106.5	....	....	....
Victoria	30.54	16.1	86.2	12.3	50.2
Duncan	50	10.6	87	11	....
Kuper Island	45.20	39.5	95	16	....
French Creek	39.79	7.5	88	11	47.9
Nanaimo	40.36	28.5	90.3	7.3	48.9
Carmanah	112.86	10	70	18	....
Alberni	71.59	36	94.2	12.9	49.3
Clayoquot	146.56	nil.	87	18	48.9
Cape Scott	137.76	"	81.5	16	46.3
Bella Coola	36.20	46	91.5	0	44.9
Point Atkinson	63.23	20.2	....	....	....
Naas Harbour	58.16	17.9	....	....	....
Port Essington	121.10	68.5	....	....	....
Rivers Inlet	105.66	42.8	84.1	16.1	46
Masset	30.4	59.4	78	20	46.2
Port Simpson	71.26	34	74	15.6	46

ALTITUDES.

Altitudes given by various authorities are as follows :—

Fraser River, at Lillooet	700	feet.
Seton Lake	802	"
Pemberton Meadows	695	"
Pavilion	1,357	"
Pavilion Mountain	3,500 to 5,000	"
Fraser River, at Big Bar	1,200	"
" Alkali Lake and mouth of Riskie Creek	1,350	"
Riskie Creek, Chilcotin	2,170	"
Chilcotin Valley, average	2,625	"
Midway	1,850	"
Anarchist Mountain	3,500	"
Grand Forks	1,750	"
Greenwood	2,400	"
Phoenix	3,900	"
Keremeos	1,000	"
Princeton	1,650	"
Okanagan (District)	1,150 to 1,500	"
Griffin Lake	1,000	"
Craigellachie	1,450	"
Sicamous	1,300	"
Notch Hill	1,708	"
Kamloops	1,193	"
Nicola Lake	2,170	"
Spence's Bridge	996	"
Lytton	687	"
Ashcroft	1,508	"
Bridge Creek	3,086	"
Lac la Hache	2,682	"
70-Mile House	3,300	"
Cache Creek, about	1,500	"
Hat Creek	1,686	"
Clinton	2,973	"
Bonaparte Valley, "The Mound"	2,144	"
Bridge Creek and Lac la Hache	1,800	"
Soda Creek Crossing	1,690	"
Fraser River, at Alexandria	1,420	"
" Quesnel	1,490	"
Quesnel	1,700	"
" Forks (Bullion)	2,775	"
Barkerville	4,180	"
Stuart Lake	1,800	"
Tobacco Plains	2,300	"
Goldstream Lake	1,510	"
Victoria	Practically sea level.	
Duncan	"	"
Nanaimo	"	

TABLE OF CONTENTS.

——o——

LIST OF ILLUSTRATIONS.

VICTORIA, B. C.
Printed by RICHARD WOLFENDEN, V.D., I.S.O., Printer to the King's Most Excellent Majesty.
1906.

VICTORIA, B. C.

ed by RICHARD WOLFENDEN, V.D., I.S.O., Printer to the King's Most Excellent Majesty.
1906.

CHALLENGE
OF THE
MOUNTAINS

CONTENTS

ISSUED BY THE
CANADIAN PACIFIC RAILWAY COMPANY
D-07

Canadian Pacific Railway Hotel System

BANFF SPRINGS HOTEL

Some places of interest
near Banff

Buffalo Park	Museum
Lake Minnewanka	Cave and Basin
Spray Falls	Hot Sulphur Springs

Bankhead Coal Mines

The Observatory on Sulphur Mountain

Canadian Pacific Railway Hotel System

LAKE LOUISE FROM THE CHALET

Some places of interest
near Lake Louise

Mount Lefroy and Glacier	Lake Agnes
Victoria Hanging Glacier	Mirror Lake
Valley of the Ten Peaks	Moraine Lake
Saddleback Lookout	Paradise Valley

MOUNT STEPHEN HOUSE, FIELD, B.C.

Some places of interest
near Field

Drive to Emerald Lake	Natural Bridge
Yoho Road Drive	Monarch Mine Cabins
Cathedral Mountain	Mount Stephen
Fossil Beds	Burgess Pass

EMERALD LAKE FROM CHALET

Some places of interest
near Emerald Lake

Lookout Point	Takakkaw Falls
Twin Falls	Summit Lake
Yoho Glacier	Wapta Glacier

Yoho Valley

GLACIER HOUSE, GLACIER, B.C.

Some places of interest
near Glacier

The Great Glacier   Glacier Crest

Mount Abbott   Observation Point

Cougar Valley   Caves of Nakimu

Lake Marion   Mount Sir Donald

The Loops of the Selkirks

THE CHALLENGE
OF THE MOUNTAINS

⚜

"The joy of life is steepness overcome,
And victories of ascent, and looking down
On all that had looked down on us."
—Tennyson.

IN no other country in the world is there such an attractive district to the tourist and the lover of Alpine scenery as in the Provinces of Alberta and British Columbia, along the line of the Canadian Pacific Railway. It is a wonderful land of natural beauty, mountain peaks, rushing rivers, peaceful lakes, stupendous glaciers, remarkable natural phenomena of caves, hot springs, curious formations of rock and ice, interesting flora and animal life, all combined making a holiday district of unequalled attractiveness. It is a land whose boundaries would include fifty Switzerlands, where it has been estimated only one mountain peak out of thousands has ever yet been climbed; for it is the newest of the world's great natural playgrounds, and only that portion contiguous to the railway has yet been fully explored.

New and interesting discoveries are constantly recorded of unknown peaks, beautiful lakes, charming valleys, also new forms of bird and plant life. The Canadian Government has set aside 5,732 square miles as a national park, and the Canadian Pacific Railway

have built in some of the most interesting places a
number of charming châlets and hotels at great cost,
which are furnished in the liberal manner for which
this Company has always been noted in all its departments.   During last season many thousands of people
visited this great park, and each year in ever-increasing
numbers tourists from all over the world are attracted

The Shore of Lake Superior. Wonderful scenery along the
Canadian Pacific Railway.

by this glorious mountain scenery.   Only one regret
is expressed by visitors, and that is, they unfortunately
give themselves too little time to see this charming
country.   A stay of at least several days should be
made at each of the resorts in order to fully realize

the magnificence of the surrounding mountains, which must be viewed under the various atmospheric conditions so as to see the wonderful changes in light and shadow, sunrise and sunset in the Canadian Rockies which, under favourable conditions, are scenes never to be forgotten. Unfortunately the average tourist is all too prone to stop over only between trains and thus catch but a hurried glance of these glorious

The Alpine Club of Canada making an ascent in the Canadian Rockies

peaks, which is regrettable, inasmuch as frequently the greater beauty is missed entirely, though many thousands claim travelling through these mountains without leaving the train as the most enjoyable event and greatest scenic treat of their lives.

Seekers after the grandest in the way of what Nature has provided for man's edification need not be satisfied with repeating

9

the ascents of the well-known peaks of the old world. Edward Whymper, with all the authority born of his conquest of the Matterhorn, and a lifetime spent in scaling the heights of Switzerland, the Andes, and the Himalayas, has declared the Canadian Rockies to be equivalent to "fifty or sixty Switzerlands rolled into one." Here the geologist, the botanist, the mountaineer, the naturalist, the artist, the sportsman, the health or pleasure seeker will find in these mountains a region attractive and beautiful, with many other advantages which make it unexcelled for his purpose in all the world.

The Canadian Rockies are the culminating scenic portion of the mighty Rocky Mountains called, "the backbone of America." To the northward they gradually diminish in height until the Arctic circle is reached. Southward they lack that ruggedness and glacier beauty which gives them their attractiveness to the lovers of Alpine scenery.

Every day new points of beauty are being revealed. Celebrated mountain-climbers and topographers are constantly visiting and exploring their recesses. This is particularly the case with respect to scientific men from Europe and the United States. There is no particular incentive for these men to go to Switzerland. That country has been thoroughly explored, while in the Canadian Rockies there are numbers of mountains that have never been climbed that challenge the mountaineer: and hundreds of valleys, gorges and lakes, that have never been visited. Every visitor carries a camera, and the many new scenes of grandeur which are revealed after each trip do much to spread the fame of The Canadian Rockies.

Four great ranges are crossed by the Canadian Pacific Railway, the Rockies proper, the Selkirks, the Gold Range and the Coast Mountains, the latter standing like a great bulwark along the shores of the Pacific. The traveller approaches this mighty series of ridges across a country that makes their majesty doubly imposing by reason of the contrast.

For a day or two he has traversed the prairies, a country with many beauties of its own and marvellously rich in all that man requires. As the train approaches the mountains their huge bulk seems to

Camping in the Canadian Rockies is a delightful and beneficial vacation

prohibit passage absolutely, and the clear air brings them apparently close to the train, when they are still miles away. Close by, the Kananaskis Falls of the Bow are taking a mighty plunge, the roar of which is distinctly heard from the track. The river has cut for itself a deep gorge of naked, vertical cliff, and beyond the woods that clothe the summit of the banks rise the steeps of the Fairholme Range, shutting in the view with a line of rocky precipices.

As one looks upon these peaks that seem to start out of the plain, it is difficult to realize their stupendous magnitude. Everything here is on such a gigantic scale that it takes time and effort to weigh the immensity of the great upheavals. Here are mountains that seem much higher than the diameter of their base; and their dizzy heights as one gazes upon them is awe inspiring; but one sees beyond almost interminable ranges with snow-capped tops, bearing upon their shoulders immense glaciers, the very plenitude of which seems to detract from every individual object. These mountains are tremendous uplifts of stratified rocks of the Devonian and Carboniferous ages which have broken out of the earth's surface, and heaved aloft. There are sections miles in breadth, and thousands of feet in thickness that have been pushed straight up, so that the strata of rock remain in almost as level a position as when they occupied their original beds. Other sections seem to be tilted, and stand in a more or less erect position, while others are crumbled by the crowding of other peaks. All these vast piles are doubtless worn away by the action of the elements until they now present only a fragment of their original magnitude. The strata are plainly marked on the sides of the mountains by the various colors of the

rocks that compose them, and often by broad ledges that hold the ice and snow; or when not too greatly elevated are covered with belts of trees which can gain a foothold nowhere else. On the dizzy heights of some of these peaks are piled great masses of rocks which look as though there was scant room to hold them, so sharp are the peaks on which they rest. It would require but little of the mythology of the past to picture these castellated heights as the home of the gods, and imagine them hurling the huge missiles about them for the purpose of crushing their victims below.

The Gap where the Railroad enters the Canadian Rockies

The entrance to the Rockies is by The Gap. It seems that the train has reached an *impasse,* and that there is no way by which it can surmount the lordly line of heights drawn up across its path. Suddenly, however, it takes a sharp turn and finds itself between two walls of vertical rock, **and** a passage is forced to the world of mountains beyond. It has found and

14

The Three Sisters, near Canmore,
Canadian Rocky Mountains

followed the course of the Bow River, and, keeping to the valley that stream has worn for itself in the course of ages, the track turns northward and runs between the Fairholme Range on the right and the Kananaskis mountains on the left.

Prominent among them are the Three Sisters, a trinity of noble peaks. The most distant one from the track is sharp and jagged, but on its shoulders a mantle of snow is thrown and fills up all its crevices. Round the others, to their very summits, tiers of rock run in massive spirals with curious regularity. Across the broad lower slopes they extend, till widened and softened into rolling spurs, they run right down to the River Bow, flowing like a silver streak beneath.

15

57a. Pall Mall.
London. S.W.

Immovable the Three Sisters stand, beautiful in their purity, peaceful in their solitude, steadfast in their guard. Like sentinels apart from their compeers, they seem to the traveller to hold eternal watch and ward over the wonders of the region through which he is to pass.

Cascade Mountain, Banff

Cascade Mountain, at whose base a few miles away from the railway track are the anthracite mines of Bankhead, operated by the Canadian Pacific Railway Company, which supply the country from Winnipeg to Vancouver with hard coal. The powers of the eye are greatly increased, and to one fresh from the plains, things yet far off appear quite near. However, the traveller gradualy understands his mistake, and the track, following the course of the Bow River, turns sharply to the west, just as the lowest spurs are reached, and arrives at the Canadian National Park.

BANFF, THE BEAUTIFUL

Headquarters of the Canadian National Park.

The whole of the town of Banff is the property of the Canadian Government and, under the control of the Park Superintendent, public improvements of all kinds are being constantly carried on to the great advantage of both residents and visitors. The main streets are broad and splendidly kept, the residences are in most instances tastefully designed and well maintained, and throughout the whole village there is an air of sylvan leisure and careful comfort.

Few, if any, towns are more charmingly situated. Few places have found such speedy recognition of their attractiveness, and none have better deserved the encomiums of enthusiastic visitors, than Banff, for of all the lovely resorts on the American continent, it is without a peer. Its surroundings are the mountain steeps, beside whose immense crags and peaks the works of man sink into insignificance. It is not a question of one mountain or of two, but of many, for they stretch away as far as the eye can follow them in every direction, rolling back, one behind another, in varied and sublime confusion.

The stores, while not pretentious, have from years of experience and catering to visitors gained a complete knowledge of their requirements, and few indeed will be the needs, in the way of camping equipment,

Banff from Tunnel Mountain

photography supplies, fishing tackle, and such like
necessities for tourists, that the Banff stores cannot
supply.

To the north, rises the swelling, rounded back of
Stony Squaw Mountain, with cliff-like buttresses pro-
jecting at its eastern end. Towering above this,
majestic in its strength, dominating the whole scene, is
Cascade Mountain, a huge black, timeworn pyramid.
its sides ribbed and scarred by avalanche and tempest.
A plane face looks toward the little town, and two
outward bastions, ridging back toward the centre of its
fall, have made a natural channel. marked, even in
August, by a winding trail of snow. To the west the
Bow River winds in a broad, open strath, the Sawback
range flanking it at the northern side, with Mount
Edith, a splendid dolomite peak. its symmetrical upper
cone glistening virgin white in its mantle of everlasting

18

snow, almost concealed, despite its superior height, by intervening mountain masses. The Bourgeau and Sulphur Ranges are contrasts, both of them, to the craggy and precipitous peaks north of the river, for they are rounding and hummocky in outline, with but a few rock terraces protruding, till near the summit outbulging bastions break the contours, revealing the rugged strength underlying the harmony of many hued forest with which they are clothed. Eastward lies Tunnel Mountain, a knob shaped hill, with a precipitous face to the south, and with a zigzagging carriage road traceable up its eastern side. Because of its ease of access,—many a visitor climbs it as an appetizing walk before breakfast—and the magnficent view, make it the first and favorite trip of every tourist. Opposite it rise the up-tilted terraces of Mount Rundle, almost 10,000 feet high, its sides furrowed and trenched by snowslides. From the valley it appears to have two summits, and so it is sometimes called Twin Peaks.

Looking down the Bow Valley, Banff

A Mountain Split in Two.

The northern one is some thousand feet or more lower than the other. It is evident that time was when Tunnel was merely a shoulder of Rundle, but some tremendous cataclysm of nature split the huge mountain and Tunnel tilted northward—its rocky ribs being plainly discernible in the lateral stratification—and the sleepless, tireless Bow River forced and fought itself through the opening, boring its way towards the limitless plain to the eastward. Above the murmur of pines can be heard, rising and falling on the wind, the noise of the boiling river, as it tears through the rapids, and its roar as it leaps over Bow Falls.

It is a scene possessing almost every element of beauty, and many of sublimity. Over-arched, as it is in summer, with a sky that in its deep azure outrivals that of Italy, lit with the brilliant sunshine characteristic of Western Canada, and possessing an exhilarating atmosphere, full of ozone, purified by frost and forest, is it any wonder that overworked business men absorb its quiet peace like a sponge, and declare it to be the most invigorating spot on the Continent, or that pilgrims in search of the beautiful, pronounce the views superior to those of Zarmatt or the Engadine?

The Museum.

The Canadian Government maintains at Banff, a museum of very great interest to visitors, as it contains many splendidly preserved specimens of the animals. fishes and birds to be found within the Park; a carefully mounted and classified herbarium are also among its chief attractions. Indian relics and specimens of Indian workmanship, many of them of extraordinary interest, are also to be seen. The official in charge has for years taken a record of temperature, and the meteorological charts will repay examination by the weatherwise.

The Bow Falls at Banff

The Bow Falls.

Another of the sights that is sure to claim early attention from the visitors is the Bow Falls, situated beneath the Banff Springs Hotel. Almost as soon as the Bow passes under the Banff bridge, it eddies and rushes as if preparing for its final leap. Soon it begins to foam and boil. Jagged black rocks, with their softer tissues worn away by the rushing stream, stand up here and there out of the roaring flood, dripping and glistening like natural fangs. Churned to a whiteness like that of milk, it roars and hisses through the trench it has worn at the base of Tunnel Mountain, leaps down to small ledges, and then hurls itself a stream 80 feet wide, in a deafening cataract of wonderful beauty. It is not, of course, comparable with the

Falls of Niagara or the Yellowstone, but among the lesser falls of the Continent it has few rivals.   Comfortable rustic seats are placed at various points within view, and at all hours of the day can be seen visitors quietly reading, or gazing at the panorama of beauty of which the Falls form so striking a centre.

Banff Hot Springs.

The Banff Hot Springs undoubtedly possess wonderful curative value for rheumatic and kindred ailments and the cures recorded almost stagger belief.

It may be of interest to give an analysis of the hot sulphur water effecting such marvellous cures.   Mr. McGill, assistant analyst of the Canadian Government, who made a full examination of the Banff water supplies, reports:

"The water is very free from organic impurities and gives no albuminoid nitrogen. * * *   Each gallon contains dissolved sulphuretted hydrogen to the amount of 0.3 grains (equivalent to 0.8 cubic inch).

Hoodoos, near Banff

22

" The dissolved solids are as follows:—

Chlorine (in chlorides)........... 0.42 grains.
Sulphuric Acid (SO³)........... 38.50 "
Silica (SiO₂).................... 2.31
Lime (CaO).................... 24.85
Magnesia (Mg°)............... 4.87
Alkalies (As Soda, Na²O).......: 0.62 "
Lithium ....A decided trace.

" The temperature of the spring is 114.3 degrees Fahrenheit."

Banff Springs Hotel

Banff Springs Hotel
of the Canadian Pacific Railway Hotel System.

Located on a rocky elevation on the south bank of the Bow River near the mouth of the Spray, this splendid hotel commands a view perhaps unrivalled in America. Many important improvements and additions have been made for the comfort and convenience

23

of guests. In the refinement of its appointments and the completeness of detail marking the whole establishment, the Banff Springs Hotel ranks among the finest summer hotels to be found anywhere. The excellence of the cuisine, and the perfection of the waiting—a characteristic of the Canadian Pacific service—are enhanced by the magnificence of the outlook from the dining hall and the music rendered during dinner by an orchestra. In the evenings, after the day excursions, when the guests are lounging in the roomy rotunda, basking in the warmth of the huge log fires in the big open fire-places on either side, a charming concert is given by the orchestra.

On Tunnel Mountain Drive, Banff

Tunnel Mountain.

The drive on which is the finest in the park—distance seven miles. A spiral drive known as the Corkscrew, leads along the side of the mountain at an altitude of over 5,000 feet, the return being made down the further side on a steep grade passing the barracks of the Mounted Police and through the town.

The Lithia Spring.

On the way down to Banff from the Hot Springs, another spring is passed locally known as the lithia spring. It is as yet unimproved, though its curative properties for kidney trouble have a wide reputation in the Canadian West. Analyst McGill reports that the quantity of Lithium in the spring is at least one hundred times as great as in some of the so-called lithia waters placed on the market. Many of the Banff citizens bottle it for private use.

The Basin, Banff

The Cave and Basin.

A delightful drive for about a mile up the valley of the Bow River along a winding road between tall pines at the base of Sulphur Mountain and the Cave and Basin are reached.

The cave itself is covered in by a natural roof of rock and is fed by water from the springs still higher up the mountain. It is not much larger than a good-sized room, but the curious deposits of sulphur about its roof and wall make it well worth a visit. Adjoining it is a natural basin, at which the Government has erected bathing houses, and so popular is this resort that at almost any hour of the day can be heard the splash of waters and the joyous shouts of the bathers.

Bankhead Coal Mines.

One of the most popular drives in the Park, and a little more than half way to the Lake Minnewanka, where the interesting operations of an anthracite coal mine may be seen.

Buffalo at Banff

A large corral of 2,000 acres, in which is a magnificent herd of fifty-six buffalo and calves—the last remnant of the countless thousand bison which once roamed the adjacent plains. Bands of elk, moose, antelope, deer and Angora goat, amongst which are some fine specimens, have also been added to the Park, which is one mile east of the railway station, on the way to Lake Minnewanka.

Lake Minnewanka, near Banff

Lake Minnewanka.

Distance nine miles—drive skirting Cascade Mountain, and following Devil's Head River until the precipitous sides of Devil's Head Canon are crossed by a rustic bridge. The lake is 16 miles long, with a width of from one to two miles. On it is placed a launch, which can be chartered by visitors at the rate of $1.00 per head for parties of five and over. The sail usually occupies three hours. Fishing tackle, boats, etc., may be procured, this being a favorite resort for anglers. A cluster of Hoodoos (natural concrete pillars) and the Devil's Gap, on the way to Ghost River, are amongst the points of interest in this locality.

The Loop

A beautiful drive around the Bow Valley in full view of Bow Falls—distance about seven miles—skirting the base of Mount Rundle, to the banks of the Bow River.

The Observatory.

The Government Observatory on the summit of Sulphur Mountain (8,000 ft.) is reached by a bridle path by way of Hot Springs, and is four miles from the Banff Springs Hotel. There are shelters en route, and from the summit magnificent views of the entire Bow Valley are to be had.

The Observatory on Sulphur Mountain

Attractions of Banff.

It is simply impossible to properly enumerate the many attractions of this delightful spot. The carriage drives along excellent roads with new beauties of scenery unfolding with every turn of the road are delightful.

LAKE LOUISE AND —————
LAKES IN THE CLOUDS

"Lakes of gray at dawn of day,
In soft shadows lying,
Lakes of gold with gems untold,
On thy bosom glowing.
Lakes of white,
At holy night,
Gleaming in the moonlight."

The beautiful Lake Louise

Thirty-four miles westward from Banff is Laggan (the station for Lake Louise and Lakes in the Clouds). Two and a half miles distance from the station by a fine carriage road and Lake Louise (altitude 5,645 ft.)

—the most winsome spot in the Canadian Rockies—is reached. Of the beauty of this remarkable lake there is no divided opinion; every visitor to its shores sings its praises, and it is acknowledged by the most competent judges to be one of the great masterpieces in the world's gallery of Nature. As a gem of composition and coloring it has no rival. At every hour of the day the view is ever-changing with the shadows. This is especially true of the early morning and evening hours. Walter Dwight Wilcox, F.R.G.S., in his charming book, "The Rockies of Canada," describes the colorings of Lake Louise as follows: " It is impossible to tell or paint the beautiful colors, the kaleidoscopic change of light and shade under such conditions. They are so exquisite that we refuse to believe them even in their presence, so subtle in change, so infinite in variety, that memory fails to recall their varying moods. I have seen twenty shades of green and several of blue in the waters of Lake Louise at one time. Sometimes in the evening when the quantity of light is rapidly diminishing, and the lake lies calm, or partly tremulous with dying ripples, marked vertically by the reflections of cliffs and trees, there is a light green in the shallowest water of the east shore, a more vivid color a little farther out, and then a succession of deeper shades merging one into another by imperceptible change, yet in irregular patches according to the depth of water to the deep bluish green and the blue of the middle lake. The eye wanders from place to place and comes back a few moments later to where the brightest colors were, but no doubt they are gone now and the mirror surface is dulled by a puff of air, while the sharp reflections have been replaced by purple shadows, or the obscure repetition of the red brown cliffs above the water. It may be that a day, a year or possibly a century will pass before these identical glories of color will come again."

Lake Louise lies at an elevation of 5,645 feet and is shut in on every side by rocky, snow-capped heights, offering a picture o f perfect peace. Mr. Edward Whymper has compared it to Lake Oeshinen in Switzerland, but has declared it "is more picturesque and has more magnificent environments." It is about a mile and a half long and half a mile broad, while its depth is over 200 feet.

Monument to Sir James Hector at Laggan

Two miles across the boulder covered glacier. lake there begins to rise southward the forefront of the great glaciers where the ice slants away upward until it reaches a depth possibly five hundred feet of solid blue and green, to where it is fed by continuous avalanches from the endless groups of enormous heights beyond. At the upper end of t h i s brown, rises a stern b l a c k

On the trail to Lakes in the Clouds

31

wall to a height fully half a mile, over which the avalanches thunder. This wall is five miles away, but looks to be but one, because of the clearness of the atmosphere.

Above this black avalanche-wall there gradually rises, like the roof of the universe, the pure white snow field on Mount Victoria to a height of ten or twelve thousand feet. Joining with Victoria in forming this ice field are the towering heights of Lefroy, Beehive, Whyte, Niblock, St. Piran, Castle Crags, and many other lofty peaks. To the east an upright mountain forms a perpendicular wall of several thousand feet.

From Lake Louise the ascent altitude of 6,280 to Mirror Lake and Lake Agnes is made easily on horseback or afoot. Lake Agnes, the higher of the two, with an feet, is about two and a quarter miles from the hotel by a good trail.

One of the Chinese Waiters at
Lake Louise Chålet

Lake Louise Châlet

Lake Louise Chalet.

Charmingly situated on the very verge of the water in the midst of the evergreen wood, the Canadian Pacific Railway has built a lovely châlet which has since been enlarged to a great hotel. It is open from June to September, and at it Swiss guides, horses, and packers can be hired for excursions near or far. It affords most comfortable accommodation and conveyances to meet every train. The rates are $3.50 a day, and by pre-arrangement the round trip can be made from Banff at single fare, tickets being issued on presentation of certificate signed by the manager of the Banff Hotel. Telephonic communication exists between the station and the châlet and telegrams may be sent to any part of the world.

mall,
London, S.W.

Lakes in the Clouds.

Mirror Lake is another of these beautiful gems which has no visible outlet, its waters escaping through an underground channel to Lake Louise 1,000 feet below. The waters of this lake rise or fall as the inflowing stream pours its flood into the lake more or less rapidly than they are carried off. Lake Agnes is much frequented by those who revel in the wild chaos of erratic Nature, and at this charming lake is found scenes which aspires to the ideal in beauty, and the grand in sublimity. On the side, like sentinels, stand Mounts Whyte and Niblock, grim and silent; and the irregular peaks running back tell of violent irruption in that great and terrible day of upheaval far back in the misty ages of the earth's infancy. A little way down the valley nature smiles, not broadly but none the less sweetly; for here among the mosses are found the forget-me-nots, the wood anemones, the blue bells of the Scottish Highlands, the ferns, the Alpine eidel-weiss, (the bridal flower of the Swiss mountaineer) and the heather that reminds the sons and daughters of Bonnie Scotland of their native hills. It is an Alpine garden, and the eternal hills seem worthy guardians of this spot of peerless beauty.

Wild Flowers of the Canadian Rockies.

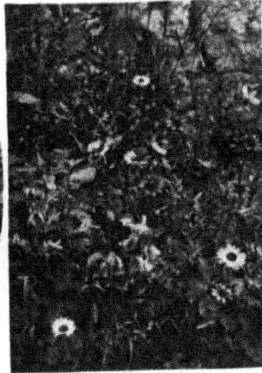

Orchid
Avalanche Lily

Harebell
Asters and Columbines

Among the many flowers found in the Lake Louise region are moss campion, alpine campion, alpine dandelion, crepis, star thistle, erigeron, arnica, arctic saxifrage, stonecrop and alpine willows, and harebells. romanzoffia, grass of parnassus, pentstemon, anemones, large thistle, chives, shooting-star.

ALPINE CLUB OF CANADA.

The Second Annual Meet of the Alpine Club of Canada will be held in Paradise Valley, during the second week in July, 1907. The President, Mr. Wheeler, expects about two hundred climbers to be in attendance. The camp will be very easy of access, being but twelve miles distant from Laggan and nine from Lake Louise. The usual route, is by the trail of nine miles from Lake Louise. Paradise Valley is bounded on the east and west by some splendid glacier mountains, such as Mt. Temple, (11,607 feet above sea level); Mt. Lefroy, (11,115); Mt. Aberdeen, (10,450); Mt. Hungabee, (11,305); and Mt. Victoria, (11,600), is nearby. The most difficult and most dangerous mountain in the whole region is Mt. Hungabee (Indian for Chieftain) which has only been climbed once, and that by one of the most strenuous Alpinists in America — Prof. H. C. Parker, of Columbia University, New York. The glacier which feeds Paradise River is packed in the lap of Mt. Hungabee and is said to be one of the most dangerous glaciers in the Rockies.

The camp will be situated on a lovely Alpine meadow right at the base of

Victory

36

this glacier. The meadow which is studded with Lyall's larch, covers about a square mile. Excursions will be made over the mountain passes to contiguous valleys and lakes, and a round dozen of neighboring peaks will be ascended by members of the club. Mount Aberdeen has been chosen as the official climb; that is to say, those who qualify for active membership will climb this mountain. Swiss guides will be in attendance and also a dozen or so experienced mountain climbers. No neophyte will be allowed to climb without proper mountaineering equipment or unless in fit physical condition. Preparations will be made to accommodate two hundred persons in addition to the large staff of guides and outfitters. During the Alpine Meet, the President, A. O. Wheeler, is in command of the club, and everything is managed with military precision.

Since the Yoho camp of 1906, many new members have been added to the club, and those persons who are thinking of joining the club, ought to make application to the Secretary, Mrs. H. J. Parker, 160 Furby street, Winnipeg, Manitoba, at an early date. Active members must have climbed at least 10,000 feet above sea level. Graduating members have the privilege of qualifying under the auspices of the club at the Annual meet.

Many prominent people will take part in the climb of Mount Aberdeen, and this year's meet of the Alpine Club of Canada will be a time of rare enjoyment.

In a high place

37

Paradise Valley.

To the east of Laggan run two mountain valleys, both of which are noted for their exquisite scenery. Paradise Valley, the nearer to Lake Louise, lies between Mt. Sheol and Mt. Temple, while the Valley of the Ten Peaks, as its name implies, is lined by ten great peaks, and holds at its head, Moraine Lake.

Its entrance to Paradise Valley is under the shadows of Mt. Sheol, that rises to nearly 10,000 feet. The traveller as he gazes into the valley spread at his very feet, cannot but be struck by the wondrous beauty laid out before him, and the immensity of the scale and the perfection of the symmetry of Nature's work.

Paradise Valley, near Lake Louise

Moraine Lake and Valley of the Ten Peaks

The valley of the Ten Peaks extends parallel to Paradise Valley on the other side of Mt. Temple. In it is Moraine Lake, two miles long and half a mile wide, in which there is trout fishing. The Government have recently constructed a splendid carriage road from Lake Louise to Moraine Lake.

A great glacier has found its way down the heights at the head of the lake and has forced its course between and around the peaks. For a third of the distance from the lake to the summit the ice is entirely covered by a picturesque mass of rocks, piled in such disorder as chance directed the ice should have them. It is a picturesque and awe-inspiring sight, the effect of which is magnificent in the extreme.

39

An interesting feature about this glacier is that it seems to be advancing. For some reason that cannot be explained, the glaciers, not only in the Canadian mountains but the world over, have of late years been receding, and the Moraine Lake ice-river is, therefore, an exception to the usual rule. Its force is tremeudous, and it is most impressive to note how the woods have fallen before its resistless force.

Abbot Pass pierces the divide and by it are reached Lakes O'Hara and Oesa, the latter of which is at so great an altitude that its waters are released from the grip of the frost for barely five weeks a year, and has, therefore, received a name that means in the Indian tongue the Lake of Ice. North of Lake O'Hara lie the Wiwaxy Peaks, to the south the Ottertail and the Prospectors' Valleys, lead on into a maze of mountains.

Soon after leaving Laggan the track quits the valley of the Bow and turns south-west to cross the divide. A fine view is obtained of the valley of the Bow extending in a north-westerly direction to the Bow Lakes, while, overtopping the Slate and Waputekh ranges that the railway skirts, loom up the enormous buttresses of Mt. Hector, named after Sir James Hector, who as a member of the Palliser expedition of 1858, was one of the first to explore that pass. Into the solitudes over which it broods, few have yet penetrated, but it is known to be a land rich in beauties and full of marvels, where ice-bound crags and splendid glaciers shut in valleys of great beauty and lakes of infinite charm.

Six miles from Laggan the summit of the Rockies is reached, and the Great Divide is passed, 5,296 feet above sea level. It is marked by a rustic arch spanning a stream, under which the waters divide by one of

The Great Divide

those curious freaks with which nature occasionally diverts herself. For the two little brooks have curiously different fates, though they have a common origin. The waters that deviate to the east eventually mingle with the ice-cold tides of Hudson Bay, while the rivulet that turns to the west finally adds its mite to the volume of the Pacific.

This is the region of mighty avalanches. It is said that by actual count, and without the aid of a glass, eighty distinct glaciers can be seen. In some of this region the scenery is almost terrible.

Stephen, the most elevated station on the Canadian Pacific Railway line, takes its name from the first president of the Company, Lord Mount Stephen, while the next on the westward slope, Hector, recalls Sir James Hector. Of the latter the Kicking Horse River also preserves the memory, for the "kicking horse" was one that inflicted upon him serious injuries during the Palliser expedition. The story is a curious one, as it shows on what chances the success of an exploration may depend. The expedition was encamped on the banks of the Wapta, where a pack horse broke three of the leader's ribs by a kick. He lay unconscious for hours till his Indians thought him dead and prepared to bury him, but as they bore him along he regained his senses. When he recovered he went to inspect his grave, that had been dug some little way from the camp, and then fired by curiosity determined to discover where led the valley in which it had been intended to leave him for ever. He explored it further and found it a practicable way of crossing the mountains. Thus was the Kicking Horse River brought to light and received the name of a vicious animal, which all unintentionally had led to so important a discovery.

But soon all eyes are centred on Cathedral Mt., 10,204 feet high, that rises on the south side of the track, just before Field is reached. It is happily named, for its summit bears a wonderful resemblance to some noble

Pack Horses in the Canadian Rockies

42

by HUSBAND WOLFENDEN, V.D., I.S.O., Printer to the King's Most Excellent Majesty.
1908.

Kicking Horse Canyon

ruin of Gothic architecture. From the very verge of
the rise, where the gradual slope has given place to a
precipice, springs a great crag, like the shattered tower
of a cathedral. The eye can almost trace the windows,
their tracery gone, their mullions in pieces; the buttresses remain, but battered out of all shape and proportion, while the truncated shaft of an arch juts up
behind, solitary and desolate, speaking eloquently of
the noble fane that seems to have been demolished.
The illusion is made all the more realistic by a long,
low line of crags that extend along the summit of the
mount, the perpendicular sides of which might well be
the unroofed, half fallen nave of a cathedral.

At Field the prospect widens, and the Kicking Horse River for a short distance flows across broad, level flats, that are only covered when the water is high.   The place itself is a prosperous little village, but is dwarfed into insignificance by the splendid mountains that hem it in.   On one side is Mt. Burgess,

Hotel at Field, B.C.

on the other Mt. Stephen, one of the grandest of all the Rockies.   Field is the gateway of the wonderful Yoho Valley, and the headquarters for mountaineers of the more ambitious type.   The Yoho Valley is now included within the confines of the National Park.

44

Looking from the shoulder of Mt. Burgess or Mt. Stephen the valley seems narrow, the river a mere stream, and the dwellings in the village dolls' houses. From below Mt. Stephen fills all the view; so rounded, so symmetrical that the spectator hardly realizes at first that he has before him a rock mass towering 10,000 feet above sea level and 6,500 feet above the valley. But as he gazes its majesty bears in on him and he is filled with a sense of awe and wonder. One great shoulder is thrown forward, a mountain in itself, and then the dome swells, g e n t l y, e a s i l y, till it r e a c h e s t h e c l o u d s. Sometimes, indeed, the mist settles on it and obscures half its bulk, sometimes t h e sun lights u p i t s c r e v i c e s a n d touches its peak with gold, sometimes a cloud lies like a m a n t l e across i t s face, but with it all it dominates everything and seems to defy man and nature. There is nothing broken or

Field from Emerald Lake Road

45

rugged in its outlines, no suggestion of wildness or desolation; it impresses by its sheer bulk and massiveness and forces the admiration of the most careless.

To practised climbers the ascent of Mt. Stephen presents no insuperable difficulties, and, indeed, the trip to the summit and back from Mt. Stephen House has been made in eight hours. Swiss guides are stationed at the hotel, and will help the ambitious to accomplish the feat. The lower slopes of the mountain have one spot well worth visiting, the Fossil bed, where for 150 yards the side of the mountain for a height of 300 or 400 feet has slid forward and broken into a number of shaly, shelving limestone slabs.

From the top of Mt. Stephen a magnificent view is obtained, that well repays the toil and difficulty of the ascent. The Van Horne range is seen beyond the Kicking Horse Valley to the west, the Emerald group occupies the north, while on the east the peaks that line the Yoho Valley, Mts. Habel, Collie, Gordon, Balfour, and many another are in full view. Across the river to the south a number of fine mountains are in sight, Mts. Assiniboine, Goodsir, The Chancellor and Vaux. For miles and miles the tourist can see over valleys and peaks, and he realizes the immensity, as well as the beauty of the Rockies.

As a centre for the numerous expeditions to be made from Field, the Canadian Pacific Railway has built there a comfortable hotel and has since been called upon to enlarge it twice. It is planned cunningly, and has splendid accommodations, including a billiard room and suites of rooms with private baths. Moreover, at the livery, carriages, pack and saddle horses, mountaineering outfits and Swiss guides can be engaged at reasonable rates.

Mount Burgess and Emerald Lake

Emerald Lake.

From Field a delightful drive of seven miles round
the spurs of Mt. Burgess to Emerald Lake, another of
those charming tarns that spangle the mountain side.

Emerald Lake Chalet

The road leads through a splendid spruce forest. In one place the road has been cut straight as an arrow for a mile in length. Snow Peak Avenue this stretch is called, and the effect of the narrow way with the mighty trunks standing bolt upright on either hand, with a glimpse of the mountains at the end of the vista, is curious and unique. At Emerald Lake is a charming châlet operated by the Canadian Pacific Railway, where tourists may find first-class accommodation, and rest at the very entrance to the Yoho Valley. The lake, apart from its beauty, is a favorite resort for anglers, as the trout are many and gamey, and heavy are the creels that have been filled from its waters.

Natural Bridge.

One of the most interesting of the short excursions to be made from Field is a walk of two and a half miles to the Natural Bridge, spanning the Kicking Horse River. This is caused by the action of the water of the river itself on the soft limestone rock. Once upon

48

a time the bed of the river extended up to the rocks that now bridge it, and its waters poured over it in headlong fall. Gradually, however, the soft stone was eaten away, and a hole was formed 'n the very rock. Once the way was found nothing could stop the flood, and day by day it enlarged the outlet, until now it has carved a tunnel for itself, and the rocks that once faced a waterfall remain to bridge a rapid. But the end is not yet, and some day the river will win. The rocks will be hurled down from the position they have held so long, and will lie as mere boulders in the bed of the stream.

'Natural Bridge, near Field

The Yoho Valley.

Emerald Lake is half way to the Yoho Valley, one of the most beautiful mountain vales in all the world. From the Châlet by the lake the tourist may take a pony or can walk around the lake and up the mountain beyond. He passes mighty glaciers, their surface lit up and decked with many hues in the sunlight, and

49

charming cascades, their waters leaping a scanty thread-like line, 800 feet or more. Thick timber shuts in the summit of the pass, but parts asunder to grant a glimpse of Summit Lake, a stretch of water, 1,800 feet above Emerald Lake.

Takakkaw Falls, Yoho Valley

A short walk brings one to the Look-Out Point, where a superb view of the celebrated Takakkaw Falls, the highest cataract in America, is obtained. In the course of ages the water has worn for itself a regular semi-circle in the face of the cliffs, and as the trees stand well away on either side, its white foam stands out magnificently against the brown, wrinkled surface of the rock. As it begins its fall, it sparkles in the sunlight; but soon it grazes a narrow ledge, widens out and unravels into a fleecy foaming tangle, till at length, all spray, it reaches the valley, and joins the Kicking

Horse River. Eight times as high as Niagara (1,200 feet), it compares with anything in the Yosemite Valley, and fed by the melted snows of the glacier, it is at its best in summer.

All up the valley other cascades are seen or heard. The hills are crowned with glaciers and the water melted from them seeks the shortest way to the valley, even at the cost of

Twin Falls, Yoho Valley.

a plunge of hundreds of feet, and among them the Laughing Falls charm particularly. Its leap is only 200 feet, but its waters seem to laugh with glee as they go, and its milk-white flood smiles delightfully through the dark evergreens around it. Further up the valley on the left branch of its forked stream are the Twin Falls, an almost unique phenomenon and as beautiful as it is unexpected. Two streams plunge side by side into the abyss. Every waterfall is beautiful and no one can help marvelling at the ever-varying, ever-constant flow of a cascade with its wondrous force and grace. But when there are two falls leaping side by side, when there is life and motion in two separate

57a, Pall Mall.

cascades, when the light plays across them and the rainbow tinges their spray, but remains still for not two seconds together, then indeed the spectator is entranced and he lingers long, loath to tear himself from a sight that appeals to his deepest sense of beauty.

But there is sterner scenery than any the waterfalls present along the Yoho Valley. A great glacier too, far larger even than the famous Illecillewaet Glacier of the Selkirks, overhangs the right hand fork of the valley. The Wapta Glacier, as it is named, is part of the great Waputekh ice field guarded by Mt. Gordon. Mt. Balfour and the broken crags of Trolltinderne (The Elfin's Crown).

At the fork of the Yoho Valley another shelter has been provided for visitors, and there will be many that take advantage of it. It is possible to make the trip round the valley from Emerald Lake in a day, but all who can will spare another day or two.

The return to Field may be varied by crossing the Burgess Pass. From this lofty trail Emerald Lake is seen thousands of feet below, with the Emerald Range rising beyond, while on the other hand Mts. Cathedral, Stephen and Dennis and the Ottertail Range excite admiration. From this eminence a zig-zag path leads down by easy stages to Mt. Stephen House.

FIELD **TO GLAC**IER.

Field left behind, the train has to descend the western slope of the Rockies to the valley of the Columbia. To reach it the course of the Kicking Horse River is followed through some of the finest mountain scenery in the world.

The track runs between the Ottertail and Van Horne ranges. The highest of the range, Mt. Goodsir, a victim to the prowess of Professor Fay, of Tuft's College, stands miles from the railroad, but its hoary head is seen towering above its sisters. The Van Horne Range, just across the narrow valley, is less severe in its outline; its slopes are ochre-hued, and its summit is an alternating succession of crest and trough. To the southeast the Beaverfoot Mountains, a splendid line of peaks, stretch in regular array as far as the eye can reach, and between them and the Ottertails rises the mass of Mt. Hunter.

At Leanchoil, the canon of the Kicking Horse is entered. Straight up and down the rocky sides extend a well that seems impregnable. Thousands of feet in the air they rise, and their summit is lined with a number of peaks, perpetually covered with snow, to which no names have yet been given. The cleft is a bare stone's throw across, and through it river and railway find their way. Ledges have been blasted in the face of the rock; jutting spurs have been tunnelled through; from side to side the track has been carried; and always below is the river foaming and roaring, breaking itself against the sides of the canon. The effect is marvellous and stupendous, and the ingenuity of man had to fight a great battle with the forces of Nature.

All of a sudden there is a wonderful change. The descent is completed and the track emerges in the broad valley of the Columbia. One of the resting places of the mountains has been reached and the travellers gaze with pleasure upon the thriving little town of Golden.

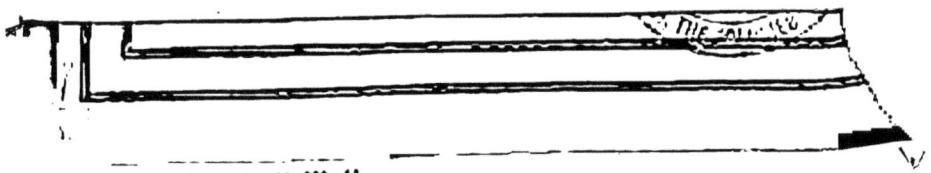

53

57a. Pall Mall.

One of the principal difficulties in constructing this part of the line was caused by the mountain torrents, which rush down these mountain sides in deep narrow gorges over which the railway must cross. The largest of these bridges crosses Stony Creek, a noisy stream flowing in a narrow V-shaped channel, 300 feet below the rails. This is said to be one of the highest railway bridges in the world.

Stoney Creek Bridge

Rogers' Pass was named after Major A. B. Rogers, by whose energy it was discovered in 1883, prior to which time no human foot had trod the summit of this central range. Here is a vast amphitheatre, where seven or eight thousand feet above the valley half a dozen glaciers may be seen at once, and so near that their green fissures are disincly visible. Here one may behold the never-to-be-forgotten spectacle of the rising sun as it gilds the mighty battlements; or look upon the green valleys, and see the snowstorm trailing along the crests with perhaps a peak or two standing sphynx-like and serene above the clouds.

Mt. Macdonald, over 6,000 feet above the railway.

Mt. Hermit, which takes its name from the cowled figure that with a dog appears on the western spurs, has regular strata, running in parallel rows across its front, to which undulating waves just marked by snow give grace and lightness. Mount Macdonald rises with precipitous walls far above the railway.

GLACIER

Nestled in a niche of the narrow valley a few rods from the railway, and surrounded by the beautiful evergreen trees that everywhere thrive in this region, is a charming hotel, the Glacier House, which has become so popular that the Canadian Pacific Railway Company has found it necessary to enlarge the original structure, erect new buildings, and increase the capacity of the annex, so that now over one

Glacier House, Glacier, B.C.

On the Trail, near Glacier, B.C.

hundred guests can be accommodated. A Surgeon-General in the Army wrote recently in the guests' book at the hotel: "My wife and I have travelled for nearly forty years all over the world, and are both agreed the scenery at Glacier House is the finest we have seen in Europe, Asia, Africa or America." The first to attract the tourist is the Great Glacier of the Selkirks, which crowds its tremendous head down the mountain gorge within thirty minutes' walk of the hotel. At the left Sir Donald rears his mighty head more than a mile and a half above the railway. This monolith was named after Sir Donald A. Smith (now Lord Strathcona and Mt. Royal), who was one of the chief promoters of the Canadian Pacific Railway. A mountain rivulet rushes down the abruptly rocky sides of the mountain opposite the hotel, and a trail has been cut up the steep incline to a spot beside the rushing stream, where a rustic summer house has been erected. The effect is novel and pleasing. The waters from this stream have been utilized to supply the hotel and fountains that play in the foreground. All the streams here are simply ice water from the glaciers. A tower has been

erected near the annex of the hotel, on which is a large telescope commanding a view of the great glacier and surrounding objects. As one alights here a feeling of restfulness comes over him. Everything conspires to a feeling that all the cares and rush of the business world are shut out by the great mountain. The trees, the streams, and even the mountains speak of peace and quiet. The mighty rushing winds never reach this secluded spot. The ever-green trees are restful to the eye, and

Mountain Game.

the mid-day sun is tempered by their emerald sheen. Let the visitor step abroad, inhale the vitalizing air, look at the mighty glaciers, where are stored the snows of centuries; gaze upon the wild rage of the mountain torrent as it takes its mad plunge from the rocks among the clouds; look away and see the ice-bound peaks of these mighty ranges as they stand sphynx-like, serene and grim, as if man beheld a type of the eternal.

At the west end of this range stands the pyramid form of Cheops, and between two rugged peaks is the Mount Bonney Glacier, known also as the Purity; and to the right is the amphitheatre of the Cougar Range. In the background may be seen the picturesque Asulkan Glacier, and the two sharp peaks furthest south are Castor and Pollux. It is said that from the summit of Sir Donald, 120 glaciers may be seen.

The Great Glacier

The Great Glacier is a mile and a half from the hotel, but among such gigantic surroundings looks much nearer. Its slowly receding front with crevasses of abysmal depths cutting across its crystal surface is only a few hundred feet above the level of the railway. Good trails have been made to it, and its exploration is not difficult, although it is not wise to traverse some portions of it without a guide to pilot the way among yawning bergschrunds that slash its surface. This Great Glacier is said to be greater than all those of Switzerland combined. It is the centre of a group of glaciers embracing more than 200 square miles, and the hoary head seen from the hotel is one of several outlets. The great ice peaks and glaciers are truly an interesting study. Solemn, stately, and serene, smiling not in the beautiful sunshine; and still unmoved when the fierce blasts of the tempests strike. At times they clothe themselves in thick clouds waiting only the bright rays of noonday sun to step forth armored in glittering silver, or robed in the gorgeous colors of evening; and in the mysterious silent night the moon and the stars look down to see their faces in the glassy surface. The last rays of departing day linger upon the lofty spires; and when the night has passed and the moon has sunk behind the grand old peaks, they catch the first gleam of returning light, and their gilded tops herald the coming morn. The elements combine to pay tribute to such matchless beauty. The sun steals through the sparkling fountains which flutter over the crystal surface in summer, and the hues of the rainbow betray the sun's warm kiss. In winter the hoar frosts gather as a mantling shroud over the silent forms only to add new beauty in the resurrection of spring. Good-bye, grand old glaciers! For untold ages you have lifted your hoary heads among the clouds! For unnumbered ages you will still remain! " Men may come, and men may go," but you keep your silent vigils unmoved by the lapse of Time!

Above the Snow Line.

Those interested in glaciers and glacial phenomena should ask for a copy of a little handbook, " Glaciers," published by the Canadian Pacific Railway and kept for gratuitous circulation at the company's agencies and hotels.

The Illecillewaet Glacier, like nearly every other observed glacier in the world, is receding. It is reckoned the sun drives it back on the average 35 feet a year, and recovers this much from the bonds of ice. However, after the ice is gone, the moraine remains, and it will be many centuries before the great rocks carried down by the glacier are reduced to dust, and the land thus reclaimed supports renewed vegetation.

From Glacier House other expeditions of great interest may be made. One trail leads first to the shores of Marion Lake, 1,750 feet above, and two miles distant from the hotel, where a shelter is erected. Splendid views are obtained on the way of the range from Eagle Peak to Sir Donald, and a path strikes off for Observation Point, where another shelter is built for those who would dwell on the glories of Rogers' Pass to the north-east and the Illecillewaet Valley to the west.   Mt. Abbott is a day's climb, but it is an easy one, and should be undertaken by all, for from it a splendid view is obtained of the Asulkan Valley.

From Observation Point an extremely fine view is obtained, adown the Illecillewaet Valley, down the precipitous sides of which the track has had to make a descent of 522 feet in seven miles.   The feat taxed to the utmost the skill of the engineers, and they accomplished it by means of the famous Loops of the Selkirks.

The course the railway has to follow to gain the valley has been called the Loops of the Selkirks.   First, the track crosses a valley leading from Mt. Bonney glacier.   Then it touches for a moment the base of Ross Peak.   It doubles back to the right for a mile or more, and so close are the tracks that a stone might be tossed from one to the other.   Next it sweeps around and reaches the slope of Mt. Cougar on the other side of the Illecillewaet, but it has to cross the stream once more before it finally finds a way parallel to the general trend of the valley.   The line has made a double " S " in its course, and has cut two long gashes on the mountain side, one above the other.

Albert Canon

The Illecillewaet River is, of course, of glacial origin,

and takes its rise from the Great Glacier of the Selkirks; it is, therefore, at first a pea-green color from the glacial mud, but afterwards, as it flows through the valley, it clarifies itself and in the end is perfectly pure. Caribou are found all the way down the valley to the Columbia in considerable numbers.

Twenty-two miles from Glacier, the Illecillewaet River runs through the Albert Canon, a gorge so marvellous that several of the regular trains stop for a few minutes to allow passengers to see its wonders. The Illecillewaet issues from an exceedingly narrow pass, through which the river must pass. The canon widens a little, but it still remains deep, abrupt and narrow. From its brink rocks torn, rent and split can be seen 300 feet straight below. It is but 20 feet across, and in the gloom the white foam of the flood can be made out, while the noise of its fury is redoubled by the closeness of its confinement.

More mountains there are, and we shall not lose sight of them all when the waters of the great Pacific dash at our feet; for in the mighty upheaval the deep waters of the sea were no barrier, as is seen by the ever admire, but he may here find Nature as it passes from the great Creator, untarnished by the hand of man. Succeeding generations of the children of men will gaze upon these majestic mountains, whose peaks of eternal ice tower above the clouds that would hide the sun; and will look with awe at the wild canons and mountain torrents; and will behold with ecstacy the many scenes of Edenic beauty, too sacred to remain in the gaze of the multitude, but "sought out of all those who have pleasure therein."

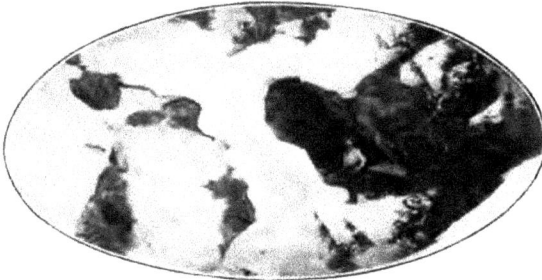

Entrance to the Caves of Nakimu

The Great Caves of Nakimu.*

These great caves which were recently discovered by Charles H. Deutschman are situated about six miles from Glacier, B.C., at the head of a beautiful valley, the altitude being 1,980 feet from the track and above the snow line. The wonderful caverns are formed by the action of water for ages upon the solid rock, and are a series of chambers with large entrances, the ceilings being polished strata of rock varying in height. The main chamber is about 200 feet in height, with a varying width of from 150 to 200 feet. The walls

* "Nakimu"—Indian for grumbling caves.

64

sparkle with the quartz crystals, and myriads of minia-
ture lights are reflected from the darkness. In other
parts the walls are smooth as marble, the harder
portions of the formation showing like the rounded
rafters of a cathedral dome. Recesses are abundant
where the eddying waters found a softer and more
yielding rock. A natural bridge marks the point where
other streams in ages past have worn two other passages in the mountain. Vast bowls of water are all
that remain to show where former waterfalls existed.
None are deep, however, and flint-like ledges afford an
easy method of progress. No evidence has so far been
discovered that any portion of these caverns have ever
been used as the habitation of human beings.

Revelstoke to Kamloops.

Revelstoke is an important centre; from it there is
water communication with the rich Kootenay and
Boundary districts. It is on the Columbia River,
which has made a great bend since the train crossed it
at Donald and, flowing now south instead of north, is
much increased in size. Twenty-eight miles below

Steamer Rossland.

Revelstoke it ex-
pands into the
Arrow Lakes,
which fill the trough
between the Selkirk
and Gold ranges as
they run north and
south. A branch line
runs down to
Arrowhead, and
from there well-
appointed Canadian

Pacific Railway steamboats carry travellers to Nakusp and Robson, from which the Slocan, Kootenay, Boundary and Rossland districts are reached.

Down Arrow Lake the steamer plies to Nakusp and Robson, passing near the head of the lake the famous Halcyon Hot Springs. This is a favorite summer resort, having a good hotel, while opposite is Halcyon Peak, 10,400 feet high, and several fine waterfalls. A spur of the Canadian Pacific Railway connects it with Sandon on Slocan Lake, in the centre of the silver-lead district and with Rosebery, to join the steamer that plies down the lake to Slocan City. Here again the rails begin and communicate with Robson at the end of the Lower Arrow on the west,.and with Nelson on an arm of Kootenay Lake on the east.

The Arrow Lake steamer has also come the full length from Robson, 165 miles, through splendid mountain scenery, while from Robson trains run over a short but important line to Trail and Rossland through one of the richest mining regions in the world. Yet another branch from Robson has been constructed through the Boundary district to Midway and opens up another prosperous mining locality.

The Crowsnest branch of the Canadian Pacific Railway ends at Kootenay Landing, and from there to Nelson there is communication by Canadian Pacific Railway steamer. A steamboat line has been established from Nelson up Kootenay Lake to Lardo, whence an isolated branch of railroad runs 32 miles north to Gerrard, and a steamer plies across Trout Lake to Trout Lake City, a matter of 17 miles, so that every part of Southern British Columbia may be reached by the Canadian Pacific Railway and its connections.

The thriving town of Revelstoke stands in the broad valley of the Columbia, over which a bridge half a mile long has been built.

As Craigellachie is passed a monument may be seen which marks the spot where the last spike was driven into the great line that joins the Atlantic and the Pacific. The work had been begun from both ends of the railroad, and it was on Nov. 7, 1885, that, with fitting ceremonial, the last strokes were put to the truly stupendous task—five years before the stipulated time.

The chain of lakes passed, the valley closes in until Sicamous Junction is reached. Sicamous is at an altitude of only 1,300 feet above sea level, and is remarkable as a sporting resort and as the gateway to a splendid ranching and farming district. From it can be visited by the Okanagan branch, Okanagan Lake, down the 70 miles of which plies the Canadian Pacific Railway steamers to Penticton, from which the mining towns to the south may be reached by stage. The whole region of the Okanagan is a land with a balmy climate where fruit grows to perfection, and at Vernon and at Kelowna on the lake shore Lord Aberdeen, late Governor-General of Canada, has splendid farms. The

Hotel at Sicamous, B.C.

67

names, Peachland and Summerland, given to places not far from Penticton, are suggestive and fully justified.

Shuswap Lake is a most beautiful sheet of water. It runs up the valleys between the mountains wherever its waters can find a level, and its long arms have been compared to the tentacles of an octopus. Each of them is many miles long and at places as much as two miles broad, but they often narrow down to a few hundred yards, and at one such spot the railway crosses the Sicamous Narrow by a drawbridge. It then follows the south shore of the Salmon Arm, crossing the Salmon River.

At Tappen the Salmon Arm is left and the track strikes boldly out for Shuswap Arm, though in so doing a way has to be cut through the forest and Notch Hill, 600 feet above the lake level, has to be passed. From this elevation a charming view is obtained. On every side the lake extends silvery arms that wander along among rounded hills and thick woods.

Shuswap Lake gradually narrows into the south branch of the Thompson River, and steadily downhill along its banks runs the line. The country is an excellent ranching district and has been long settled from the Pacific Coast.

Near Kamloops

Kamloops is a thriving little town, and an air of activity is given to the place by the numerous sawmills and the steamboats, that ply on the lake. It draws much profit from the mining fields, being a supply point for them, and from the ranching district to the south, communication being by stage.

The Thompson and Fraser Canons.

Nicomen is a little mining town where, on the opposite side of the river, gold was first discovered in British Columbia. The discovery was doubtless the clue to the finding of the rich gold fields of Caribou, as miners always prospect up stream to find the lode from which the placer came. We are now in the Thompson Canon, whose gold gorge narrows and deepens till the

Cariboo Bridge, Spuzzum, B.C.

scenery is wild beyond description. At Lytton, a small trading town, the canon widens to admit the Fraser, which comes from the north, between two ranges of mountain peaks.

Harrison Springs Hotel

The old Government road to Cariboo is in evidence all along the Fraser and Thompson valleys. Since the building of the railway the use of the waggon road has been discontinued except in some places where local interests make it convenient. At Spuzzum it crosses the river on a suspension bridge 110 feet above low water; yet it is said that in 1881 the river rose to such a height that it was only by the greatest exertion that the bridge was saved from destruction by driftwood.

For fifty-four miles between Lytton and Yale, the river had cut through this lofty range of mountains, thousands of feet below their summits. On this section of fifty-four miles, a construction army of 7,000 men worked.

During the building of this road, men were suspended by ropes hundreds of feet below the tops of the cliffs to blast a foothold. Supplies were packed in on the backs of mules and horses; and building materials often had to be landed on the opposite bank of the stream and taken across at great expense. It is estimated that portions of this work cost $300,000 per mile. Below the town of Lytton the river is spanned by a cantilever bridge 530 feet long, the centre span being 315 feet. The difficulty of its construction was great, owing to the fact that the site could only be approached from one end. ·One half the materials were sent across the river on a steel cable one and one-fourth inches in diameter. Several pieces of the structure weighed over five tons each. It is claimed that in this respect the bridge is without a rival.

North Bend is now reached, which is certainly a place whose memory will long linger in the recollection of those who have ever seen it from the car windows.

The Pacific Coast.

At Yale he feels the balmy air of the Pacific. At Spence's Bridge he saw a curious Indian cemetery, with rudely carved birds perched even on the Cross, the totem intruding on the Christian symbol. All down the canons he has seen occasional natives fishing for salmon or washing for gold, and at Agassiz he finds a fine Government experimental fruit farm, while

Steamer Princess Victoria, Seattle, Victoria and Vancouver Service

71

57a, Pall Mall,
London, S.W.

five miles away to the north is Harrison Lake with its hot sulphur springs, the visitors at which stay at Harrison Springs Hotel.

At Mission Junction he can, if so disposed, change to the branch line, that runs to the international bonndary and there joins the Northern Pacific Railroad. By this route he reaches Seattle and makes connection with the Shasta route for San Francisco and all the Pacific states. The main line, however, keeps on past Westminster Junction, where a branch line leads to Westminster, and arrives at the terminus of the Canadian Pacific Railway at Vancouver.

There he finds his long journey ended and himself on the shores of Burrard Inlet, one of the finest harbors on the Pacific. If the inducements of Vancouver and the splendid service of the Canadian Pacific

Canadian Pacific Vancouver Hotel

Railway Hotel, Vancouver, do not tempt him to stay, he can embark at the very railway station on steamships that will take him to the ends of the earth. The Canadian Pacific Railway Company's Empresses will transport him swiftly and comfortably to Japan or China, the Canadian-Australian line runs regularly to Honolulu, Fiji, Australia, and New Zealand, while if such long journeys do not suit his pleasure, he can sail by a Canadian Pacific Railway steamer to Victoria on Vancouver Island, or take longer coasting trips to the golden Yukon, or to Seattle.

Station and Offices, Vancouver, B.C., Canadian Pacific Railway

Vancouver has a fine harbor, landlocked, well-lighted and safe, to which resort, besides the liners already mentioned, freighters from all parts of the world. They bring silks and teas from the Orient; they take away the lumber and canned fish of British Columbia and the wheat and flour of the Canadian

West; and they make the port one of the most important of the Pacific Coast.

The city, though only nineteen years old and burnt to the ground in 1886, now numbers over 50,000 and is the centre of flourishing industries. Industries there are in plenty, and Vancouver has everywhere the appearance of a rapidly progressing community. Its well-built, wide streets add to the impression, and the extremely picturesque surroundings of the city make it pleasant as a residence and delightful to visit. Stanley Park is its crowning glory, in the depths of which the Douglas fir and giant cedar are secn·iu all their magnificence and nature is allowed to display her unspoiled beauty.

A few hours steam from Vancouver is Victoria, the capital of British Columbia. Across the Straits of Georgia daily plies the fast new Canadian Pacific Rail-

New Empress Hotel, Victoria, B.C.

74

way steamer " Princess Victoria," passing through a world of small islands, comparable to the Thousand Islands of the St. Lawrence, though with infinitely finer timber. Victoria itself is a city of lovely homes and the seat of the Provincial Government, its Parliament buildings being one of the handsomest piles on the continent. This city is of singular beauty and has a population of over 30,000. There is now nearing completion a palatial hotel by the Canadian Pacific Railway Company, which will be completed during the coming summer. Beacon Hill Park, 300 acres in extent, is no less beautiful than Stanley Park.

Farewell, old mountains! Your vales with their beautiful verdure, and your sunny slopes shut in from the fierce winds, and fiercer business of the outside world, have spoken of earthly peace, and given glimpses of Edenic beauty too rarely seen on earth! Your snowy crests, reaching above the clouds into the purer atmosphere of the heavens, have been an inspiration, speaking to the inner consciousness with a "voice as of a trumpet," ever pointing to the Infinite! Your great glaciers with their enduring ice have been a monitor of the Eternal. . Grand old mountains! Your frown is terrible!

Canadian Pacific Ry. Co., Atlantic Service

EMPRESS OF BRITAIN

One of the palatial Royal Mail steamships of the Canadian Pacific Railway Company's Atlantic Service. Length 570 feet, breadth 65 feet, displacement 20,000 tons 18,000 horsepower, and makes the passage between Liverpool and Quebec in less than a week.

EMPRESS OF JAPAN—PACIFIC SERVICE, CANADIAN PACIFIC RAILWAY CO.

TO JAPAN AND CHINA

"Empress of India," "Empress of Japan," "Empress of China," "Tartar" and "Athenian."

Sailing between Vancouver and Victoria. B.C., and Yokohama, Kobe and Nagasaki, Japan, and Shanghai and Hong Kong, China.

THE SHORTEST AND SMOOTHEST ROUTE ACROSS THE PACIFIC

The Canadian Pacific Railway

THE WORLD'S HIGHWAY BETWEEN THE ATLANTIC AND THE PACIFIC

SPECIAL ATTENTION IS CALLED to the PARLOR, SLEEPING and DINING CAR SERVICE—so important an accessory upon a railway whose cars run upwards of THREE THOUSAND MILES WITHOUT CHANGE.

These cars are of unusual strength and size, with berths, smoking and toilet accommodation correspondingly roomy. The Transcontinental Sleeping Cars are fitted with double doors and windows to exclude the dust in summer and the cold in winter. The seats are well upholstered, with high backs and arms.

The upper berths are provided with windows and ventilators. The exteriors are of polished red mahogany and the interiors are of white mahogany and satinwood.

No expense is spared in providing the **DINING CARS** with the choicest viands and seasonable delicacies, and the bill of fare and wine list will compare favorably with those of the most prominent hotels.

OBSERVATION CARS, specially designed to allow an unbroken view of the wonderful mountain scenery, are run on transcontinental trains during the Summer Season (from May to about October 15th).

THE FIRST CLASS DAY COACHES are proportionately elaborate in their arrangement for the comfort of the passengers; and for those who desire to travel at a cheaper rate, **TOURIST CARS,** with bedding and porter in charge, are run at a small additional charge; **COLONIST SLEEPING CARS** are run on transcontinental trains without additional charge. The colonist cars are fitted with upper and lower berths after the same general style as other sleeping cars, but are not upholstered, and the passenger may furnish his own bedding, or purchase it of the Company's agents at terminal stations at nominal rates.

The entire passenger equipment is **MATCHLESS** in elegance and comfort.

First Class Sleeping and Parlor Car Tariff

FOR ONE DOUBLE BERTH, LOWER OR UPPER, IN SLEEPING CAR BETWEEN		TOURIST CAR TARIFF
Halifax and Montreal	$ 4	....
St. John, N.B., and Montreal	2	....
Quebec and Montreal	1	....
Montreal and Toronto	2	....
Montreal and Chicago	5	....
Montreal and Winnipeg	8	$4 00
Montreal and Calgary	13	6 50
Montreal and Banff	14	7 00
Montreal and Revelstoke	15	7 75
Montreal and Vancouver	18	9 00
Ottawa and Toronto	2	....
Ottawa and Vancouver	17	8 75
Fort William and Vancouver	15	....
Toronto and Chicago	3	....
Toronto and Winnipeg	8	4 00
Toronto and Calgary	12	6 00
Toronto and Banff	13	6 50
Toronto and Revelstoke	14	7 25
Toronto and Vancouver	17	8 50
Boston and Montreal	2	....
Boston and Vancouver	19	....
New York and Montreal	2	....
Boston and St. Paul	7	....
Boston and Chicago	5	....
Montreal and St. Paul	6 00	....
St. Paul and Winnipeg	3 00	....
St. Paul and Vancouver	12 00	6 00
Winnipeg and Vancouver	12 00	6 00

Between other stations rates in proportion.

Rates for full section double the berth rate. Staterooms between three and four times the berth rate.

Accommodation in First Class Sleeping Cars and Parlor Cars will be sold only to holders of First Class transportation, and in Tourist Cars to holders of First or Second Class accommodation.

57a, Pall Mall.
London. S.W.

Canadian Pacific Hotels

While the sleeping and dining car service of the Canadian Pacific Railway furnishes every comfort and luxury for travellers making the continuous overland through trip, it has been found necessary to provide comfortable well managed hotels at the principal points of interest among the mountains, where tourists and others might explore and enjoy the magnificent scenery.

ALGONQUIN HOTEL—ST. ANDREWS, N.B.
(Open from June to September)

This popular Atlantic Seaside Resort, is situated on a peninsula five miles long, extending into Passamaquoddy Bay. Good deep sea and fresh water fishing may be enjoyed; the roads are perfect, making driving and cycling most enjoyable. The facilities for yachting and boating cannot be surpassed, and there are golf links that have no superior in Canada.

The hotel, on which a large expenditure has recently been made, in improvements, offers every modern accommodation for tourists.

Rates, $3.50 per day and upward. Special rates to those making prolonged visits.

McADAM STATION HOTEL—McADAM JUNC., N.B.,

offers the visitor in search of sport a choice of routes through the whole provinces. It gives him, too, an outing at a summer retreat, free from the heat and crowds of the fashionable resorts, whence the hunting and fishing grounds are easily accessible.

The rates are from $2.50 per day upwards.

THE CHATEAU FRONTENAC, QUEBEC,

In the quaintest and historically the most interesting city in America, is one of the finest hotels on the continent. It is fireproof, and occupies a commanding position overlooking the St. Lawrence, its site being, perhaps, the grandest in the world. The Chateau Frontenac was erected at a cost of over a million of dollars. Great taste marks the furnishing, fitting and decorating of this imposing structure, in which comfort and elegance are combined to an unequalled extent.

Rates, $4.00 per day and upward, with special arrangements for large parties and those making prolonged visits.

THE PLACE VIGER, MONTREAL,

is a handsome structure in which are combined a hotel and station. The building, which faces Place Viger, is most elaborately furnished and modernly appointed, the general style and elegance characterizing the Chateau Frontenac, at Quebec, being followed.

Rates, $3.50 per day and upward, with special arrangements for large parties or those making a prolonged stay.

CALEDONIA SPRINGS HOTEL—CALEDONIA SPRINGS, ONT.,

is situated at the famous Caledonia Springs, so well-known all over the American Continent.

Rates, $3.00 per day and upward.

THE ROYAL ALEXANDRA—WINNIPEG, MAN.,

a newly completed 300 room house situated at the Railway station, furnished with every modern convenience, including Cafe and Grill Room. European and American plan. Rates:—American plan, $4.00 per day up; European plan, $2.00 per day up.

MOOSE JAW HOTEL—MOOSE JAW, SASK.,

in the Canadian North-West, at the junction of the Soo-Pacific road with the main line of the C.P.R. The hotel is appointed in the most modern style and is elegantly furnished.

Rates, $3.00 per day and upward, with reductions to those making prolonged visits.

BANFF SPRINGS HOTEL—BANFF, ALBA.,
(Open from May to October)

In the Canadian National Park, on the eastern slope of the Rocky Mountains, is 4,500 feet above sea level, at the junction of the Bow and Spray Rivers. A large and handsome structure, with every convenience that modern ingenuity can suggest, costing about half a million dollars.

Rates, $3.50 per day and upward, according to the rooms. Special rates by the week or month will be given on application.

Canadian Pacific Hotels.—*Continued.*

THE LAKE LOUISE HOTEL—LAGGAN, ALBA.
(Open from June to October)

This quiet resting place in the mountains is situated on the margin of Lake Louise, about two miles distant from the station at Laggan, from which there is a good carriage drive and an excellent base for tourists and explorers desiring to see the lakes and the adjacent scenery at their leisure.

The rates are $3.50 per day and upward.

MOUNT STEPHEN HOUSE—FIELD, B.C.,

is a magnificent mountain hotel, several times enlarged, fifty miles west of Banff in Kicking Horse Canon, at the base of Mount Stephen, the chief peak of the Rockies, towering 8,000 feet above. This is a favorite place for tourists, mountain climbers and artists, and sport is plentiful. Emerald Lake, one of the most picturesque mountain waters, being within easy distance. The newly-discovered Yoho Valley is reached from Field.

Rates, $3.00 per day and upward, with special arrangements for parties making prolonged visits.

EMERALD LAKE CHALET—NEAR FIELD, B.C.,
(Open from June to October)

is a Swiss Chalet Hotel, situated on the margin of Emerald Lake, near Field, and affords splendid accommodation for those wishing to remain at the Lake or who intend visiting the famous Yoho Valley, to which excellent trails lead from this point.

Rates, $3.00 per day and upward. Special rates to those making prolonged visits.

GLACIER HOUSE—GLACIER, B.C.,

is situated in the heart of the Selkirks, within forty-five minutes' walk of the Great Glacier, which covers an area of about thirty-eight square miles.

The hotel, which has recently been enlarged several times to accommodate the ever-increasing travel, is in a beautiful amphitheatre surrounded by lofty mountains, of which Sir Donald rising 8,000 feet above the railway is the most prominent. The dense forests all about are filled with the music of restless brooks, which will irresistibly attract the trout fisherman, and the hunter for large game can have his choice of "big horn, mountain goat, grizzly and mountain bear." The main point of interest, however, is the Great Glacier. One may safely climb upon its wrinkled surface or penetrate its water-worn caves.

Rates, $3.50 per day and upward, with special arrangements for parties making prolonged visits.

HOTEL REVELSTOKE—REVELSTOKE, B.C.,

at the portal of the West Kootenay gold fields and the Arrow Lakes, situated between the Selkirk and Gold Ranges, is complete in all details.

Rates, $3.00 per day and upward.

HOTEL SICAMOUS—SICAMOUS, B.C.,

a fine structure, built on the shores of the Shuswap Lakes, where the Okanagan branch of the Canadian Pacific Railway leads south to the Okanagan Valley and the contiguous mining country. The hotel has all modern appointments and conveniences.

Rates, $3.50 per day and upward, with reductions to those making prolonged visits.

HOTEL VANCOUVER—VANCOUVER, B.C.,

is at the Pacific Coast terminus of the Railway. This magnificent hotel, lately much enlarged, is designed to accommodate the large commercial business of the place, as well as the great number of tourists who always find it profitable and interesting to make here a stop of a day or two. It is situated near the centre of the city, and from it there is a glorious outlook in every direction. Its accommodations and service are perfect in every detail, and excel those of the best hotels in Eastern Canada or the United States.

Rates, $3.00 per day and upward, with special terms for those making prolonged visits.

The new Empress Hotel at Victoria, now in course of construction, and which will be one of the grandest on the Continent, will be opened for guests during the coming Summer.

Enquiries as to accommodation, rates, etc., at any of the Canadian Pacific Hotels will be promptly answered by addressing managers of the different hotels, or communicating direct with

The Manager-in-Chief of C.P.R. Hotels, MONTREAL.

57a, Pall Mall,
London, S.W.

AGENCIES

Adelaide... SOUTH AUS..Australasian United Steam Nav. Co. [Ltd.] ....
Antwerp......BELGIUM..H. Debenham, Agent....................................33 Quai Jordaens
Auckland...........N.Z..Union S.S. Co. of New Zealand [Ltd.]................
Baltimore.........MD..A. W. Robson, Passr. and Ticket Agent..............127E Baltimore St.
Bellingham......WASH..W H. Gordon, Passenger Agent........ ....1225 Dock St.
Berlin........GERMANY..International Sleeping Car Co.....................71 Unter den Linden
Bombay..........INDIA..Ewart Latham & Co. Thos. Cook and Son..........18 Esplanade Rd.
Boston..........MASS.{ F. R. Perry, Dist. Passr. Agent.....................362 Washington Pt.
                     { G. A. Titcomb, City Passr. Agent
Brisbane ....QD..The British India and Queensland Agency Co. [Ltd.]
Bristol..........ENG..F. W. Forster, Agent.....................18 St. Augustine's Parade
Brussels....BELGIUM{ International Sleeping Car Co.........................Nord Station
                   { Thos. Cook & Son...........................41 Rue de la Madeleine
Buffalo...........N.Y..R. A. Burford, City Passenger Agent ....233 Main St.
Calcutta.........INDIA{ Thos. Cook & Son............................9 Old Court House St.
                     { Gillanders, Arbuthnot & Co ....
Canton.........CHINA..Jardine, Matheson & Co............................
Chicago...........ILL..A. C. Shaw, Gen. Agent. Passr. Dept.............232 South Clark St.
Cincinnati........OHIO..R. L. Thompson, G. A., P. D....Sinton Hotel Block. 15E Fourth St.
Cologne....GERMANY{ International Sleeping Car Co ....Central Station
                  { Thos. Cook & Son...1 Domhof
Colombo....CEYLON..Bois Brothers & Co., Thos. Cook & Son...................
Detroit..........MICH..A. E. Edmonds, City Passr. Agent..................7 Fort Street W.
Duluth.........MINN..M. Adson, Gen. Passr. Agt., D.S.S. & A. Ry.........Manhattan Bldg.
Frankfort...GERMANY..International Sleeping Car Co ....1 Kaiserstrasse
Glasgow.....SCOTLAND..Thomas Russell, Agent..........................67 St. Vincent St.
Halifax..........N.S..J. D. Chipman, City Passr. and Frt. Agent ....107 Hollis St.
Hamburg...GERMANY..Thos. Cook & Son, Tourist Agents....................39 Alsterdamm
Hamilton.........ONT..W. J. Grant, Commercial Agent............Cor. King and James Sts.
Hobart......TASMANIA..Union S.S. Co. of New Zealand [Ltd.]
Hong Kong ....D. W. Craddock, General Traffic Agent. China. etc.
Honolulu.........H.I..Theo. H. Davies & Co. [Ltd.]............................
Kobe..........JAPAN..J. Rankin, Agent ....14A Maye-Machi
Liverpool........ENG..J. J. Gilbertson, Agent.................................24 James St.
                    { Allan Cameron, Gen. Traffic Agt. } 62-65 Charing Cross S W. and
London..........ENG.{ F. W. Flanagan, Gen. Passr. Agt. } 67-68 King William St. E.C.
                    { H D. Annable, Gen. Freight Agt. }
London..........ONT..W. Fulton, City Passr. Agt...........................161 Dundas St.
Los Angeles.......CAL..F. A. Valentine, Travelling Passr. Agent.......Room 349, Wilcox Bldg.
Madrid..........SPAIN{ International Sleeping Car Co ... 18 Calle de Alcala [Equitable Bldg.]
                     { Thos. Cook & Son ....5 Carrera de S. Geronimo
Melbourne..........AUS..Union S.S. Co. of New Zealand [Ltd.].............
Minneapolis......MINN..W. R. Callaway, General Passr. Agent, Soo Line...
Montreal..........QUE{ E. J. Hebert, Gen. Agt. Passr. Dept.............Windsor St. Station
                     { A. E. Lalande, City Passr. Agent....................129 St. James St.
Moscow..........RUSSIA..International Sleeping Car Co.....................Hotel Metropole
New York..........N.Y.{ E. V. Skinner, Assistant Traffic Manager..................458 Broadway
                      { International Sleeping Car Co.......................281 Fifth Avenue
Niagara Falls......N.Y..D. Isaacs.....................................Prospect House
Nice...........FRANCE{ International Sleeping Car Co......................Avenue Massena
                     { Thos. Cook & Son............................16 Avenue Massena
Ottawa............ONT..George Duncan, City Passr. Agent ....42 Sparks St.
Paris ....FRANCE{ International Sleeping Car Co.....................3 Place d'Opera
                     { Hernu, Peron & Co, [Ltd.] Ticket Agents .... 61 Boulevard Haussman
                     { Thos. Cook & Son...................................1 Place d'Opera
Philadelphia.......PA..F. W. Huntington, Gen. Agent, Passr. Dept ....629.681 Chestnut St.
Portland...........ME..H. A. Snow, Ticket Agt., Main Central Rd...............Union Depot
Portland..........ORE..F. R. Johnson, Freight and Passr. Agent..............142 Third St.
Quebec...........QUE..Jules Hone, City Passr. Agent.......30 St. John St., cor. Palace Hill
Rome...........ITALY{ International Sleeping Car Co.....................Place San Silvestro
                    { Thos. Cook & Son........................54 Piazza Esedra di Termini
Sault Ste. Marie..MICH..W. J. Atchison, City Passr. Agt.; W. C. Sutherland, Depot Ticket Agent
St. John..........N.B..W. B. Howard, District Passr. Agent......................8 King St.
St. Paul........MINN..L. M. Harmsen, City Ticket Agent, Soo Line..........379 Robert St.
St. Petersburg....RUS..International Sleeping Car Co..................5 Perspective Newsky
San Francisco.....CAL..E. E. Penn, C.P.A.; J. H. Griffin, D.F.A.77 Ellis St., James Flood Bldg.
Seattle.........WASH..A. B. Calder, G.A.P.D... ....Mutual Life Bldg., 609 First Avenue
Shanghai........CHINA..A. R. Owen.....................................
Suva...........FIJI..Union S.S. Co. of New Zealand [Ltd.] ....
Sydney...........AUS..Union S.S. Co. of New Zealand [Ltd.]................
Tacoma.........WASH..J. O'Grady Passr. Agent..........................1113 Pacific Avenue
Toronto..........ONT..C. B. Foster, Dist. Passr. Agent.............71 Yonge St., cor. King
Vancouver........B.C..E. J. Coyle, Asst. Gen. Passr. Agent; W. R. Thomson, Ticket Agent
Victoria..........B.C..Geo. L. Courtney, Dist. Freight and Passr. Agent..58 Government St.
Warsaw........RUSSIA..International Sleeping Car Co.........................Hotel Bristol
Washington.......D.C..Wm. Linson, C.F & P.A...Bond Bldg., 14th St. and New York Avenue
Winnipeg.........MAN..A. C. Smith, City Ticket Agt........Cor. Main St. and Portage Avenue
Yokohama.......JAPAN..Wm. T. Payne, General Traffic Agent for Japan, etc............14 Bund

MESSRS. THOS. COOK & SONS, Tourist Agents, with offices in all parts of the world
are also Agents of the Canadian Pacific Railway, and can supply tickets and information.

Mont Louis
Douglastown
New Carlisle
Sayabec
Rimouski
Matapedia
Bathurst
Riviere du Loup
Mouth of St. Francis
Van Buren
Caribou
Presque Isle
Woodstock
Houlton
Levis
Tring Jc.
St. Francis
Moosehead Lake
Vanceboro
Mc Adam Jc.
Fredericton
St. John
Kentville
Digby
Nova Scotia
Point du Chene
Painsec Jc.
Cape Tormentine

HANDBOOK
of
BRITISH COLUMBIA CANADA

SPLENDOR SINE OCCASU

HANDBOOK
of
BRITISH
COLUMBIA
CANADA

OFFICIAL GOVERNMENT BULLETIN.

HAND BOOK

OF

BRITISH COLUMBIA,

CANADA.

ITS POSITION,
ADVANTAGES,
RESOURCES,
And CLIMATE.

MINING,
LUMBERING,
FISHING.

FARMING,
RANCHING,
FRUIT GROWING.

INDEX.

BRITISH COLUMBIA, CANADA.

BRITISH COLUMBIA, Canada's Maritime. Province on the Pacific Ocean, is the largest in the Dominion, its area being variously estimated at from 372,630 to 395,610 square miles, or about 250,000,000 acres. It is a great irregular quadrangle, about 700 miles from north to south, with an average width of about 400 miles, lying between latitudes 49 degrees and 60 degrees north. It is bounded on the south by the Strait of Juan de Fuca and the States of Washington, Idaho and Montana, on the west by the Pacific Ocean and Southern Alaska, on the north by Yukon and Mackenzie Territories, and on the east by Athabasca and the Province of Alberta. From the 49th degree north to the 54th degree the eastern boundary follows the axis of the Rocky Mountains, and thence north the 120th meridian.

The Province is traversed from south to north by four principal ranges of mountains—the Rocky and Selkirk ranges on the east, and the Coast and Island ranges on the west. The Rocky Mountain range preserves its continuity, but the Selkirks are broken up into the Purcell, the Selkirk, the Gold and the Cariboo mountains. Between these ranges and the Rockies lies a valley of remarkable length and regularity, extending from the international boundary line along the western base of the Rockies northerly 700 miles. West of these ranges extends a vast plateau or table land with an average elevation of 3,500 feet above sea level, but so worn away and eroded by water courses that in many parts it presents the appearance of a succession of mountains. In others it spreads out into wide plains and rolling ground, dotted with low hills, which constitute fine areas of farming and pasture lands. This interior plateau is bounded on the west by the Cascade, or Coast, Range, and on the north by a cross range which gradually merges into the Arctic slope. It is of this great interior plateau that Professor Macoun says: "The whole of British Columbia, south of 52 degrees and east of the Coast Range, is a grazing country up to 3,500 feet and a farming country up to 2,500 feet, where irrigation is possible."

The Coast Range is a series of massive crystalline rocks, averaging 6,000 feet in height, and a mean width of 100 miles, and descends to the Pacific Ocean. The Island Range, supposed to have been submerged in past ages, forms the group of islands of which Vancouver and the Queen Charlotte are the principal.

One of the most noticeable physical features of British Columbia is its position as the watershed of the North Pacific slope. All the great rivers flowing into the Pacific, with the exception of the Colorado, find their sources within its boundaries. The more important of these are: the Columbia (the principal waterway of the State of Washington), which flows through the Province for over 600 miles; the Fraser (750 miles long); the Skeena (300 miles); the Thompson, the Kootenay, the Stikine, the Liard and the Peace. These streams, with their numerous tributaries and branches drain an area equal to about one-tenth of the North American continent. The lake system of British Columbia is extensive and important, furnishing convenient transportation facilities in the interior. The area of lake aggregates 1,560,000 acres.

Many of the smaller streams are not navigable, but these furnish drive-ways to the lumbermen and supply power for sawmills, and electric plants, and water for irrigation. The water power is practically unlimited and so widely distributed that no portion of the Province need be without cheap motive power for driving all necessary machinery.

HISTORICAL.

In 1537 Cortez discovered California and for nearly half a century the Spaniards were the only navigators of the North Pacific. Sir Francis Drake was the first of the British to visit the Pacific Coast, in 1578, when he raided the Spanish settlements and set up the British flag at Drake's Bay near San Francisco and took possession of the country in the name of Queen Elizabeth, calling it New Albion. In 1592 Juan de Fuca discovered the Strait which bears his name, and other captains Juan Perez, Quadra, Behring and others, visited the coast at intervals until 1778, when Captain James Cook cast anchor in Nootka Sound, while on a mission to discover a north-east passage to the Atlantic. After sailing north to the Arctic Ocean and naming many sounds,

AMONG THE ISLANDS, GULF OF GEORGIA.

inlets and islands, Cook's ships sailed to the Sandwich Islands, where he was killed in a fight with natives. His vessels, the Resolution and the Discovery, returned to England, however, and the reports of their crews respecting the great opportunities for fur trading aroused so much attention that several expeditions were outfitted in England and in China and India for the North Pacific trade. For several years merchant adventurers, British, Spanish and Dutch, visited the coast as rival fur traders, but it was only in 1788 that Captain Meares established a

permanent settlement on Nootka Sound, where he built a ship called the North West America. The following year a Spanish force under Don Estevan Martinez seized the settlement in the name of his sovereign, confiscated the British ships and imprisoned the crews. These arbitrary acts nearly caused war between Britain and Spain, but the affair was finally settled by arbitration, Spain abandoning the territory and paying an indemnity of $210,000. Subsequently in 1792 and following years, Captain George Vancouver made a survey of the coast and established the existence of Vancouver Island, which had been a matter of dispute since the days of Juan de Fuca. The Mainland was for many years "No Man's Land," and it is due to the North-West Fur Company and the Hudson's Bay Company that that vast territory was brought to the notice of the world.

Alexander Mackenzie, who was the first man to cross the continent north of the Mississippi, reached the shore of the Pacific at the mouth of the Bella Coola River in July, 1793.

In 1800 David Thompson, travelling overland from Red River, near the present site of Winnipeg, reached the Bow River, near the present site of Calgary, and subsequently crossed the mountains and discovered the river which bears his name.

Sir Alexander Mackenzie, Simon Fraser and David Hearne also made extensive explorations and added materially to the knowledge of the great North West and the Pacific Coast.

In 1849 the Island of Vancouver was granted to the Hudson's Bay Company for a period of ten years. A government was established and Richard Blanchard was sent from England as Governor. He resigned in 1850 and was succeeded by James (afterwards Sir James) Douglas. An Assembly was called and held its first meeting at Victoria in August, 1856. While Vancouver Island was thus constituted a Crown Colony, the Mainland, known as New Caledonia, remained practically unknown, and inhabited only by Indians and a few fur traders. Gold was discovered on the Fraser River in 1857. and miners began to crowd into the country, making the establishment of some form of government a necessity. Therefore the whole of the Mainland, west of the Rocky Mountains, was created a Crown Colony under the name of British Columbia.

In 1866 the two colonies were united by Act of the Imperial Parliament and on July 20th, 1871. British Columbia became a Province of the Dominion of Canada. British Columbia entered Confederation upon the condition that within two years the construction of a railway should be begun which would connect it with the Eastern Provinces. This road is now the Canadian Pacific Railway. It was completed in 1885 and gave Canada and the Empire a great highway from the Atlantic to the Pacific.

The Provincial Government consists of a Lieutenant-Governor, appointed by the Dominion Government, an Executive Council, or Cabinet, of five members (who are elected members of the Legislative Assembly), and a Legislative Assembly of forty-two (including the Cabinet Ministers), elected by the constituencies into which the Province is divided.

The estimated revenue and expenditure of British Columbia for the fiscal year 1906-7 is as follows:

Revenue............. ....$2,647,976
Expenditure 2,912,916

The excess of expenditure over revenue is provided for, however, by a surplus of $268,265 remaining from the previous year.

RESOURCES.

With the exception of nickel (which has not yet been discovered in quantity) all that the other provinces of Canada boast of possessing in the way of raw material is here in abundance. British Columbia's coal measures are sufficient to supply the world for centuries; it possesses the greatest compact area of merchantable timber in North America; the mines have produced over $226,000,000 and may be said to be only in the early stages of development; the fisheries produce an average annual value of $5,500,000, and, apart from salmon fishing, their importance is only beginning to be realised; there are immense deposits of magnetite and hematite iron of the finest quality which still remain undeveloped; the agricultural and fruit lands produced $6,482,366 in 1905, and less than one-tenth of the available land is settled upon, much less cultivated; the Province has millions of acres of pulpwood as yet unexploited; petroleum deposits, but recently discovered, are among the most extensive in the world, and much of the territory is unexplored and its potential value unknown. With all this undeveloped wealth within its borders can it be wondered at that British Columbians are sanguine of the future? Bestowed by prodigal Nature with all the essentials for the foundation and maintenance of an empire, blessed with a healthful, temperate climate, a commanding position on the shores of the Pacific, and encompassed with inspiring grandeur and beauty, British Columbia is destined to occupy a position second to none in the world's commerce and industry.

TRADE AND TRANSPORTATION.

The trade of British Columbia is the largest in the world per head of population, amounting to close upon $300. What may it become in the future when the resources of the Province are generally realized and actively developed? In 1904 the imports amounted to $12,079,088, and the exports totalled $16,536,328. The leading articles of export are fish, coal, gold, silver, copper, lead, timber, masts and spars, furs and skins, fish oil, hops and fruit. A large portion of the salmon, canned and pickled, goes to Great Britain, Germany, Eastern Canada, the United States, Hawaiian Islands, Australia and Japan; the United States consumes a large share of the exported coal, and immense quantities of lumber are shipped to Great Britain, South Africa, China, Japan, India, Australia and South America. The valuable furs—seal, sea-otter and other peltries—are sent to Great Britain and the United States. China also buys a considerable amount of lumber, timber and furs. Valuable shipments of fish oil, principally obtained from dog fish, are consigned to the United States and Hawaii. A large inter-provincial trade with Alberta, Saskatchewan, Manitoba and the Eastern Provinces is rapidly developing, the fruit grown in British Columbia being largely shipped to the Prairie Provinces, where it finds a good market. With the shipping facilities offered by the Canadian Pacific Railway and its magnificent fleets of steamships running to Japan, China, New Zealand, Australia and Hawaii, backed by her natural advantages of climate and geographical position, British Columbia's already large trade is rapidly increasing. The tonnage of vessels employed in the coasting trade is 8,661,740 tons, and of sea-going vessels carrying cargoes to and from the ports of the Province, 2,545,741 tons. The Canadian Pacific is the principal railway in the province. It has two main lines, the Canadian Pacific Railway and the Crow's Nest Pass Railway, and several branches and steamboat connections on the inland lakes, besides its large fleet of ocean-going and coasting steamers. The railway mileage of the province is 1,544 miles, being one mile of track to each 250 square miles of area.

The Canadian Pacific Railway Coast Service employs a fleet of steamers, many of them model ships of their class, which ply between coast points, from Victoria, Vancouver, Seattle, Nanaimo and Ladysmith to northern British Columbia and Alaskan Ports. The Empress liners, world-famed for their speed, comfort and safety, make regular trips from British Columbia ports to Japan and China, while the Canadian-Australian liners give splendid service to Hawaii, Fiji, New Zealand and Australia.

The Canadian Pacific Railway Company recently acquired the Esquimalt and Nanaimo Railway, running from Victoria to Wellington, on Vancouver Island, a distance of 78 miles, together with the land grant of the E. & N., some 1,500,000 acres, and will probably extend the road to meet the growing requirements of the Island's trade.

CLIMATE.

Varied climatic conditions prevail in British Columbia. The Japanese current and the moisture-laden winds from the Pacific exercise a moderating influence upon the climate of the coast and provide a copious rainfall. The

VIEW OF LONG LAKE, OKANAGAN VALLEY.

westerly winds are arrested in their passage east by the Coast Range, thus creating what is known as the " dry belt " east of those mountains, but the higher currents of air carry the moisture to the loftier peaks of the Selkirks, causing the heavy snowfall which distinguishes that range from its eastern neighbor, the Rockies. Thus a series of alternate moist and dry belts are

57a. Pall Mall.
London. S.W.

formed. The climate of British Columbia, as a whole, presents all the conditions which are met with in European countries lying within the temperate zone, the cradle of the greatest nations of the world, and is, therefore, a climate well adapted to the development of the human race under the most favourable conditions. As a consequence of the purity of its air, its freedom from malaria, and the almost total absence of extremes of heat and cold, British Columbia may be regarded as a vast sanitarium. People coming here from the east invariably improve in health. Insomnia and nervous affections find alleviation, the old and infirm are granted a renewed lease of life, and children thrive as in few other parts of the world.

The climate of Vancouver Island, and the coast generally, corresponds very closely with that of England; the summers are fine and warm with much bright sunshine, and severe frost scarcely ever occurs in winter. On the Mainland similar conditions prevail till the higher levels are reached, when the winters are cooler. At Agassiz, on the Lower Fraser, the average mean temperature of January is 33 degrees, and of July 64 degrees; the lowest temperature on record at this point is—13 degrees, and the highest 97 degrees. There are no summer frosts, and the annual rainfall is 67 inches, 95 per cent. of which falls during the autumn and winter.

GARDEN. VICTORIA.

To the eastward of the Coast Range, in Yale and West Kootenay, the climate is quite different. The summers are warmer, the winters colder and the rainfall rather light—bright, dry weather being the rule. The cold winter is, however, scarcely ever severe, and the hottest days of summer are made pleasant from the fact that the air is dry and the nights are cool.

Further north, in the un developed parts of the province, the winters are more severe.

The great diversity of climate and the unique climatic. conditions existing in the mountains, valleys and along the coast, to which, if is added the scenic beauty of the landscape, give to life in British Columbia an undescribable charm. There is scarcely a farm house in all the valley regions that does not look out upon great ranges of majestic mountains, more or less distant. The floral beauty of the uncultivated lands and the wonderfully variegated landscape are a source of constant delight. Each one of the numerous valleys appeals to the observer with some special charm of scenic beauty, and presents distinct qualities of soil and climate, bounded by mountains stored with precious and economic minerals, watered by lakes and streams of crystalline purity, and clothed with a wealth of vegetation which demonstrates the universal fertility. These impress one with the great extent of the province and its inexhaustible resources. And this great natural wealth is so evenly and prodigally distributed that there is no room for envy or rivalry between one district and another, each is equally endowed, and its people firmly convinced that their's is the "bonanza" belt, unequalled by anything on top of the earth.

DISTRICTS OF BRITISH COLUMBIA.

British Columbia is divided into the following districts:

Kootenay (East and West)	15,000,000	acres
Yale	15,500,000	,,
Lillooet	10,000,000	,,
Westminster	4,900,000	,,
Cariboo	96,000,000	,,
Cassiar	100,000,000	,,
Comox (Mainland)	4,000,000	,,
Vancouver Island	10,000,000	,,

Each of these great districts would require a separate and detailed description in order to set forth its particular advantages of soil, climate, mineral and timber resources, and diversity of scenery, but space forbids more than brief mention.

THE KOOTENAYS.

Kootenay District (or, "The Kootenays") forms the south eastern portion of British Columbia, west of the summit of the Rocky Mountains, and is drained by the Columbia and Kootenay rivers. East Kootenay contains a large extent of agricultural land, much of which requires irrigation, but suited to fruit growing and all kinds of grain and vegetables. Most of the land is well timbered and lumbering is, next to mining, the principal industry. There are considerable areas of fertile land in West Kootenay and a good deal of it is being utilized for fruit growing. The fame of the Kootenay mines is world-wide, the mountains being rich in gold, silver, copper and lead, and the eastern valleys underlaid with coal and petroleum. British Columbia mining has reached its highest development in Kootenay, and, as a consequence, many prosperous cities and towns have been established. The development of the Crow's Nest coal fields and the revival in metalliferous mining has caused a rapid increase in population, especially in East Kootenay, where it is estimated to have more than doubled since 1901.

YALE.

Lying west of the Kootenays is the splendid Yale District, rich in minerals and timber and possessing the largest area of agricultural land in the province. It includes the rich valleys of the Okanagan, Nicola, Similkameen, Kettle River, and North and South Thompson, and the Boundary, and has been appropriately named "the Garden of British Columbia." The main line of the Canadian Pacific passes nearly through the centre of Yale from east to west, while the Okanagan branch and the lake steamers give access to the southern portions. New branch lines are projected, and some are in course of construction, which will serve to open up a very large mining and agricultural area. Cattle raising on a large scale has been one of the chief industries, but many of the ranges are now divided into small parcels which are being eagerly bought by fruit growers and small farmers. The district is very rich in minerals and coal, but development has been delayed by lack of transportation facilities—a drawback which will soon be removed.

LILLOOET.

In natural features Lillooet resembles Yale. It is largely a pastoral country. well adapted to dairying, cattle raising and fruit growing. Placer and hydraulic mining is carried on successfully, and quartz mining is making fair progress, but railway communication is needed to insure success.

WESTMINSTER.

One of the richest agricultural districts of the province is Westminster, which includes all the fertile valley of the Lower Fraser. The climate is mild, with much rain in winter. The timber is very heavy and the underbrush thick. Westminster is the centre of the great lumbering and salmon canning industries. Its agricultural advantages are unexcelled in the province, heavy crops of hay, grain and roots being the rule, and fruit growing to perfection and in profusion. A great deal of the land in the Fraser valley has been reclaimed by dyking.

CARIBOO AND CASSIAR.

The great northern districts of Cariboo and Cassiar are practically unexplored and undeveloped, although in the early days parts of them were invaded by a great army of placer miners, who recovered about $50,000,000 in gold from the creeks and benches. Hydraulic mining on a large scale is being carried on by several wealthy companies at different points in the district with fair success, and individual miners and dredging companies are doing well in Atlin. Recently large deposits of gold and silver quartz were found in Windy Arm, east of Atlin, and give promise of rich returns. Large coal measures have been located on the Telqua River and at other points, and copper ore is found in many localities. The country is lightly timbered and promises in time to become an important cattle raising and agricultural district, as there are many fertile valleys, which, even now, despite the absence of railways, are attracting settlers. In the southern part of Cariboo, along the main wagon road, are several flourishing ranches which produce cattle. grain and vegetables, finding a ready market in the mining camps.

COMOX.

The northern portion of Vancouver Island and a portion of the opposite mainland is known as Comox District. It is very rich in minerals and timber, and there is considerable fertile land. The deeply indented coast line and the adjacent islands afford fine opportunities for the fishing industry, which is now being developed on a considerable scale.

VANCOUVER ISLAND.

Not the least important portion of British Columbia is Vancouver Island, which, from its great wealth of natural resources and its commanding position on the Pacific Coast, should become one of the richest and most prosperous districts of the province. Coal mining and lumbering are the chief industries, and fishing, quartz mining, copper smelting, shipbuilding, whaling and other branches are being rapidly developed. The Esquimalt & Nanaimo Railway, running from Victoria to Wellington, serves a section of country which it would be difficult to surpass anywhere in the world for beauty of scenery and natural wealth. There are prosperous agricultural communities along the railway and in Comox District, and several mines are being developed. There is quite a large area of agricultural land, but it is heavily timbered and costly to clear by individual effort. The Esquimalt & Nanaimo Railway Company is considering a scheme for clearing portions of its land grant (which consists of about 1,500,000 acres), and it is expected that the economic methods adopted will enable the company to sell the land to settlers at moderate prices.

MINING.

British Columbia has been pertinently named "The Mineral Province," a title justified by the fact that the province produces annually more than 28 per cent. of the total mineral output of Canada, and that in 1904 the production of gold, silver, copper, lead and coal almost equalled the total production of those minerals by all the other provinces. With the prospect of a marked increase in the production of coal and coke in the near future, and the development o fthe iron and zinc deposits more than a probability, the day does not seem far distant when British Columbia will hold her title "The Mineral Province," by undisputed right.

Established as an industry in 1858, gold mining progressed rapidly. The output in 1858 was $705,000, in 1863 it had increased to nearly $4,000,000. The fame of British Columbia's gold fields had reached the ends of the earth and adventurers crowded from all quarters to share in the golden harvest. After 1868 the output of the placers decreased, but they continued to produce an average considerably over $1,000,000 per annum until 1882, when the industry gradually declined until hydraulic and dredging operations again placed it upon a substantial footing. The output for six years past has averaged close to $1,000,000 annually, with several companies operating on a large scale in the northern districts of the province.

The silver-lead mining was established in Slocan district about 1886, and in 1887 the first output of silver and lead is recorded. It amounted to 17,690 ounces of silver, valued at $17,331, and 204,800 pounds of lead, valued at $9,216. Access to the mineral zone, which includes Rossland and many other mining camps, was difficult, and it was not until the construction of railways that lode mining assumed important proportions. In 1894 the province began the export of copper, the output for that year being 324,680 pounds, of the value of $16,243. In ten years (1904) the production reached 35,710,125 pounds, worth $4,578,037. Lode gold mining had also a small beginning, the first record of production being 1,170 ounces, worth $23,404, in 1893, which had increased to 222,042 ounces, worth $4,589,608 in 1904, and 238,660 ounces, valued at $4,993,102 in 1905.

HYDRAULIC MINE AT KAMLOOPS, B.C.

It will thus be seen that mining is indeed in its infancy in British Columbia. The recent advances in the prices of copper, silver, and lead has given a new impetus to all branches of the industry. Mines that have lain idle for months are being operated ; smelters, closed for lack of ore, are again in full blast and increasing their capacity ; new discoveries are being developed into shipping mines—in fact, from all reports, the world is waking up to the importance of British Columbia's mineral resources, and the business is progressing more rapidly and on more substantial lines than ever before.

Coal mining, the oldest established branch of the industry, has kept pace with the requirements of the market, the output varying yearly with the demand, and averaging for ten years a little over a million long tons. A demand for coke was caused by the establishment of smelters, and in 1895 a few ovens were built at Comox, Vancouver Island, from which about 1,000 tons of coke were produced. The manufacture of coke has increased proportionately with the increase of the smelter industry, the Vancouver Island and Crow's Nest Pass collieries producing 238,428 long tons in 1904.

MINERAL RESOURCES.

Gold is distributed all over British Columbia. There are few places where "colors" may not be found for the seeking, and the metal is met with in paying quantities in almost every section. In 1904 the following districts and divisions contributed to the total production of gold : Cariboo, Quesnel, Omineca, Cassiar, Atlin, Skeena, East Kootenay, Fort Steele, Windermere, Golden, West Kootenay, Nelson, Slocan, Trail Creek, Revelstoke, Trout Lake, Lardeau, Lillooet, Yale, Grand Forks, Greenwood, Osoyoos, Similkameen, Vernon, Ashcroft, Kamloops, New Westminster, Nanaimo, Alberni, West Coast Vancouver Island and Victoria. A considerable portion of the gold is found incorporated with silver, copper and lead ores, from which it is separated at the smelters and refineries.

Silver, which for the most part is found in conjunction with lead and copper, is also widely distributed, the districts contributing to the grand total being : Cassiar, East Kootenay, Fort Steele, Golden, Windermere, West Kootenay, Ainsworth, Slocan, Nelson, Trail Creek, Trout Lake, Lardeau, Revelstoke, Arrow Lake, Lillooet, Yale, Greenwood, Grand Forks, Osoyoos, Ashcroft, Kamloops, Similkameen, Victoria, Alberni, Quatsino, Nanaimo and New Westminster. About 80 per cent. of the silver produced is obtained from silver-lead ores, the remaining 20 per cent being chiefly found associated with copper. Half the silver produced in 1904 came from Slocan District, about one-fourth from Fort Steele district, while the other districts mentioned above accounted for the rest. The lead production was chiefly in East Kootenay, Slocan and Ainsworth.

The range of copper is almost, if not quite, as wide as that of the more precious metals, the discovery of large bodies of ore being constantly reported from as yet undeveloped parts of the province. The chief sources of copper production at present are Boundary, Rossland, Coast, Yale, Kamloops, Nelson, Nanaimo, Alberni and Victoria Districts.

Coal is found in commercial quantities in many sections. The only working mines are at Ladysmith, Nanaimo and Comox, on Vancouver Island, and at Fernie, Morrissey, Michel and Carbonado in East Kootenay, but there are extensive beds of coal at several points, viz.: Quatsino Sound, Alert Bay, Port McNeil, Port Rupert, and Sooke on Vancouver Island ; a large deposit of anthracite on Queen Charlotte Islands, and on the mainland in East Kootenay, Nicola, Similkameen, Tulameen, Kamloops, Bulkley River Valley, Telqua River, Omineca and Peace River. The wide distribution and great extent of those numerous coal measures, surrounded as they are by a country of endless agricultural and mineral resources, gives assurance of prosperity to future generations for centuries to come and must be considered one of the most important assets of the province.

TYEE COPPER CO.'S SMELTER. LADYSMITH VANCOUVER ISLAND.

Large deposits of iron ore have been discovered in various localities on the mainland and on Vancouver and other islands, but none of them have been developed in a commercial sense. About 20,000 tons have been taken from Texada Island to supply a small iron furnace established at Irondale, Washington, which ceased operations in 1901, but has now resumed work. The only place on the mainland where iron has been mined in any quantity, and only to the extent of 3,000 to 4,000 tons is at Cherry Creek, near Kamloops, the magnetite being shipped to Nelson for use as a flux in lead smelting. At Bull River, Grey Creek and Kitchener, in East Kootenay, are iron deposits of considerable extent, as well as near Trail, West Kootenay. Iron also exists in large bodies at Sechelt, and near Fort George. The principal deposits occur on Vancouver Island, and are of large extent and conveniently situated for manufacturing purposes. The growing demand of all the country west of the Rocky Mountains for manufactures of iron and steel and the increasing Oriental trade should be an inducement to capital to establish an iron industry in this province, where all the necessary elements are found in abundance and so closely grouped as to ensure economic production. It may be added that the iron ores of Vancouver Island are of exceptionally high grade, and almost wholly free from sulphur and phosphorus. The principal deposits are on the Gordon River, Bugaboo Creek, and Barkley Sound, all within forty miles of Victoria, and on Quatsino Sound on the west coast.

Besides those mentioned above, British Columbia has deposits of almost every known economic mineral. Amongst these may be mentioned plumbago, platinum, cinnabar, molybdenum, chromic iron, manganese, asbestos, mica, asphaltum, gypsum, schulite, aquerite, pyrites, osmiridium and palladium. Several of these have been found in workable quantities, while others are mere occurrences, the extent of which has not yet been ascertained.

Much attention is now being given to the petroleum fields of South-east Kootenay, where a large area of oil-bearing strata is known to exist. Several companies are at work boring and otherwise developing their properties, and the reports of the progress are encouraging, leading to the hope of the establishment of a new and important industry. Specimens of oil from the Flat Head Valley and other localities are of superior quality and singularly free from impurities.

Marble, granite, sandstone, lime, brick and fire clay, cement and pottery clay are well distributed, and are being utilized to meet local demands. Considerable lime and cement is now being manufactured for domestic use and exportation and the trade is increasing satisfactorily. A form of slate is found on one of the Queen Charlotte Islands which cuts easily, hardens with exposure, and takes a fine polish.

The amount and value of mineral products for 1903, 1904, and 1905 were :

	Customary Measure.	1904.		1905.		1906. Approximate.	
		Quantity.	Value.	Quantity.	Value.	Quantity.	Value.
Gold placer	Ounces	55,765	$1,115,300	48,465	$969,300		$6,070,000
„ lode ..	„	222,042	4,589,608	238,660	4,933,102		
Silver ...	„	3,222.481	1,719,516	3,439,417	1,971,818		2,2 0,000
Lead ...	Pounds	35,6 6,214	1,421,874	56,580,703	2,399.022		690,000
Copper ...	„	35,710,128	4,578,037	37,692,251	5,876,222		8,690,000
Coal ...	Tons	1,253,628	3,760,884	1,384,312	4,152,936		4,580,000
Coke ...	„	238,428	1,192,140	271,785	1,358,925		1,050,000
Other materials		—	600,000	—	800,000		1,100,000
			$18,977,359		$22,461,325		$26,390,000

The tonnage of ore mined in 1904 was 1,461,609 tons (exclusive of coal), an increase of 14 per cent. over 1903, and, counting the increased production of 1903 over 1902, an increase in two years of 46 per cent. The number of mines shipping in 1904 was 142, as against 125 in 1903, and the number of men employed, 3,306. These shipping mines were located as follows : Skeena, 2 ; Fort Steele, 2 ; Windermere, 5 ; Ainsworth, 12 ; Nelson, 17 ; Slocan, 48 ; Trail, 13 ; other Kootenay divisions, 10 ; Lillooet, 1 ; Boundary, 20 ; Ashcroft and Kamloops, 2 ; Coast, including Vancouver Island, 10. During the year there were 26 mines working which did not ship, so that the actual number of working mines was 168.

SMELTING AND REFINING.

The smelting industry has fairly kept pace with the mining development In the early days of mining several smelting plants were installed before there was ore mined or blocked out with which to supply them and, consequently, some heavy losses were sustained by too sanguine promoters. These costly lessons made capital over-cautious, and for some years practically all the ore mined was sent abroad for treatment. The development of mining on business principles, which followed the " wild-catting " period common to all new mining countries, eventually restored confidence, and smelting soon grew to be an important and profitable industry. Expert metallurgists and chemists and skilled mechanics experimented till the most economic methods and processes were devised for treating the different classes of ores, and to-day British Columbia has eleven smelters and one refining plant, with a combined daily capacity of about 7,500 tons of ore. These plants are distributed as follows : Grand Forks (Granby smelter); Greenwood, and Boundary Falls, in the Boundary district ; Trail, Canadian Smelting Works (including lead and copper smelting works, a lead refining plant, and a sheet lead and lead pipe manufacturing plant) ; Nelson, Hall Mines smelter, and Pilot Bay, in West Kootenay ; and at Marysville in East Kootenay. There are two smelters on Vancouver Island, at Crofton and Ladysmith, one at Van Anda, Texada Island, and a zinc smelter at Frank, just over the boundary of British Columbia in Alberta. A good indication of the present healthy state of the mining industry lies in the fact that some of the smelters which had lain idle for some time have resumed operations, while others are increasing, in some instances doubling, their capacity.

MINING OUTLOOK.

The progress made in mining during 1905 has been very satisfactory —more so, perhaps, than in any previous year, for the industry now seems more firmly established on legitimate business lines than ever before in its history. Mining on paper and on the Stock Exchange has given place once for all to actual hard-headed work, with the result that confidence in British Columbia mines is greatly strengthened, and capital is beginning to loosen its purse-strings and aid in their development. An incentive to increased production and the opening of new ore bodies is the increase in prices of copper, silver and lead, which will easily make a million dollars difference in the value of this year's output of those metals. Another hopeful feature of the situation is the almost universal tendency of prospectors and mine owners to accept reasonable prices or grant practicable working privileges on mutually favorable terms. In the past capital was often driven away by the inflated notions of men who saw a bonanza in every prospect hole and demanded the price of a working mine for an outcropping of ore.

This year will also be marked by the addition of zinc to the list of British Columbia's metallic products, as a good deal of attention has been given to this metal and a serious effort is being made to turn it to account. It occurs in the form of zinc blend in many of the galena ores of Slocan District, and there is one mine which may be distinctly called a zinc mine. Zinc is also found on Quatsino Sound and on the coast near Vancouver.

SYNOPSIS OF MINING LAWS.

The mining laws of British Columbia are very liberal in their nature and compare favorably with those of any other part of the world. The terms under which both lode and placer claims are held are such that a prospector is greatly encouraged in his work, and the titles, especially for mineral claims and hydraulic leases, are absolutely perfect. The fees required to be paid are as small as possible, consistent with a proper administration of the mining industry, and are much lower than those of the other Provinces of Canada or the mineral lands under Dominion control.

The following synopsis of the mining laws of British Columbia is not applicable to Yukon Territory:

A free miner is a person, male or female, above the age of 18 years, who is the holder of a valid free miner's certificate, which costs $5 for a full year, or a proportionate sum for any shorter period, but all certificates expire on May 31st. A free miner may enter on Crown lands and also on other lands where the right to enter has been reserved, and may prospect for minerals, locate claims and mine. Claims may not be located on Indian reserves nor within the curtilage of any dwelling. Should a free miner neglect to renew his certificate upon expiry, all mining claims held by him under its rights, if not Crown granted, revert to the Crown, unless he be a joint owner, in which case his interest or share reverts to his qualified partners or co-owners. It is not necessary for a shareholder in an incorporated mining company, as such, to possess a free miner's certificate.

A mineral claim is a rectangular piece of ground not exceeding 1,500 feet square. The claim is located by erecting three posts, as defined in the Act. In general, location of a claim must be recorded within a period varying according to distance from a registrar's office from the date of location. A mineral claim, prior to being Crown granted, is held practically on a yearly lease, an essential requirement of which is the doing of assessment work on the claim annually of the value of $100, or, in lieu thereof, payment of that amount to the mining recorder. Each assessment must be recorded before the expiration of the year to which it belongs, or the claim is deemed abandoned. Should the claim not meantime have been relocated by another free miner, record of the assessment work may be made within 30 days immediately following the date of expiry of the year, upon payment of a fee of $10. A survey of a mineral claim may be recorded as an assessment at its actual value to the extent of $100. If during any year work be done to a greater extent than the required $100, any additional sums of $100 each (but not less than $100) may be recorded and counted as assessments for the following years. When assessment work to the value of $500 has been recorded the owner of a mineral claim is, upon having the claim surveyed and on payment of a fee of $25, and giving certain notices, entitled to a Crown grant, after obtaining which further work on the claim is not compulsory. The Act includes, too, liberal provisions for obtaining mill and tunnel sites and other facilities for the better working of claims.

There are various classes of placer claims severally defined in the "Placer Mining Act" under the heads of creek, bar, dry, bench, hill and precious-stone diggings. Placer claims are 250 feet square, but a little variation is provided for under certain conditions. They are located by placing a legal post at each corner and marking on the initial post certain required information. Locations must be recorded within three days if within 10 miles of a recorder's office; but if further away another day is allowed for each additional 10 miles. Record before the close of each year is requisite for the retention of placer claims. Continuous work, as far as practicable, during working hours, is necessary, otherwise a cessation of work for 72 hours, except by permission of the Gold Commissioner, is regarded as an abandonment. The Commissioner, however, has power to authorize suspension of work under certain conditions and also to grant rates to facilitate working of claims. No special privileges are granted to discoverers of "mineral" claims, but those satisfying the Gold Commissioner that they have made a new "placer" discovery are allotted claims of extra size.

No free miner may legally hold by location more than one mineral claim on the same lode or vein, and in placer diggings he may not locate more than one claim on each creek, ravine or hill, and not more than two in the same locality, only one of which may be a creek claim.

In both mineral and placer Acts provision is made for the formation of mining partnerships, both of a general and limited liability character; also for the collection of the proportion of value for assessment work that may be due from any co-owner.

Leases of unoccupied Crown lands are granted for hydraulic mining or dredging, upon the recommendation of the Gold Commissioner, after certain requirements have been complied with. An application fee of $20 is payable. Leases may not exceed 20 years' duration. For a creek lease the maximum area is ½ mile and the minimum rental $75; hydraulic lease, area 80 acres, rental $50, and at least $1,000 per annum to be spent on development; dredging lease, area 5 miles, rental $50 per mile, development work $1,000 per mile per annum, and a royalty payment to the Government of 50c. per ounce of gold mined.

Mineral or placer claims are not subject to taxation unless Crown granted, in which case the tax is 25c. per acre per annum; but if $200 be spent in work on the claim in a year this tax is remitted. A tax of 2 per cent. is levied on all ores and other mineral products, the valuation being the net return from the smelter; that is, the cost of freight and treatment is deducted from the value of the product, but not that of mining. These taxes are in substitution for all taxes on land, and personal property tax in respect of sums so produced, so long as the land is used only for mining purposes. A royalty of 50c. per 1,000 feet is charged on all timber taken from the land for mining uses.

Applications for coal or petroleum prospecting licences must, after the publication of certain notices, be made to the Gold Commissioner, accompanied by the plans of the land and a fee of $100, which sum will be applied as the first year's rent. Limit of land a licence will cover is 640 acres. Extension of lease for a second or third year may be granted. Upon proof of discovery of coal, royalty of 5c. and a tax of 10c. per ton of coal mined, 9c. on coke, and 12½c. per barrel of petroleum, is payable. After proof that land covered by lease has been worked continuously, lessee may, within three months of expiry of lease, purchase said land at $10 per acre.

Fees payable are: For a free miner's certificate, $5 per annum; records' $2.50 each; leases under "Placer Mining Act," $5, etc., etc. Incorporated companies pay for a free miner's certificate $50 per annum where the nominal capital is $100.000 or under, or $100 where it exceeds that sum.

MINERS' WAGES.

The current wages paid in and about the mines are as follows: Miners, $3 to $3.50 per day (12 to 14 shillings); helpers, $2.00 to $2.50 (8 to 10 shillings); laborers, $2.00 to $2.50 per day (8 to 10 shillings): blacksmiths and mechanics, $3.00 to $5.00 per day (12 to 20 shillings). Board is usually $7.00 (28 shillings) per week at mining camps.

ASSAY OFFICES.

The Provincial Government Assay Office at Victoria purchases gold from the miners at its full value less charges of assaying, which usually amount) to less than one-half of one per cent. The Dominion Government also maintains an assay office at Vancouver, where gold is bought on the same terms.

TIMBER.

Next to her great industry of minerals, the most readily available, if not the most important, of British Columbia's natural resources is her immense timber reserve. This province may now be said to possess the greatest compact area of merchantable timber in North America. The total forest area of Canada is estimated at 1,657,600,000 square acres (exceeding that of the United States and Europe combined), and of this British Columbia has 182,750,000 acres. This immense extent of forest and woodland is not, of course, all of present commercial value, as much of it is covered with small trees, only fit for fuel and domestic purposes, which would not be considered as "timber" by loggers, who choose only the largest and best trees. As far north as Alaska the coast is heavily timbered, the forest line following the indentations of the shore and the river valleys and fringing the mountain sides. The Douglas fir, the most widely distributed and valuable tree found on the Pacific Coast, grows as far north as 51 deg., where it is supplanted by the cypress, or yellow cedar, red cedar, hemlock and spruce. The fir is very widely distributed, being found from the coast to the Rocky Mountains. On the coast it attains immense proportions, sometimes towering to a height of 300 feet with a base circumference of 30 to 50 feet. The best average trees are 150 feet clear of limbs and five to six feet in diameter. The fir is the staple of commerce, prized for its durability and strength. The great bodies of this timber are found on Vancouver Island, on the coast of the mainland and in the Selkirk and Gold Mountains. Next to the Douglas fir in importance are the cypress and red cedar, both of which are of great value and much in demand. Red cedar shingles are the standard, and are finding an increasing market in Eastern Canada. The white spruce is also much sought after by certain builders for use in the better class of buildings. Hemlock is abundant in the province and possesses qualities which should make it more valued than it is. The western species is different and much superior to the eastern hemlock, and is as serviceable in many ways as more expensive lumber. There are many other trees of commercial value which are manufactured into lumber, including white pine, tamarac, balsam, yew, maple and cottonwood.

The trees indigenous to the province are: White fir, western white fir, mountain balsam, large-leaved maple, vine maple, red alder, arbutus, western birch, canoe birch, western dogwood, red cedar, American larch, mountain larch, western larch, white spruce, western black spruce, black spruce, white-marked pine, scrub pine, white mountain pine, yellow pine, western crab-apple, balsam, poplar, cottonwood, aspen, cherry, Douglas fir, western white oak, lance-leaved willow, willow, western yew, giant cedar, yellow sypress or cedar, western hemlock, Alpine hemlock.

There are between 100 and 150 sawmills in the province, big and small, and a large number of shingle mills, planing mills, and sash and door factories, representing—with logging plants, logging railways, tug boats, etc., and exclusive of the value of lands purchased and leased as timber limits—about $15,000,000 of capital invested in this industry.

The lumber cut of 1904 aggregated 325,271,568 feet, cut off 703,433 acres under lease from the Provincial Government; but in addition to this there was cut on Dominion lands within the Canadian Pacific Railway Belt 22,760,222 feet, the product of 17 mills out of 27 which manufactured timber cut on these limits. Ten of these mills did not operate during the season of 1904. Despite this drawback, however, the cut of 1904 considerably exceeds that of any former year, the figures standing thus: Cut on Provincial lands, 325,271,598 feet; cut on Dominion lands, 22,760,222 feet. Total, 348,031,790 feet.

57a, Pall Mall,
London. S.W.

ROSS-MACLAREN MILL, BURRARD INLET.

BRITISH COLUMBIA TOOTH PICKS.

57a, Pall Mall,
London. S.W.

In 1905, the acreage under lease has increased to 770,362 acres and the total cut on Provincial lands aggregated 450,385,544 feet, an increase of 135,311,986 feet over 1904. The quantity cut on Dominion lands within the Railway Belt for the year ending June 30, 1905, totalled 23,328,175 feet, making the grand total for 1905, 473,713,719 feet.

OAK TREES, BEACON HILL PARK, VICTORIA, B.C.

Previous to 1905, twenty-one year leases of timber lands were granted on certain conditions, but this law has been repealed, and now all timber is cut by special licences, which are granted for an area not exceeding 640 acres and are renewable annually. The annual fee for a timber licence is:—West of the Cascade Range, $140; east of the Cascade Range and in Atlin District, $115. The number of special licences granted in 1905 was 2,200, representing 1,400,000 acres.

The principal markets for the coast mills are Great Britain, Australia, New Zealand, Japan, China, Africa and South America, while the interior, or mountain mills, ship most of their product by rail to Alberta, Saskatchewan and Manitoba, and an increasing quantity to Ontario, Quebec, and the Middle and Eastern States. The foreign shipments have averaged considerably over 50,000,000 feet annually, with an average value of over $700,000. Figures as to the value of lumber used in the local markets and exported by rail are not available, but in 1904 these shipments amounted to 138,000,000 feet, and would be worth about $2,500,000.

British Columbia cedar shingles are in high favor in Eastern Canada and the Atlantic States, as well as in the Middle West. The output of shingles for 1905 is estimated at 250,300,000, and these are reported to have been all sold, so that the mills began the new year with no stocks on hand. The future of the shingle business, as well as that of the lumbering generally, is very promising.

A few years ago the lumber industry was confined almost wholly to the coast districts, where the big trees attracted capital, but as population increased in the interior and in the Prairie Provinces, the demand for lumber became greater and sawmills were set up in many localities in the mountains to supply the new market. There are at present about 45 mills in the interior, with a combined output of about 280,000,000 feet annually, representing an investment of nearly $10,000,000. They pay out for wages and supplies $2,500,000 annually. These mountain mills look almost entirely to the prairie country for their market. The outlook for the lumber business is favorable and the companies are making extra large cuts of logs this winter.

PULP AND PAPER.

As a field for the manufacture of paper pulp and paper, British Columbia stands without a rival. Possessing as she does her full share of the enormous timber reserve of Canada, her geographical position gives her a decided advantage over other places, for her pulp wood borders the ocean or the numerous rivers and streams which furnish easy and cheap communication with deep water harbors. With transportation charges at a minimum and an unlimited supply of the raw material of the very best quality, British Columbia should be in a position to supply the greater half of the world with wood pulp, or, better still, with paper of every grade and quality and in every form in which paper is used in the industrial arts. While the pulp and paper mills of Eastern Canada may find markets in the Eastern States and Europe, British Columbia should absolutely control the rapidly developing markets of Asia and Australia. An important factor in favour of this industry is the density of the British Columbia forest. Another important point is the mildness of the coast climate, which permits of work being done the year round.

In order to encourage the establishment of pulp and paper mills the Provincial Government, a few years ago, passed a law providing for the granting of special leases to individuals or companies desiring to embark in this enterprise. The result has been the formation of several companies, at least two of which are now engaged in preliminary work and promise to be in active operation before the close of another year.

•

FISHERIES.

The coast of British Columbia, embracing all the sea-front which lies between the 49th and 55th parallels of north latitude, presents an ideal field for the prosecution of a great fishing industry in all its branches. The coast is indented by innumerable bays, sounds, inlets and

Fishing Fleet at Mouth of Fraser River B.C.

DAY'S SALMON CATCH.

other arms of the sea, so that the actual shore line exceeds 7,000 miles, while thousands of islands shelter the inshore waters from the fury of ocean storms. This vast maze of water is alive with all kinds of fish, from the mighty whale to the tiny sardine, but until very recently commercial fishing has been practically confined to the taking of salmon. The fertility of the soil, the wealth of the mines and the quality and quantity of the timber have all served to divert attention from the fisheries, and it is but lately that their importance has begun to be recognized. The salmon, swarming in myriads to the mouths of the rivers during the spawning season, forced men to appreciate their value, and as they proved an easy prey, salmon canning was established as one of the great industries of the province

A Winter Day's Fishing on Okanagan Lake.

To convey an idea of the importance of the fisheries, it is only necessary to quote from the report of the Department of Marine and Fisheries. Taking an average year British Columbia produced in sea fish, and exclusive of seal and sea otter, a value of $4,999,417, of which $3,753,892 was derived from salmon. No account is taken of the fresh water fishes which are found in great abundance in the inland waters. The total yield of Canadian fisheries in 1904 fresh and salt water, was $23,506,439, so that British Columbia's contribution to the output was about 21 per cent. of the whole, and of that 21 per cent. 75 per cent. was for salmon. These figures, while proving beyond question the great commercial value of British Columbia's salmon fisheries, suggest the immense possibilities which await the development of the numerous other branches of the fishing business. The product of the Atlantic Coast fisheries, long recognized as a prime factor in the economic fabric of Canada, will one day be equalled and probably surpassed by those of British Columbia. Prince Edward Island, Nova Scotia, New Brunswick and Quebec produced fish to the value of $14,787,126 in 1904, as against British Columbia's $4,999.417, but it must be borne in mind that the Atlantic fisheries have reached almost their maximum of development, while those of the Pacific have been seriously attacked in only one branch—salmon. It is therefore but reasonable te expect very different results when British Columbia's fisheries reach the same stage of development, for, practically inexhaustible, the only limit to their output should be the lack of markets, a contingency not likely to arise.

The principal food fishes of the North Pacific are :—Salmon (five varieties, viz.: sockeye, spring, or tyee, cohoe, humpback and dog), halibut, cod (several varieties), herring, sturgeon, bass, oolachans. smelts, perch, trout, skill. sardines, anchovies, shad, oysters, clams, crabs, shrimps, and prawns. Whales are very plentiful along the coast and in Behring sea, and a whaling company recently organized, with headquarters at Sechart, Barkley Sound, is doing a profitable business. Dog fish, a species of shark, which prey upon the salmon and other fish, are valuable for their oil and the manu· facture of guano, and several companies are taking them in large quantities.

Halibut are caught in great numbers off the coast and their exportation to the Eastern markets has become an important industry, second only to salmon canning. In 1903 the total catch of halibut on the Pacific Coast, from California north, was about 25,000,000 pounds, of which British Columbia supplied over 10,000,000.

Herring of excellent quality are taken on the east coast of Vancouver Island, the present centre of the industry being Nanaimo. They are pronounced equal to the Atlantic fish by experts, engaged by the Dominion Government to instruct the British Columbia fishermen in the best methods of curing and packing. The catch of herring is increasing annually and promises to become a very important branch of the fishing business. Cod fishing has not been given much attention, but seems to offer good opportunities for profit if carried on systematically. The oolachan, a fish of the smelt family, swarms to the rivers in the early summer and is caught in large quantities by the Indians, with whom it is a staple food. It is a delicious fish, delicate in flavor, and should afford profitable business if canned or otherwise preserved for export.

There are many other sea products, which might be turned to account with advantage. Very little has been done in the minor branches, of the fishing industry, yet, there is little doubt, that canning crabs, clams, sardines, smelts, prawns, shrimps, etc., could be made to pay handsomely, while giving employment to a large number of people.

Sealing was at one time a leading source of profit in British Columbia but the business has fallen off considerably of late, owing to restrictions imposed by the Behring Sea Award and the decreasing number of seals. The average catch for five years ending 1903 was 26,300 skins, as compared with an average of 62,600 skins for a corresponding period ten years ago. The value of 1904 catch was $219,690. The Victoria sealing fleet of 1904 consisted of 38 schooners, employing 637 men, of whom 299 were whites and 338 Indians, and the catch totalled 14,646 skins.

Apart from the commercial aspects of British Columbia's fisheries they offer exceptionally good sport to the amateur fisherman and angler. All the numerous rivers, creeks and lakes, as well as the sea, teem with fish, so that the gentle art may be enjoyed at all seasons and in every part of the province.

The sockeye salmon, the kind most prized for canning, appear in greater numbers every fourth year. These are called "big years," and fishermen and canners make special preparations for them. In 1901, a big year, 1,247,212 cases of salmon were put up in British Columbia, worth $5,986,000, and containing 12,500,000 fish, weighing 60,000,000 pounds. The present year (1905) was another big one and, although the early run disappointed the hopes of the canners, the fish came in sufficient numbers to fill 1,167,460 cases, valued at $5,750,000, compared with 465,894 cases in 1904. There were about 80 canneries operated in 1905, employing over 20,000 fishermen and others, 168 steamers and other vessels, and 500 small boats. The capital invested in the salmon industry is about $4,000,000.

AGRICULTURE.

The traveller whose knowledge of British Columbia is gained from a trip through the province by railway and steamboat departs after having spent many delightful hours in Nature's picture gallery, in which she has collected her most precious treasures in bewildering profusion. He has quite failed to catch the details of her mighty work and carries away with him the impression that the principal asset of the country is its scenery, and it is with a feeling of relief that he looks out upon the more homelike scenes of the Fraser delta. The cultivated fields, the cosy farm houses surrounded by orchards, the fat cattle knee deep in the lush meadows, come as a balm to eyes surfeited with the sublimity of all that has passed before them. "Here," he whispers to himself, "reign peace and plenty created by the hand of man, an essential to national prosperity lacking in the mountains."

Here and there since crossing the Rockies he has seen cultivated patches, stretches of bench and bottom lands which might be utilised for farming and ranching, but the mountains dominated all else and he can only vaguely surmise as to the agricultural possibilities of the valleys separating the numerous ranges. He is therefore prepared to accept the statement that British Columbia is "a sea of mountains," in which mining and lumbering must furnish the only occupations for its population. The creation of this false impression is not far to seek. The Canadian Pacific Railway seeking the shortest path to the Pacific, let no barrier of nature, however formidable, stand in its way, but pushed its main line through regions the most unpromising, from an economical point of view. Thus the traveller catches the merest glimpses of the rich agricultural valleys which intersect the mountain ranges from north to south, and which are capable of supplying a population of many millions with all the products of farm, ranch, orchard and dairy.

To form a just estimate of the extent and importance of the agricultural areas of British Columbia one must make many excursions to the north and south of the main line of the Canadian Pacific Railway—over its branches and steamboat connections—and even then, if he trusts to what he may be able to see from the car window or the deck of a lake steamer, his knowledge will be far from complete. In the Shuswap and Okanagan Valleys, for instance, for every acre of arable land within sight of the railway or lake, there are thousands hidden away behind the beautiful grass-covered hills which border the highway of travel, and the same may be said of Kootenay, Boundary, Arrow Lake, Similkameen and other districts. The agricultural capabilities of the many sections of Southern British Columbia are, as a matter of fact, only beginning to be realized. So far they have been practically ignored, for the mineral seeking prospectors who first invaded the country had no eye for aught save the object of their quest. Now, however, branch lines of railway and lake steamers are enabling a new class of men to enter and explore this land of promise and many have embarked in fruit-growing, mixed farming and dairying.

The agricultural and pastoral lands are not restricted to a small proportion of the total acreage, for Professor Macoun, after personal investigation on the ground, says ; " The whole of British Columbia, south of 52 degrees and east of the Coast Range, is a grazing country up to 3,500 feet, and a farming country up to 2,500 feet, where irrigation is possible." This is a most important statement and its truth is being confirmed by the practical experience of settlers who have established themselves in the country. Within the boundaries thus roughly defined by Professor Macoun the capabilities of the soil are practically unlimited. All of it that is not too elevated to serve only for grazing purposes will produce all the ordinary vegetables and roots, much of it will grow cereals to perfection, while everywhere the hardier varieties of fruits can be successfully cultivated. As far north as the 52nd degree it has been practically demonstrated that apples will flourish, while in the southern belt the more delicate fruits, peaches, grapes, apricots, etc., are an assured crop. Roughly estimated, the extent of these fertile lands may be set down at one million acres, but this figure will probably be found far below the actual quantity capable of cultivation when the country has been thoroughly explored. The anticipation of such a result is justified from the fact that at several points in the mountains, even in the most unpromising looking localities, where clearing and cultivation has been attempted it has proved successful. In several instances also, bench land, pronounced only fit for pasturage by "old timers" has been broken and cropped with very satisfactory results. The agricultural lands just mentioned are located as follows :

	Acres.
Okanagan	250,000
North and South Thompson Valleys	75,000
Nicola, Similkameen and Kettle R'r Valleys	350,000
Lillooet and Cariboo	200,000
East and West Kootenay	125,000

West of the Coast Range are several extensive tracts of arable land of the richest quality, notably the Lower Fraser Valley, Westminster District, Vancouver Island and adjacent islands in the Gulf of Georgia. These sections of the province are recognized as agricultural districts and are fairly well settled, but much of the land is still wild and untilled. North of the main line of the Canadian Pacific Railway, on the Pacific slope, and but partially explored, are vast areas of agricultural and grazing lands, which will be turned to profitable account when the country is a few years older. Much of this northern region is fit for wheat growing, and all of it will produce crops of the coarser cereals, roots and vegetables, except the higher plateaux, which will afford pasturage to countless herds of cattle, horses and sheep. Some of these districts, best known and in which settlements have been established, are Chilcotin, Nechaco, Blackwater, Bulkley, Ootsa, Kispyox, Skeena and Peace River Valleys, and they are estimated to include some 6,500,000 acres. That this is a conservative estimate is clear from the fact that the late Dr. Dawson and Professor Macoun credited that portion of Peace River Valley lying within British Columbia with 10,000,000 acres of wheat land.

DIVERSIFIED FARMING.

The advantages of diversified farming over special farming are many and important, and there is scarcely a district in British Columbia in which diversified farming may not be carried on more profitably than any special branch of the industry. Large areas which require irrigation and are now used for grain growing and stock raising will at no distant day be supplied with water and will afford men of moderate means the opportunity to acquire

57a, Pall Mall,
London, S.W.

A VIEW FROM MOUNT TOMLIE, VICTORIA, B.C.

FARM HOME, VANCOUVER ISLAND.

57a, Pall Mall.
London. S.W.

homes and pursue general farm work under conditions similar to, but more advantageous and profitable than the Eastern Provinces.

Irrigation, though far from general, has already wrought a change in agricultural methods in those districts in which it has been introduced, but so far farming under this system does not appeal to the average easterner. Many who have had no experience with irrigation entertain the feeling that it is suited to special farming only. When they learn the use of water, applied where and when it is needed, and come to understand that there is nothing intricate or difficult to be learned in respect to it, they quickly appreciate its advantages. The productive value of land in British Columbia which has good water facilities, is easily four times as great as land in Eastern Canada. The milder climate contributes to this in a measure, but the great advantage of irrigation lies in being able to control the elements, or, in other words, in being independent of them in the conduct of farm work. Diversified farming is essentially practicable where irrigation is required. It enables the farmer to gratify his fancy with respect to crops, and at the same time realise from the land the greatest possible returns. By studying the needs of his locality and adjusting his products to the demand, he derives a continuous income without fear of failure from drought or excessive rain. The general farmer may combine stock raising, which includes dairying, in a small way, hay and grain, poultry, hogs and sheep, with a great variety of small fruits and vegetables. The farmer who understands how to reduce his product to compact form, making his alfalfa or hay field support a few cows, which will yield with their increase a considerable annual return each, a few sheep and hogs. which find ready sale at all seasons, a small band of hens and turkeys, always saleable at good prices, can easily wait for his fruit trees to come to bearing—he will never find it necessary to confine himself to a special branch. Thousands of men who are struggling for a meagre livelihood on exhausted fields elsewhere may find prosperous homes here with profitable occupation in a climate and amidst scenes of beauty and grandeur unequalled in the world.

AGRICULTURAL OPPORTUNITIES.

The opportunities for profitable diversified farming are practically unlimited. The demand for every product of the farm is great and ever increasing, the present supply being wholly inadequate for the local market.

Under a system of small land holdings, with diversified field culture, every object of cultivation is highly profitable, because produced by labor that might otherwise be unproductive.

DAIRYING.

Dairying pays handsomely, especially in cases where the farmer is not obliged to employ skilled labor to do the milking and butter-making. The local demand for butter is constantly increasing with the population and the prices secured are far higher than in the East. In 1904 the creameries of the province produced 1,119,276 pounds of butter, which was sold at an average of 26½ cents per pound, or $296,608, little more than twenty-five per cent. of the value of butter imported. The production for 1905 was 1,398,778 pounds from the creameries, while the dairy product was about 400,000 pounds, 1,800,000 pounds in all, valued at $480,000. Quite a large proportion of the imported article was forwarded to Yukon, but that fact only serves to show the great possibilities for dairying in British Columbia. The province possesses many elements necessary to constitute it a great dairying country, the products of which should include cheese and condensed milk. There are extensive areas of pastoral land in the interior, while increased cultivation in the lower country will form the necessary feeding ground. With a plentiful supply of good water and luxuriant and nutritious grasses, there is every required facility added. The coast climate is most favorable to the dairying industry. Clover, one of the most valuable plants in cultivation, is practically a weed in British Columbia west of the Cascade Range. Once it gets established in the soil it is almost impossible to get it out. Lucerne, or afalfa, is succeeding admirably. In Okanagan Valley, Thompson River Valley, and many other points, three heavy crops of this nutritious fodder are produced annually.

There are fifteen co-operative and private creameries established in the province, all doing well and earning satisfactory dividends. The Provincial Government aids the establishment of co-operative creameries by loaning the promoters one-half the cost of the creamery building, plant and fixtures, repayable in eight instalments with interest at five per cent., the first of such instalments to be paid at the expiration of three years, and the other seven annually thereafter.

Cheese making has scarcely been attempted on a commercial basis, although there is a good field for that branch of dairying, and the Government is prepared to assist the establishment of factories on the same terms as those in which aid is granted to creameries.

POULTRY RAISING.

Poultry raising is an important branch of general farming which is gradually developing in British Columbia, but not to the extent which its importance warrants. The home market is nowhere nearly supplied either with eggs or poultry, large quantities being imported from Manitoba, Ontario, California, Washington, Oregon. In 1904 the value of eggs and poultry imported amounted to over $400,000, and good prices prevail at all seasons, the average wholesale prices for eggs on the coast being : Fresh eggs, 30 cents per doz.; case eggs, 22 cents per doz.; while the retail price for fresh eggs averaged 37½ cents per doz., ranging from 25 cents to 70 cents. Fowls bring from $5 to $8 per doz.; chickens, $4 to $7; ducks, $5 to $11; geese, $1 to $1.50 each, and turkeys from 22 to 30 cents per pound.

A practical poultry raiser who has made a success of the business on Vancouver Island says : " I have no hesitation in saying that there are good profits in the business, conducted on a strictly commercial basis. In fact, I know of no other branch of agriculture which is so profitable, having in view the amount of capital to be invested and the expense of conducting it. . . . . Properly managed, in any number, poultry ought to reap a profit of at least $1 per head per annum."

Every portion of British Columbia is suitable for poultry raising. In the coast districts hens, ducks and geese can be bred to great advantage, and the drybelts and uplands are particularly well adapted to turkeys.

With such facts before them it is a matter for surprise that many farmers in British Columbia send to the nearest store for their eggs and fowls. Eggs and chickens are by-products on every well-conducted Eastern farm, and they add considerably to the annual income, as well as providing agreeable and healthful variety to the family's bill of fare.

GRAIN GROWING.

Wheat is grown principally in the Fraser Valley, Okanagan, Spallumcheen, and in the country around Kamloops in the Thompson River Valley, and is manufactured at local mills, at Enderby, Armstrong and Vernon. Until the northern interior of the province is brought under cultivation through the construction of railways the wheat area will not be increased. Wheat is only grown on the Mainland Coast and Vancouver Island for fodder and poultry feeding.

Barley of excellent quality is grown in many parts of the province.

Oats are the principal grain crop, the quality and yield being good, and the demand beyond the quantity grown. Rye is grown to a limited extent, and is used for fodder.

The average yields of grain and prices are as follows:

Wheat, bushels per acre ...... 25 02 ; Price per ton $33 15
Oats,   „   „   ...... 39 05 ;   „   „ 27 00
Barley,   „   „   ...... 33 33 ;   „   „ 28 00

These averages are very much exceeded in many cases, and according to nature of soil and local conditions. In the matter of oats as high as 100 bushels to the acre is not an uncommon yield.

ROOT CROPS.

Potatoes, turnips, beets, mangolds, and all the other roots grow in profusion wherever their cultivation has been attempted. Sixty-eight tons of roots to a measured acre is recorded at Chilliwack, and near Kelowna, on Okanagan Lake, 20 acres produced 403 tons of potatoes, which sold at $14 per ton. The Dominion census places the average yield of potatoes at 162.78 bushels to the acre. The average price of potatoes is $14 to $16 per ton, while carrots, turnips, parsnips and beets sell at an average of about 60 cents per bushel.

HOP CULTURE.

The Okanagan, Agassiz and Chilliwack districts are well suited to hop growing and produce large quantities, unexcelled in quality. British Columbia hops command good prices in the British market and most of the crop is sent there, though recently Eastern Canada and Australia are buying increasing quantities. The yield of hops averages 1,500 pounds to the acre and the average price is 25 cents per pound.

FODDER CROPS.

Besides the nutritious bunch grass which affords good grazing to cattle, horses and sheep on the benches and hillsides, all the cultivated grasses grow in profusion wherever sown. Red clover, alfalfa, sainfoin, alsike, timothy and brome grass, yield large returns—three crops in the season in some districts and under favourable circumstances. Hay averages about 1½ tons to the acre and the average price was $17.25 in 1904.

SPECIAL PRODUCTS.

Tobacco growing has proved successful in several districts, notably in Okanagan, where a leaf of superior quality is produced. Tobacco of commercial value will grow in almost any part of Southern British Columbia, and there is no reason why the farmers of the province should not cultivate it in a small way for their own use, as is the custom in many parts of Quebec and Ontario.

The importance of apiculture is beginning to be recognised and a considerable quantity of delicious honey of home production is found in the local markets. As the area of cultivation extends, bee-keeping should become a profitable adjunct of general farming.

The Coast districts and many of the lowlands of the interior are well-suited to cranberry culture, which is being tried in a small way, but with success, by settlers on the West Coast of Vancouver Island.

Celery, another vegetable luxury, is grown in limited quantities, but the soil and climate warrant its cultivation on a more general scale. Celery properly grown and packed would command good prices, and an unlimited market.

Sugar beets grow to perfection in several localities, but their cultivation on a large scale has not been attempted.

Indian corn, melons and tomatoes are profitable items in the output of the small farmer, and are successfully grown in all of the settled districts.

LAND CLEARING.

Companies are being formed with the object of taking contracts to clear land and prepare it for cropping at a fixed price per acre, according to the density of the forest growth. These companies, wherever possible, calculate to utilise the timber and cordwood so that the cost of clearing will be reduced to a minimum.

The Canadian Pacific Railway Company, which owns a large tract of agricultural land on Vancouver Island, is formulating a systematic plan for clearing portions of it, by the most approved methods, before offering it for sale to settlers. The company's work in this direction will be the first instance of clearing land on economic principles.

IRRIGATION.

As already observed, a very considerable percentage of the agricultural lands in the interior districts requires irrigation in order to insure crops. Generally speaking, there is abundant water within reach, but there are sections where the height of the land above water level or distance from the source of supply stands in the way of individual attempts at irrigation, but the work may be accomplished by co-operation and with the expenditure of capital. The supplying of water to these higher plateaux is, however, a matter for future consideration, as there is sufficient land capable of irrigation at comparatively small cost to meet the requirements for some years to come. In Okanagan, Similkameen and Kamloops districts companies have purchased large tracts of land, formerly used as cattle ranges, which they are subdividing into small holdings of ten acres and upwards, and constructing reservoirs and ditches, which will provide an unfailing supply of water. These companies are already reaping the reward of their enterprise, as the land is being rapidly sold to actual settlers, who are planting orchards and engaging in mixed farming. The example set by the Canadian Pacific Railway in Alberta in wresting over 2,000,000 acres from barren sand and low-producing grain fields, and making them yield millions of bushels of wheat, is one which cannot be overlooked by British Columbians, who, witnessing the transformation which is taking place on their eastern border, cannot fail to profit by the lesson. It is therefore safe to predict that the next few years will witness the reclamation of many hundreds of thousands of acres of bench lands from pasturage to flourishing orchards and farms, the homes of thousands of prosperous settlers.

DYKING.

British Columbia, although generally accepted as a country of high altitudes, includes large tracts of alluvial lands, which are overflown at certain seasons, and therefore require dyking in order to make them available for cultivation. These lowlands are located on the Lower Fraser, at Canal Flats (the head waters of the Columbia River), in West Kootenay, and on the northwest coast of Vancouver Island. The Government of British Columbia early recognized the importance of reclaiming the rich alluvial meadows in the Fraser River Valley, and to that end established a system of dykes, which has rendered over 100,000 acres fit for cultivation. These reclamation works represent an expenditure of $981,000 up to November, 1904. The Government undertakes the redemption of dyking debentures issued by the municipalities benefited and payable in forty years. In West Kootenay from the international boundary a tract of meadows extends to the south end of Kootenay Lake, a distance of about 35 miles, comprising about 40,000 acres. These lands have been partially reclaimed by dyking, and are very productive, but the greater portion is still a vast haymeadow. Fronting the west and north coast of Vancouver Island is a very large body of land, which could be made available for mixed farming and dairying by inexpensive dyking and drainage. The extent of this land is estimated at over 150,000 acres.

LIVE STOCK.

Cattle raising on a large scale was once one of the chief industries of the province, and many of the large ranches are still making money, but the tendency of late has been for smaller herds and the improvement of the stock. The efforts of the Dairymen's and Live Stock Association have proved successful in this direction. The Association imports and sells to its members every year a certain number of young pure-bred stock, purchased in Eastern Canada by a special agent, who visits the principal stock markets in the interests of the farmers. In 1904 the Association imported and distributed 43 cattle, principally Shorthorns; 10 mares (Clydes); 13 sheep, Hampshire Downs; 14 pigs, Yorkshire; 33 fowls, White Leghorn. The bulls sold from $100 to $150 and up to $500; the mares averaged about $300.

While the province is capable of raising all the beef, mutton and pork required for home consumption, a very large quantity is imported, the money sent abroad annually amounting to about $3,000,000. The parts of the province particularly adapted to cattle raising are the interior plateaux and the Fraser River Valley, though there is scarcely a district in which the keeping of a few head will not pay well, for the high prices prevailing justify

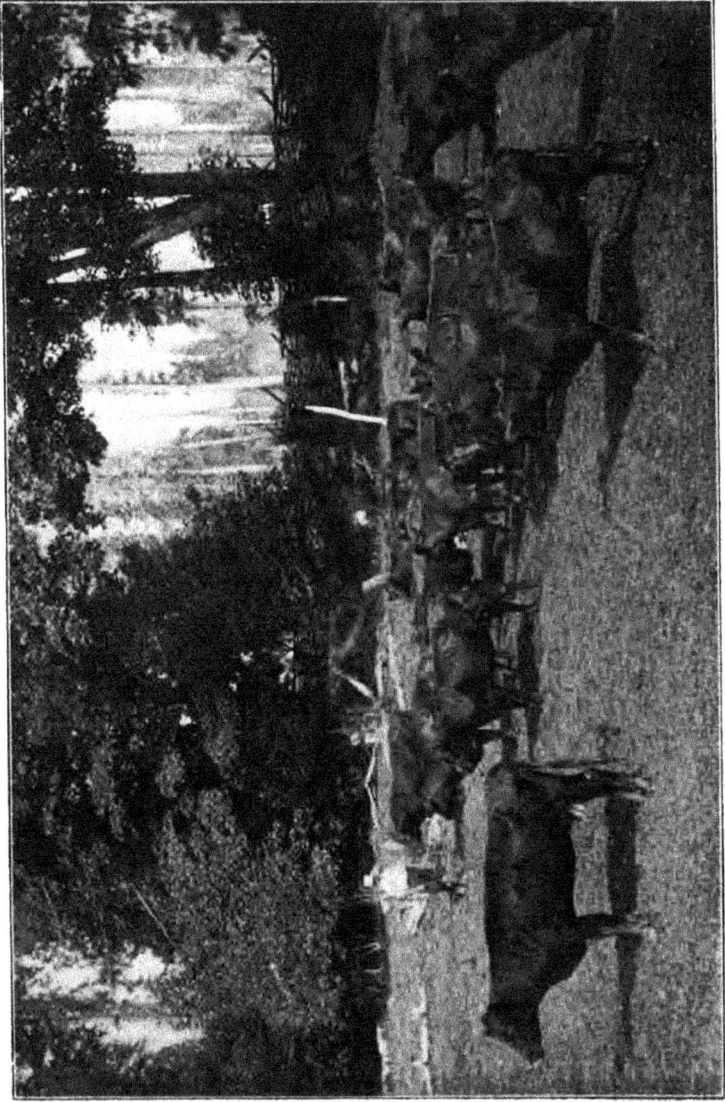

HERD OF RED POLLED CATTLE, VANCOUVER ISLAND.

stall feeding. The development of irrigation should stimulate the cattle industry, and make the province self-supporting in respect of beef.

Sheep raising is another branch of agriculture capable of great expansion. In the past the ranchers of the interior objected to sheep, as they are such close feeders, and sheep-raising was confined chiefly to Southern Vancouver Island and the Gulf Islands, where considerable numbers were produced. These are the most favourable parts of the province for sheep-raising, though they do well in many localities in the interior.

Hogs, in small farming, are probably the most profitable of live stock, owing to the general demand for pork, bacon, ham and lard, and much attention is now being given to raising them. Over $1,000,000 of hog products are imported annually, and prices are always high, so that the farmer can never make a mistake in keeping a small drove of pigs. The breeds which mature earliest are the Berkshire and Poland China. The increased production of hogs has encouraged the establishment of some small packing houses, but there is room for very extensive expansion. Hogs thrive in every part of the province, and are in demand at all seasons, especially animals weighing from 125 to 150 pounds, suitable for fresh pork.

The demand for good horses, especially heavy draft and working animals, is always increasing, and prices are consequently high. Formerly horses were raised in great numbers in the interior without much attention to their quality, and in consequence great bands of wild horses became a nuisance and a menace to the farmers and ranchers to such an extent that the Legislature had to adopt measures for their destruction. The quality of horses has been much improved of late, and, although the " cayuse, " the native pony, will always be prized for its hardihood and endurance, the tendency everywhere is for a better class of animal. The horses exhibited at the recent Dominion Exhibition at New Westminster compared favourably with those of any country in the world.

PURE BRED STOCK.

As already noted, the Dairymen's and Live Stock Association is doing splendid work in securing to the farmers of British Columbia a better class of live stock. The efforts of the Association in this direction are materially assisted by the Canadian Pacific Railway, which grants a freight rate of one-half the regular rates on all importations of pure bred stock, the only condition to granting such rate being the production of uniform record certificates in every case. The company insists that "all Record Certificates accepted by the railway must be of uniform size and appearance, and bear the seal of some central body recognized by the Dominion Department of Agriculture." While this rule protects the railway company against fraud, it acts as a double safeguard to the importer and purchaser of high-bred animals.

FRUIT GROWING.

British Columbia fruit is preferred above all others in the markets of the Middle West, where it commands profitable prices. In 1904 a small exhibit sent to England was awarded the gold medal of the Royal Horticultural Society, and last year (1905) a car lot, exhibited in London, won the first prize from all competitors, while no less than eight medals were awarded the individual exhibits which made up the collection. This goes to prove that despite the great distance British Columbia fruit has secured a prominent place in the British market, in which Oregon and California apples have heretofore sold at the highest prices.

The fruit industry of British Columbia is in its infancy, but the results so far secured are convincing as to its future importance. The actual extent of fruit growing land has not yet been ascertained, but by a conservative estimate at least one million acres south of the 52nd degree will produce all the fruits of the temperate zone. The recognized fruit districts include the southern part of Vancouver Island and the Gulf Islands, Lower Fraser River Valley, Thompson River Valley, Shuswap Lake, Okanagan, Spallumcheen, Osoyoos, Similkameen, Upper Columbia Valley, Kootenay Lake, Arrow Lake, Lower Columbia River and Grand Forks, which are all suited to the best grades of fruit, and which contain extensive areas of fruit lands. Other good fruit districts are :—West Coast of Vancouver Island, West Coast of Mainland (where patches of fruit lands are found at the heads of the numerous inlets) ; Lower Fraser Valley, Nicola, Grand Prairie, and many other localities. In some of these sections irrigation is necessary and, as mentioned elsewhere, water is being supplied where the influx of population warrants the necessary expenditure. Many localities, which are now proved to be suitable for fruit culture, were but recently "discovered," for a few years ago fruit was only raised in the settlements on the coast and along the rivers, and in quantity that failed to supply even the limited local demand. In 1891 the total orchard area of the province was 6,500 acres. In ten years it only increased 1,000 acres, but from 1901 to 1905 it jumped to 22,000 acres, and it is safe to say that that acreage will be more than doubled again before the close of 1906. Ten years ago British Columbia did not produce enough fruit to supply her own population. The following table of fruit shipments is interesting in showing the steady growth of the industry :

	By freight.	By Express.	Total.	Increase.
1902	1,469 tons	487 tons	1,956 tons	
1903	1,868 tons	676 tons	2,544 tons	588 tons
1904	2,161 tons	864 tons	3,025 tons	481 tons
1905	3,181 tons	1,176 tons	4,357 tons	1,332 tons

An increase of over 50 per cent. in four years.

These shipments by no means represent the whole fruit crop, much the greater part of which is consumed locally.

These figures may seem small compared with those of older fruit growing countries, but they show conclusively that the industry is growing steadily, and with every indication of its becoming one of the most important items in the future prosperity of the province. There has been a large increase in acreage of orchards during 1905.

The actual experience of many fruit growers is highly satisfactory to them and a temptation to every man who desires to make money pleasantly to set up in the business. In Okanagan there are instances of $500 to $600 gross profit per acre. At Kelowna nine tons of pears and 10 tons of prunes per acre are not uncommon. Near Nelson, 14 acres produced 1,000 cases of strawberries and 94 tons of roots, netting the owner $100 per acre. This land was formerly a cedar swamp. At Lytton to-day grapes, averaging four pounds to the bunch, were grown in the open. On the Coldstream Ranch, near Vernon, twenty acres produced $10,000 worth of Northern Spy apples. At Peachland one acre and a half gave a return of $700 in peaches. Tomatoes to the value of $1,500 per acre were grown on Okanagan Lake. A cherry tree at Penticton produced 800 pounds of fruit. These cases are by no means exceptional or confined to any single district, similar ones could be

CHERRY TREES IN BLOSSOM.

BING CHERRY.

cited from almost any part of the province. Apples and pears produce from 8 to 15 tons of fruit per acre, according to variety, and the average price is $26 and $30 per ton respectively. Plums, prunes, cherries and peaches invariably bear largely, and the prices are always satisfactory, if the fruit is properly picked and packed.

Fruit packing has been brought to a fine art in British Columbia, the methods used being considered perfect by experts, and other countries are following her lead in this most important matter. Careless or dishonest packing is not tolerated, offenders being severely punished.

PEACHES AND GRAPES.

Peaches are successfully grown in many parts of Southern British Columbia, and in every case the fruit has attained a good size, ripened fully and possessed an exceptionally fine flavor. Peach-growing gives promise of becoming an important industry in Okanagan, where the area of young orchards is increasing rapidly. Many of these are bearing, and peaches, from now on, will become a noticeable item in fast freight and express shipments. So far the shipments have been very small, as nearly all the peaches grown find ready sale on the spot, and there has been no surplus with which to supply even the Provincial markets. The small lots exported have been in the nature of experiments—samples with which to demonstrate the capabilities of the country.

Peaches grow to perfection in all the valleys south of the main line of the C. P. R., and as this fact becomes generally known more attention will be given to their cultivation.

Grape culture on a commercial basis can scarcely be said to be established in the province, but wherever their cultivation has been tried in the southern districts it has proved successful. The experience of Mr. Thomas G. Earle, of Lytton, who may be styled the pioneer grape grower, is that nearly every variety of grape will ripen in the "dry belt," and that in most cases they will come to maturity about two weeks earlier than in Ontario.

The fact that grapes of excellent quality and flavor can be grown in quantity, sufficient to supply the large and steadily increasing demand having been established, horticulturists in the "dry belt" will be encouraged to set out vineyards, and in time that part of British Columbia will rival Ontario's famed ;Niagara Peninsula as a producer of grapes and peaches. British Columbia grapes are as yet a novelty on the market, but their superior merits will in time win them a leading position.

OTHER FRUITS.

Nectarines, apricots, figs, almonds and several other of the less hardy fruits and nuts have been tried in a small way with success, and men of experience are not wanting who express the opinion that the sunny slopes of the lake country and the boundary will produce any fruit or vegetable which is grown for 300 miles south of the international boundary line.

LAND REGULATIONS.
Provincial Government Lands.

Crown lands, where such a system is practicable, are laid off and surveyed into quadrilateral townships, containing thirty-six sections of one mile square in each.

LORD SUFFIELD APPLE-5 YEARS OLD.

57a, Pall Mall.
London. S.W.

Any person, being the head of a family, a widow or single man over the age of eighteen years, and being a British subject, or any alien, upon making a declaration of his intention to become a British subject, may, for agricultural purposes, record any tract of unoccupied and unreserved Crown lands (not being an Indian settlement) not exceeding one hundred and sixty acres in extent.

No person can hold more than one pre-emption claim at a time. Prior record of pre-emption of one claim and all rights under it are forfeited by subsequent record or pre-emption of another claim.

APPLE BLOSSOM, ORCHARD IN OKANAGAN VALLEY.

Land recorded or pre-empted cannot be transferred or conveyed until after a Crown grant has been issued.

Such land, until the Crown grant is issued, is held by occupation. Such occupation must be a bona fide personal residence of the settler or his family.

The settler must enter into occupation of the land within thirty days after recording, and must continue to occupy it.

Continuous absence for a period longer than two months consecutively of the settler or family is deemed cessation of occupation ; but leave of absence may be granted not exceeding six months in any one year, inclusive of two months' absence.

Land is considered abandoned if unoccupied for more than two months consecutively.

If so abandoned, the land becomes waste lands of the Crown.

The fee on recording is two dollars (8s.).

The settler shall have the land surveyed at his own instance (subject to the rectification of the boundaries) within five years from the date of record.

After survey has been made, upon proof in declaration in writing of himself and two other persons of occupation for two years from date of pre-emption and of having made permanent improvement on the land to the value of two dollars and fifty cents per acre, the settler on producing the pre-emption certificate obtains a certificate of improvement upon payment of a fee of $2.

After obtaining the certificate of improvement and paying for the land, the settler is entitled to a Crown grant in fee simple. He pays $10 therefor.

The price of Crown lands pre-empted is $1 (4s.) per acre, which must be paid in four equal instalments, as follows : First instalment two years from date of record or pre-emption, and yearly thereafter, but the last instalment is not payable till after the survey, if the land is unsurveyed.

Two, three or four settlers may enter into partnership with pre-emptions of 160 acres each, and reside on one homestead. Improvements amounting to $2.50 per acre made on some portion thereof will secure Crown grant for the whole, conditions of payment being same as above.

The Crown grant reserves to the Crown a royalty of five cents per ton on every ton of merchantable coal raised or gotten from the land, not including dross or fine slack, and 50 cents per M. on timber. Coal and petroleum lands do not pass under grant of lands acquired since passage of Land Act Amendment of 1899.

No Crown grant can be issued to an alien who may have recorded or pre-empted by virtue of his declaring his intention to become a British subject, unless he has become naturalized.

The heirs of devisees of the settler are entitled to the Crown grant on his decease.

Crown lands may be purchased to the extent of 640 acres, and for this purpose are classified as first, second and third class, according to the report of the surveyor. It has not, however, been the policy of the Government for some time past to sell lands, except when required for special purposes.

Lands which are suitable for agricultural purposes, or which are capable of being brought under cultivation profitably, or which are wild hay meadow lands, rank as and are considered to be first class lands. Lands which are suitable for agricultural purposes only when artificially irrigated, and which do not contain timber valuable for lumbering purposes, as defined below, rank as and are considered to be second class lands. Mountainous and rocky tracts of land which are wholly unfit for agricultural purposes, and which cannot, under any reasonable conditions, be brought under cultivation, and which do not contain timber suitable for lumbering purposes, as defined below, or hay meadows, rank as and are considered to be third class or pastoral lands. Timber lands (that is, lands which contain milling timber to the average extent of eight thousand feet per acre west of the Cascades, and five thousand feet per acre east of the Cascades, to each one hundred and sixty acres) are not open for sale.

The minimum price of first class land, $5 per acre ; second class, $2.50 per acre ; third class, $1 per acre. No settlement duties are required on such lands unless a second purchase is contemplated. In such a case, the first purchase must be improved to the extent of $5 per acre for first class ; $2.50 second class, and $1 third class.

Leases of Crown lands which have been subdivided by survey in lots not exceeding 20 acres may be obtained ; and if requisite improvements are made and conditions of the lease fulfilled at the expiration of lease, Crown grants are issued.

Leases (containing such covenants and conditions as may be thought advisable) of Crown lands may be granted by the Lieutenant-Governor in Council for the following purposes :

(a) For the purpose of cutting hay thereon, for a term not exceeding ten years.

(b) For any purposes whatsoever, except cutting hay as aforesaid, for a term not exceeding twenty-one years.

The farm and buildings, when registered, cannot be taken for debt incurred after registration ; and it is free from seizure up to a value not greater than $500 (£100 English). Cattle " farmed on shares " are also protected by an Exemption Act.

The fact of a person having a homestead in other provinces, or on Dominion Government lands in this province, is no bar to pre-empting Crown lands in British Columbia.

Twenty-one year timber leases are now subject to public competition, and the highest cash bonus is accepted, subject to the 50 cents per M. royalty above mentioned, and an annual rental in advance of 15 cents per acre. The holder must put up a saw mill capable of cutting not less than 1,000 feet of lumber per day of twelve hours for every 400 acres of land in such lease ; and such mill shall be kept running for at least six months in every year.

HOW TO SECURE A PRE-EMPTION.

Any person desiring to pre-empt unsurveyed Crown Lands must observe the following rules:—

1. Place a stake or post four or more inches square and four or more feet high—a tree stump squared and of the proper height will do—at each corner of the claim, and mark upon each of the posts his name and a description of the post, for example :—

" John Smith's land, N.E. post (meaning north-east post) ; John Smith's land, N.W. post, " and so on.

2. After staking the land, the applicant must make an application in writing to the Land Commissioner of the district in which the land lies, giving a full description of the land, and a sketch plan of it ; this description and plan to be in duplicate. The fee for recording is $2.

3. He shall also make a declaration in duplicate, before a Justice of the Peace, Notary Public, or Commissioner, in Form 2 of the Land Act, and deposit

same with his application. In the declaration he must declare that the land staked by him is unoccupied and unreserved Crown land, and not in an Indian settlement ; that the application is made on his own behalf and for his own use for settlement and occupation, for agricultural purposes, and that he is duly qualified to take up and record the land.

4. If the land is surveyed the pre-emptor must make application to the Commissioner exactly as in the case of unsurveyed lands, but it will not be necessary to plant posts.

5. Every pre-emption shall be of rectangular or square shape, and 160 acres shall measure either 40 chains by 40 chains —880 yards by 880 yards, or 20 chains by 80 chains—440 yards by 1,760 yards 80 acres shall measure 20 chains by 40 chains ; and 40 acres, 20 chains by 20 chains. All lines shall be run true north and south and true east and west.

6. When a pre-emption is bounded by a lake or river, or by another pre-emption or by surveyed land, such boundary may be adopted and used in describing the boundaries of the land.

7. Thirty days after recording the pre-emptor must enter into occupation of the land and proceed with improving same. Occupation means continuous bona fide personal residence of the pre-emptor or his family, but he and his family may be absent for any one period not exceeding two months in any year. If the pre-emptor can show good reason for being absent from his claim for more than two months, the Land Commissioner may grant him six months' leave. Absence without leave for more than two months will be looked upon as an abandonment of all rights and the record may be cancelled.

8. No person can take up or hold more than one pre-emption.

9. The pre-emptor must have his claim surveyed, at his own expense, within five years from the date of record.

10. The price of pre-empted land is $1 per acre, to be paid for in four equal annual instalments of 25 cents per acre, the first instalment to be paid two years after record.

11. After full payment has been made the pre-emptor shall be entitled to a Crown grant of the land, on payment of a fee of $10.

12. A pre-emption cannot be sold or transferred until after it is Crown granted.

USE OF WATER.

Under the provisions of the "Water Clauses Consolidation Act, 1897," and amending Acts, unrecorded water may be diverted from any natural sources for irrigation or agricultural purposes generally. The scale of fees is the same as for water for industrial purposes, and is calculated on a sliding scale. For a record fee of $10.75 per 100 miner's inches up to $110.75 for 500 inches ; $260.75 for 1,000 inches ; $560.75 for 2,000 inches ; $680.75 for 5,000 inches ; $880.75 for 10,000 inches, and so on. For industrial purposes there is an annual fee calculated according to the same sliding scale ; but no annual fee is charged on water recorded and actually used for agricultural purposes.

DOMINION GOVERNMENT LANDS.

All the lands in British Columbia within twenty miles on each side of the Canadian Pacific Railway main line are the property of Canada, with all the timber and minerals they contain (except precious metals). This tract of land, with its timber, hay, water-powers, coal and stone, is now administered by the department of the Interior of Canada, practically according to the same laws and regulations as are the public lands in Manitoba and the Territories. Government Agencies are established at Kamloops, in the mountains, and New Westminster, on the coast.

Any British subject who is the sole head of a family, or any male of the age of 18 years, may secure a homestead of 160 acres on any unoccupied land, on application to the local land agent and on payment of a fee of $10. The homesteader must reside on the land for six months in every year, and cultivate at least 15 acres for three years, when he will be entitled to a free grant or patent.

CANADIAN PACIFIC LANDS.

The Canadian Pacific Railway Company controls large areas of farming, fruit, ranching and timber lands in the Kootenay and Boundary Districts. Generally speaking the prices for agricultural lands are as follows :—

First Class Lands.—Lands suitable for agricultural purposes in their present condition, or which are capable of being brought under cultivation profitably by the clearing of the timber thereon, or which are wild hay meadow lands. Price, $5 per acre.

Second Class Lands.—Lands which are suitable for agricultural purposes only when irrigated. Price, $2.50 per acre.

Third Class Lands.—Mountainous and rocky tracts of land, unfit for agricultural purposes, and which cannot under any reasonable condition be brought under cultivation. Price, $1 per acre.

Any land in the Columbia and Western Land Grant (Boundary District) which contains timber fit for manufacture into lumber to the extent of 3,000 feet board measure, to the acre, does not come under the heading of agricultural land, but will only be disposed of under the provisions of the Company's regulations for the sale or lease of timber lands. In the remaining grants the limit for agricultural lands is fixed at 5,000 feet, board measure, to the acre.

The minimum area sold is 160 acres, and all lands must be purchased in square or rectangular parcels, viz., 160 acres must measure 40 chains by 40 chains ; 320 acres must measure 80 chains by 40 chains ; and 640 acres must measure 80 chains by 80 chains.

Land sold at $1 per acre must be paid for one-fourth cash, and the balance in three equal annual instalments.

Land sold at $2.50 per acre must be paid for one-fifth cash, and the balance in four equal annual instalments.

Land sold at $5 per acre must be paid for one-eighth cash, and the balance in seven equal annual instalments.

Interest at six per cent. is payable on all outstanding amounts of principal, and also on overdue instalments. If land is paid for in full at the time of purchase, a discount of ten per cent. will be allowed on the amount so paid in excess of the usual cash instalment, but no reduction will be allowed on subsequent payment of instalments in advance of maturity.

MOUNT STEPHEN, FIELD, ON C.P.R.

The Company has also lots for sale in the following town sites :—Elko, Cranbrook, Kimberley, Kitchener, Creston, in East Kootenay; Nelson, Proctor, Trail, Nakusp, Lemonton, Arrowhead and Revelstoke, in West Kootenay; Grand Forks, Eholt, Greenwood, Midway and Kamloops in Yale District, and at Vancouver on the coast.

The terms of payment are one-third cash, and the balance in six and twelve months.

☞ The purchaser of agricultural land will be permitted to use what timber is actually required on the land purchased by him for buildings, fences and fuel, but any timber cut for sale will be subject to the payment of dues as per the following schedule :—

Lumber, per M feet, B. M. $1 00
Shingle bolts, per cord 1 00
Firewood, per cord... 25
Fence posts, per cord.. 50
Mining props (10 ft. x 10 in. or less), per cord................................... 50
Mining props (larger), each 05
Ties, each 02
House logs (20 ft. or less), each ....,. 10
Piles, cribbing, timber, telegraph posts, per running foot.................... ½

Such dues are exclusive of all Government royalties, which must be paid by the purchaser. In the case of unsurveyed lands the purchaser must arrange his own surveys.

One half of the amount paid by new settlers for fare on the lines of the Canadian Pacific Railway in travelling to British Columbia will be applied on account of the first instalment if land is purchased from the Company in this province.

Timber leases may be secured from the C. P. R. by payment of the same schedule of dues as charged the purchaser of land, as in the foregoing.

Maps showing the Company's lands, pamphlets and regulations, containing detailed information, may be secured on application to J. S. Dennis, British Columbia Land Commissioner, Calgary, or to the local land agents at Cranbrook, Nelson, Creston, Trail and Grand Forks.

PRICES OF LAND.

Apart from the Government and railway company's lands, there is a great deal of desirable land owned by companies and individuals, the price of which varies with locality, quality of soil and cost of clearing or irrigation.

For purposes of comparison the topography and climatic conditions seem to lend themselves to a natural division of the province into the following districts :—

1. The Upper Mainland.—All the country to the eastward of the Coast Range, and including the large cattle ranges and what is known as the Dry Belt.

2. The Lower Mainland.—All that portion of the sea coast to the westward of the Coast Range, and including the rich delta lands of the Fraser River. This part of the country is generally heavily wooded with big timber and is the wettest part in the province.

3. The Islands.—All that portion including Vancouver Island and the adjacent islands. This division partakes somewhat of the characteristic of the two others, and resembles the first in the distribution of the flora and the less precipitation.

Division No. 1 includes the Boundary Country, Similkameen, Okanagan Lake, Okanagan, Shuswap Lake, Thompson River Valley (upper and lower), Nicola, Upper Fraser Valley, Chilcotin and Cariboo Wagon Road. Improved or partly cleared land in the Boundary district is held at about $50 per acre. Similkameen, $25 to $150, the latter being irrigated. Okanagan Lake, $60 to $250 for water fronts, irrigated and improved land, and from $1 to £25 for non-irrigated. Okanagan bush land, $5 to $20; partly cleared and improved, $10 to $50 and up to $100 per acre. Shuswap and Upper Thompson Valley, prices about the same as Okanagan. Land may be bought at lower rates than those quoted in Nicola, Upper Fraser Valley, Chilcotin and Cariboo. It is hard to give definite figures as the country is so extensive and conditions are so varied.

Division No. 2 includes Delta, Surrey, Langley, Matsqui, Sumas, Chilliwack, South Vancouver, Burnaby Coquitlam, Maple Ridge, Mission, Dewdney, Nicomen and Kent, and prices of land vary very much. The improved land is held at $5 to $20, while reclaimed (dyked) land sells from $40 up to $100.

Division No. 3 embraces Victoria, Esquimalt, Metchosin, Sooke, Highland, Lake Saanich, Cowichan, Nanaimo, Comox, Alberni, San Juan and Port Rupert Districts and the numerous islands of the Gulf of Georgia. As in other parts of the province, there are no fixed prices for land. They vary with locality and the estimates of the owners. Wild land, mostly heavily timbered, can be bought from $3.50 to $10 per acre, while improved land ranges all the way from $20 to $200 according to extent and value of improvements.

While some of these prices may be thought high it must be considered that a small farm well located and well tilled in British Columbia will produce more and return bigger profits than a much larger area of land in most other countries.

TAXATION.

Outside of incorporated cities, towns and municipalities, the taxation is imposed and collected directly by the Provincial Government and expended in public improvements, roads, trails, wharves, bridges, etc., in assisting and maintaining the schools, in the administration of justice.

The rates of taxation imposed by the latest Assessment Act are as follows :—

On Real Estate...........................3-5 per cent. of assessed value of $2,000
"   "   "   ....1 per cent. of assessed value over $2,000
" Wild Land .... 4 per cent.
" *On Coal Land, Class A. .... 1 per cent.
" **Coal Land, Class B. .... 2 per cent.
" On Timber Land .... 2 per cent.
On income of $2,000 or under..1½ per cent.
On income over $2,000 and not exceeding $3,000 ....1¾ per cent.
On income over $3,000 and not exceeding $4,000 .... 2 per cent.
On income over $4,000 and not exceeding $7,000 .... 3 per cent.
On income over $7,000 .... 4 per cent.

*Working Mines.   **Unworked Mines.

Discounts of 10 per cent. upwards are allowed for prompt payment of taxes and the following exemptions from taxation are granted:—

On personal property up to $500 (to farmers only).

On income up to $1,000.

On pre-empted land for two years from date of record and an exemption of $500 for four years after record.

In addition to above taxes royalty is charged on coal, timber and minerals.

EDUCATION.

The province affords excellent educational opportunities. The school system is free and non-sectarian, and is equally as efficient as that of any other province in the Dominion. The expenditure for educational purposes, amounts to $400,000 annually. The Government builds a school house, makes a grant for incidental expenses, and pays a teacher in every district where twenty children between the ages of six and sixteen can be brought together. For outlying farming districts and mining camps the arrangement is very advantageous. High schools are also established in cities where classics and higher mathematics are taught. Several British Columbia cities also now have charge of their own public and high schools, and these receive a very liberal per capita grant in aid from the Provincial Government. The minimum salary paid to teachers is $50.00 per month in rural districts, up to $150.00 in city and high schools. Attendance in public schools is compulsory. The Education Department is presided over by a Minister of the Crown. There are also a Superintendent and four Inspectors in the province, also boards of trustees in each district. According to the last educational report, there are 361 schools in operation, of which 13 are high, 65 graded and 283 common. The number of pupils enrolled in 1905 was 27,335, and of teachers, 663. The public school system was established in 1872, with 28 schools, 28 teachers, and 1,028 pupils. Its growth proves that education has not been neglected in British Columbia.

The high schools are distributed as follows:—Victoria (Victoria College), Vancouver (Vancouver College), New Westminster, Nanaimo, Nelson, Rossland, Cumberland, Vernon, Kaslo, Chilliwack, Grand Forks, Kamloops and Revelstoke. There is a provincial normal school at Vancouver and many excellent private colleges and boarding schools. Victoria and Vancouver colleges are affiliated to McGill University, Montreal, and have high school and university departments.

SOCIAL CONDITIONS.

The population of British Columbia, widely scattered and composed of many nationalities, is singularly peaceful and law-abiding. Life and property are better protected and individual rights more respected in the isolated mining communities than in some of the great centres of civilization. The province, though new as compared with older countries, enjoys all the necessaries and many of the luxuries and conveniences of modern life. There are few towns which are not provided with water works, electric lights and telephones. The hotels are usually clean and comfortable, and the stores well stocked with every possible requirement. There is little individual poverty. A general prosperity is the prevailing condition throughout the country, for none need be idle or penniless who is able and willing to work. The larger towns are well supplied with libraries and reading rooms, and the Provincial Government has a system of travelling libraries, by which the rural districts are furnished free with literature of the best description.

The spiritual welfare of the people is promoted by representatives of all the Christian denominations, and there are few communities, however small, which have not one or more churches with resident clergymen.

All the cities and larger towns have well-equipped hospitals, supported by Government grants and private subscriptions, and few of the smaller towns are without cottage hospitals. Daily newspapers are published in the larger places, and every mining camp has its semi-weekly paper.

CITIES AND TOWNS.

VANCOUVER.—The commercial metropolis, incorporated in 1886, is the largest centre of population, estimated at 45,000. The city's trade is large and ever increasing, as it is the mainland terminus of the Canadian Pacific Railway, and the home port of the Empress and Australian Trans-Pacific liners, and a wholesale distributing point for the north. The bank clearings show a steady growth of business, the figures for three years being ; 1902-3, $56,000,000 ; 1903-4, $68,000,000 ; 1904-5, $88,460,391, and the reports of import and export trade and shipping are equally satisfactory. Vancouver harbor is one of the finest in the world, land locked and sheltered from all points, and roomy and deep enough for the largest vessels. The City of Vancouver possesses many fine public buildings, business blocks and private residences, and new structures are being continually added. The churches, schools, libraries, hotels and clubs are quite equal to buildings of similar class in the older cities of the east and give one the impression of solidity and permanency. The Hotel Vancouver, owned by the Canadian Pacific Railway Company, is one of the best equipped in Canada, and is well known to world travellers. One of Vancouver's great attractions is the magnificent Stanley Park, with its groves of great towering firs and cedars, a wonder and delight to visitors. In addition to the Canadian Pacific Railway Trans-Pacific fleet of steamers, Vancouver has connections by land and sea with all important points on the coast and in the interior. The steamers of the Canadian Pacific Railway Coast Service, and other lines, ply between the city and places along the coast as far north as Alaska, and South to San Francisco. The splendid Canadian Pacific Railway steamer " Princess Victoria," the fastest boat on the Pacific, makes daily trips in the summer between Vancouver, Victoria and Seattle, Washington. Direct railway connection is made with every point on the continent, from Halifax to Mexico. The city has a very complete electric railway system with extensions to New Westminster and Lulu Island. The water supply is unlimited and of superior quality, and the sewage system is constructed on modern lines. Telephone connection is had by cable with Victoria and other cities and towns on Vancouver Island, as well as all points in the Fraser Valley, and the City of Seattle. A recently constructed water tunnel provides a water power sufficient to develop 300,000 h. p.

VICTORIA is the seat of Government and the capital of British Columbia. It is charmingly situated on the southeast of Vancouver Island, and for climate and surroundings has no rival in Canada. Victoria is the oldest town in the province, dating back to 1846 when it was known as Camosun, a Hudson's Bay Company's trading post. Victoria leaped into prominence during the gold excitement and grew rapidly in trade and population. The city is substantially built, there being many fine stone and brick blocks in the business portion, while the private houses surrounded by beautiful lawns, gardens and shrubberies are picturesque and cosy. The Parliament Building, overlooking James Bay, is one of the finest examples of architecture in America. It contains fine collections of natural history, mineral, agricultural and horticultural specimens, and is a centre of great interest to visitors. Beacon Hill Park, a natural pleasure ground, facing the Strait of Juan de Fuca, affords one of the most magnificent views in the world, the snow-clad heights of the Olympian Range and the noble dome-like Mount Baker forming the background of an enthralling picture. Victoria Arm and the Gorge form one of the most beautiful stretches of inland water imaginable, and there are many other delightful bays and inlets which lend peculiar attraction and variety to the scene. With such a wealth of natural beauty Victoria is fast becoming the Mecca of the tourist, many thousands from all parts of the world visiting Victoria every year. The Canadian Pacific Railway is building a magnificent hotel, "The Empress," near the Parliament Buildings. When completed, the two noble edifices, dominating the harbor, will make an imposing picture.

In addition to its beauty and attractiveness the city is an important business and industrial centre. It shares with Vancouver the northern trade and that of the interior, and its shipping, lumbering, mining, sealing and fishing interests are very considerable and showing evidences of increase. The development of the resources of Vancouver Island must naturally benefit Victoria, and there is a conviction in the minds of her citizens that the city is on the eve of an era of substantial progress. The volume of trade for 1904 amounted to $5,944,794, made up of $2,882,000 imports, and $2,062,794 exports. The vessels entered and cleared during the year were 2,263 and 2,273 respectively; of these 11 were foreign.

The city is growing steadily in population (estimated at 26,000), many persons of independent means choosing it as a place of residence, while new enterprises are giving employment to more laborers and artizans.

Victoria is the first port of call for the Trans-Pacific liners and northern steamers, as well as all the big freighters which round the Horn for Puget Sound points. It is the home port of the Victoria sealing fleet, the Canadian Pacific Railway Pacific Coast Service and of many coasting vessels. Daily communication is had with Vancouver, Seattle and other points, and there is a tri-weekly service to San Francisco. The distance between Victoria and Seattle is 80 miles, and Victoria and Vancouver 84 miles, the Canadian Pacific Railway steamer "Princess Victoria" making the triangular run daily during the tourist season.

The city has an electric street railway system and gas and electric light services. The business streets are paved and well kept and cement sidewalks are being laid on all the principal thoroughfares. The water works and sewage system are being extended to meet the requirements. There is telephone connection with all the principal points on the Island and lower mainland, and with Seattle.

Esquimalt, Victoria's western suburb, was until recently headquarters of His Majesty's Royal Navy's North Pacific fleet, but the ships, with the exception of one or two, have been withdrawn and Canada has undertaken the maintenance of the fortifications which are among the strongest in the Empire. Esquimalt has a fine harbor, formerly used exclusively by the navy, which will now be opened to merchant vessels.

NEW WESTMINSTER is situated on the Fraser River, about 16 miles from the mouth and 12 miles from Vancouver. It is the centre of the salmon canning industry and enjoys a big share of the lumber trade. Being the depot for a large agricultural country, New Westminster market is the most important in the province—the farmer's mart and clearing house. The city was the capital of the Crown Colony of British Columbia before Confederation, and was destroyed by fire in 1898, but, through the energy of its citizens, it has been rebuilt and greatly improved. Among the public buildings are the Penitentiary and the Provincial Asylum for the Insane. The city owns and operates an electric light plant, and has an excellent water supply, and electric street railway and telephone systems. There is an inter-urban electric railway connecting the city with Vancouver, and a branch line of the Canadian Pacific Railway connects it with the main line at Westminster Junction. An annual event of importance is the holding of a Provincial exhibition of agricultural and industrial products, which attracts exhibitors from all parts of the province. The population is about 8,000.

NANAIMO.—The "Coal City," is 72 miles from Victoria, on a fine harbor, on the east coast of Vancouver Island. Its chief industry is coal mining, but latterly it has become important as a centre of the herring fishery. It is also the chief town of an extensive farming and fruit growing country. The city has a good water system, and electric lights, telephones and gas. Nanaimo coal is shipped to California, Hawaii and China, and it is a coaling station for ocean-going steamships. The Esquimalt & Nanaimo Railway connects Nanaimo with Victoria, and there is a daily steamer service to Vancouver. The population is estimated at 7,000.

ROSSLAND.—The mining centre of West Kootenay, has grown in ten years from an obscure mining camp to a well ordered, substantial city of about 5,500. Rossland's mines are famed the world over, and their development is proving their permanency. The city is eight miles from the United States boundary on a branch of the Canadian Pacific Railway, and is provided with all the modern conveniences, waterworks, electric lights, telephones, etc. The hotels, banks, and business houses are of a substantial character and would do credit to any town of similar size.

NELSON, situated on the west arm of Kootenay Lake, has a population of 5,000 to 6,000. It is a well laid out and solidly built town, the principal buildings being of brick and stone. It is the judicial centre of Kootenay and an important wholesale business point. Its altitude, 1,760 feet above sea level, renders the climate equable and salubrious, and makes a desirable place of residence. The chief industries are mining and lumbering, and of late years fruit growing has received a good deal of attention, the shores of the West Arm being found well adapted to all kinds of fruit, which grow to perfection and ripen early. The city is lighted by electricity and has an electric street car service. Excellent fishing and shooting may be had in the neighbourhood. Nelson is connected with the main line of the Canadian Pacific Railway, the Crow's Nest Pass Railway and the Great Northern by branch lines and steamers. The Hall Mines Smelter, which handles a large tonnage of ore annually, is situated at Nelson.

KASLO is an important trade centre on the west shore of Kootenay Lake. It is supplied with good stores, hotels, churches and schools, water works, electric lights and telephones. The population is about 1,800.

LADYSMITH, on Oyster Harbor, east coast of Vancouver Island, is one of the youngest towns in the province. It is the shipping port for the adjacent Extension coal mines, and the transfer point for through freight

LOOKING OVER BURRARD INLET, VANCOUVER. B.C.

PARLIAMENT BUILDINGS, VICTORIA, B.C.

57a, Pall Mall,
London, S.W.

between the Island and the mainland. The Canadian Pacific Railway ferries freight trains from Vancouver to Ladysmith, where they are transferred to the Esquimalt & Nanaimo Railway for distribution to Island points. Most of the miners working in the Extension mines live at Ladysmith, which has a population of 2,000. Ladysmith is an important coaling station for coasters and ocean going craft, and ships load cargoes of coal for California and other foreign countries. The Tyee Copper Co. operates a smelter, and there are several minor industries which add to the prosperity of the town.

KAMLOOPS is an important business place, 224 miles west of Vancouver, on the main line of the Canadian Pacific Railway. It is beautifully situated at the confluence of the North and South Thompson rivers, both of which are navigable from this point for considerable distances. Kamloops, literally " the meeting of the waters," is one of the oldest settlements in the province, the Hudson's Bay Company having established a post there over 80 years ago, which was for a long time the centre of trade for the whole interior. The town is the distributing point for a very large agricultural, ranching and mining country, and is the chief cattle market of British Columbia. It is also the centre of a big lumbering district, and a divisional point of the Canadian Pacific Railway. The adjacent country produces some of the finest fruit grown in the province, apples attaining an immense size and superior quality. The climate is dry and bracing, with bright sunshine at all seasons, the rainfall being very light. The city is lighted by electricity, there is a good water works system, several well stocked stores, good hotels, churches, schools, and every other item which goes to make life pleasant and enjoyable. The rivers afford good fishing and the woods are full of all kinds of game, including prairie chicken, grouse and deer. The population is about 2,000 Kamloops has a steamboat service on the Thompson River and Kamloops Lake·

REVELSTOKE on the main line of the Canadian Pacific Railway, 379 miles east of Vancouver, is a railway divisional point and the gateway to West Kootenay, connection being made there with the Arrowhead branch, which gives access to the Slocan, Kootenay, Boundary and Crow's Nest countries. The town is growing rapidly, being the centre of a good mining and lumbering district. The Canadian Pacific Railway has a fine hotel at Revelstoke and there are several good stores and other business and industrial establishments. The population is about 2,500.

FERNIE, a coal town on the Crow's Nest Pass Railway, is making wonderful progress, and is rapidly assuming a metropolitan appearance. The inexhaustibility of the surrounding coal fields insures the town's stability, while the important, but minor, lumbering industry contributes largely to its present prosperity. There are 1,500 coke ovens at Fernie which supply fuel to the Kootenay and Boundary smelters. The population is 3,500.

GRAND FORKS, the chief town of the Boundary District, (pop. 2,500), is situated at the junction of the North Fork with the main Kettle River. It is the site of the Granby Smelter, the largest plant of the kind in the province, where blister copper is produced at the rate of about two and a half tons a day, besides large values in gold and silver. The city is beautifully situated in a prairie-like valley, has wide streets and good buildings, with water works, electric lights and all other conveniences. The surrounding country is well adapted to fruit growing, in which good progress is made. A local nursery has 200,000 young trees ready for distribution. The population is about 2,500.

GREENWOOD, twenty-two miles west of Grand Forks, is the centre of a rich mining district. It has several large and well stocked stores, good hotels, three banks, and all the minor industries are well represented.

HASTINGS STREET, VANCOUVER, B.C.

The British Columbia Copper Company's smelter adjoins the town, a plant with a daily capacity of 3,500 tons of ore. The population is 2,500.

TRAIL, on the Columbia River, nine miles from Rossland, is the centre of the mining industry in West Kootenay. The Canadian Smelting Works, operated by the Canadian Pacific Railway Company, covers 45 acres of ground, and is the largest plant in Canada. Power is transmitted from the West Kootenay Power Company's station at Bonnington Falls. The smelter treats silver-lead as well as copper ores, and was the first in the world to produce electrolytic lead in quantity. The company now manufactures sheet lead and lead pipe, and refines silver direct from the ore—operating the only silver and lead refinery in Canada. Apart from the business created by the smelter, Trail enjoys a prosperous trade with the adjacent mining camps, and is well equipped with all the conveniences of modern life. The population is estimated at 2,000.

CRANBROOK, a divisional point of the Crow's Nest Railway, is pleasantly situated in the fertile valley which lies between the Selkirk and Rocky Mountains. It is the principal lumbering point in East Kootenay, its four sawmills having a capacity of about 160,000 feet per day. The town has a number of good stores, banks, churches, hotels, and is very prosperous and progressive. Population 2,500.

VERNON is the centre and supply depot for the Okanagan District, and is surrounded by a splendid farming, cattle and fruit country. It is the terminus of the Shuswap and Okanagan Branch of the Canadian Pacific Railway and has steamboat connection via Okanagan Landing, 5 miles south, with all points on Okanagan Lake. The town is pretty and homelike, the climate delightful at all seasons, and its inhabitants are prosperous and energetic. The population is about 1,800.

ARMSTRONG, thirty-two miles south of Sicamous Junction, is an important lumbering and flour-milling point, it and its rival, Enderby (6 miles distant), being surrounded by wheat lands. There is a large co-operative flour mill and creamery, a large sawmill and other industries. Considerable fruit is grown in the vicinity and the fruit acreage is being increased.

ENDERBY, another prosperous and growing town, has a modern roller mill, with a daily capacity of 250 barrels, and a big sawmill which, added to the town's position in the midst of a fine farming country, assures it a good general trade.

KELOWNA, 33 miles south of Vernon, on Okanagan Lake, is a prosperous town, enjoying a good trade as the supply point for Mission Valley and Sunnyside districts. The neighborhood is being transformed into an immense orchard and vegetable garden, and shipments of fruit and vegetables are increasing very rapidly. The town has a tobacco factory, supplied by locally grown leaf, a sawmill, fruit packing house and other industrial establishments, and good stores, hotels, churches, and schools.
There are many other towns and villages of growing importance in the province, of which space precludes special mention.

HUNTING AND FISHING.

The sportsman will find a greater variety of fish and game in British Columbia than in any other part of North America ; there are, indeed, few regions that can boast of anything like the same variety of species. Whether with rifle or with smoothbore, or with rod, there is an almost bewildering choice. The three great parallel ranges of the mainland hold an immense amount of big game. In the Rockies there are bighorn sheep, goat, caribou and deer ; in the Selkirks goat and caribou, and in the Coast Range, goat and quantities of the true blacktailed deer. Grizzly bears are found in several districts, while black bear are to be found in number throughout the Province. The mule deer, miscalled blacktail, is so abundant in East Kootenay, the Boundary country, Okanagan and Lillooet as to be a very certain source of supply for the ranchers and miners to draw upon. Elk (wapiti) shooting may be indulged in by those visiting the northern end of Vancouver Island. It is believed that the elk is extinct upon the mainland, with the possible exception of the south-eastern corner of the province, but on Vancouver Island it is tolerably abundant, although unfortunately, it frequents a densely forested region, so that the hunting means hard work.

ELK AND HUNTER.

Although few persons, however keen, would visit British Columbia merely for the sake of its wing shooting, yet it is undeniable that, with the exception of Manitoba, Alberta, Saskatchewan and Athabasca, a man may find as much work for his breech-loader in the province as he would abroad anywhere. Five species of grouse and vast quantities of wild fowl, from swans to teal, abound in suitable localities. The marshes of the Columbia swarm with mallard and other choice duck in the autumn ; the Arrow Lakes and the upper valley of the Fraser form a trough much frequented by the wild geese during their migrations, and the fiords and sounds of the coast shelter great flocks of wild fowl throughout the winter—for it must not be forgotten that the winters of the Pacific are very much less rigorous than those of the Atlantic, and that a very large proportion of the birds do not go further south than Vancouver Island.

The fishing of British Columbia is so remarkably good that no one can realize the quantities of salmon and trout to be found in the streams of this province, until he has visited it. The quinnat and cohoe salmon may be taken in salt water at certain seasons in large numbers by means of a spoon bait, and a few crack fishermen have succeeded in taking the quinnat in fresh water, but as a rule British Columbia salmon do not rise to the fly. However, the trout will more than make up for the salmon's lack of appreciation. The rainbow trout is, possibly, the finest fish for his inches of all the trout family, and, happily, he is extraordinarily numerous in many of the inland waters. Where he is not found his place is taken by the black spotted trout, an excellent fish, though hardly the equal of the rainbow. Very heavy lake trout are found in all the larger sheets of water. Shuswap Lake may be mentioned as especially good and easy of access. An excellent hotel has been built at Sicamous on the very edge of this lake, at which many sportsmen reside each summer for weeks at a time in order to enjoy the fishing and shooting of the neighborhood.

ADVICE TO IMMIGRANTS.

There is no country within the British Empire which offers more inducements to men of energy and industry than British Columbia. To the practical farmer, miner, lumberman, fisherman, horticulturist and dairyman it offers a comfortable living and ultimate independence, if he begins right, preserves and takes advantage of his opportunities. The skilled mechanic has also a good chance to establish himself and the laborer will scarcely fail to find employment. The man without a trade, the clerk, the accountant and the semi-professional, is warned, however, that his chances for employment are by no means good. Much depends upon the individual, for where many fail one may secure a position and win success, but men in search of employment in offices or warehouses, and who are unable or unwilling to turn their hands to any kind of manual labor in an emergency, would do well to stay away from British Columbia unless they have sufficient means to support themselves for six months or a year while seeking a situation.

The class of immigrants whose chances of success are greatest is the man of small or moderate means, possessing energy, good health and self-reliance, with the faculty of adaptability to his new surroundings. He should have at least £300 ($1,500) to £500 ($2,500) on arrival in the province, sufficient to " look around " before locating permanently, make his first payment on his land and support himself and family while awaiting returns from his first crop. This applies to a man taking up mixed farming. It is sometimes advisable for the new comer to work for wages for a time until he learns the " ways of the country."

Settlers' effects, etc., household furniture, farming implements in use, and live stock, brought into the province by *bona fide* settlers are admitted free of duty, but most articles of domestic use may be bought in the country at reasonable prices.

The following is the authorized number of live stock allowed to be imported free of duty: Horses, 1 to every 10 acres, not exceeding 16 in all; cattle, the same; sheep, 1 to each acre, 100 in all allowed; hogs, the same.

To avoid the risk of loss the immigrant from Great Britain should pay the money not wanted on the passage to the Dominion Express Company's office in London, Liverpool or Glasgow, and get a money order payable at any point in British Columbia; or he may pay his money to any bank in London having an agency in British Columbia, such as the Bank of Montreal, Canadian Bank of Commerce, Bank of British North America, Imperial Bank, etc.

This suggestion applies with equal force to persons coming from Eastern Canada or the United States.

United States currency is taken at par in business circles.

The Provincial Government Agent at point of arrival will furnish information as to lands open for settlement, farms for sale, rates of wages, etc.

HOW TO REACH BRITISH COLUMBIA.

From the United Kingdom.—Several lines of steamships ply between British and Canadian ports, and full and reliable information regarding routes, rates of passage, &c., can be obtained at the office of the Agent-General of British Columbia, Salisbury House, Finsbury Circus, London ; the office of the High Commissioner for Canada, 17, Victoria Street, London, S.W.; the office of the Canadian Commissioner of Emigration, 11-12, Charing Cross, London, W.C.; or to the Dominion Government Agents at Birmingham, Cardiff, Liverpool, Dublin, Belfast or Glasgow.

From the United States through tickets may be bought to any point in British Columbia over any of the transcontinental railways and their branches and connections.

STRAKER BROTHERS, LTD., 44-47, BISHOPSGATE WITHOUT, E.C

4

An Address on

Nova Scotia:

By John Howard.

Agent-General for
Nova Scotia.

Illustrated.

Dominion Atlantic Railway

FRUIT-GROWING

IN THE

ANNAPOLIS VALLEY OF NOVA SCOTIA

A WORD TO THE WISE

EX-CHIEF JUSTICE SIR ROBERT WEATHERBE, who has found time amid his professional duties to plant one of the largest orchards in the Annapolis Valley, says: "We have a belt containing about 400 square miles, capable of producing an annual revenue of $30,000,000. There is no land in the world that will yield like this valley, and we should plant the whole area. There is no fear of raising more apples than are required. We can raise them more profitably than in any other part of the world."

PROF. SAUNDERS, Director of the Dominion Experimental Farm at Ottawa, says: "I know of no locality where trees bear so abundantly and continuously as in your own favoured Annapolis Valley."

In the Annapolis Valley there are about 250,000 acres of land adapted to the cultivation of fruit. Probably not more than five per cent. of this area is already set with trees. Tens of thousands of acres of choice orchard land await the incoming of capital and labour.

During the past five years the acreage of young orchards has largely increased. There is no investment open that will yield such abundant returns, for a period of fifty or one hundred years. A hundred-acre orchard means wealth.

THE ANNAPOLIS VALLEY

is traversed throughout by the DOMINION ATLANTIC RAILWAY

the Short, Quick, Cheap Route to and from

HALIFAX ST. JOHN AND BOSTON

Along this Railway's System are to be found choice Fruit Lands.

There are splendid educational facilities, and easy access to the largest cities in Canada and the United States.

"Fair Nova Scotia."

Extract from Speech by Earl Grey, Governor General of Canada, at Halifax, N.S., August 8th, 1907.

" After three years' study of Canada he was prepared to hold the field against the world—that, great as were the resources, advantages, and attractions of the sister Provinces, Nova Scotia need not be afraid to pit her charms against any of her Sister Provinces, however fair they might be.

NOVA SCOTIA:

BY

JOHN HOWARD

(Agent-General.)

HON. GEORGE H. MURRAY,
Premier of Nova Scotia.

NOVA SCOTIA:

AN ADDRESS

GIVEN AT

THE IMPERIAL INSTITUTE

BY

MR. JOHN HOWARD,

AGENT-GENERAL FOR THE PROVINCE.

ON THE 24TH MARCH, 1902.

LORD STRATHCONA & MOUNT ROYAL, G.C.M.G., &C., &C.,

High Commissioner for Canada, presiding.

REVISED 1907.

IN order to convey a comprehensive idea of Nova Scotia, it is
desirable to give a brief account of its discovery and early
history; and some of the many vicissitudes it passed through,
before it became permanently attached to the British Crown.
After the Cabots sailed from Bristol in 1497 with the intention
of reaching China by the supposed North-West passage
(having been commissioned by Henry VII. to discover the
"isles, regions and provinces of the heathen and infidels
unknown to Christians "), the first new land they encountered
was that portion of Cape Breton within view of the Sugar
Loaf Peak, near Cape North, which was sighted on the 24th
June, 1497, and called by them Prima Vista. Continuing
their voyage, they shortly afterwards arrived at Prince
Edward Island, which they named the Island of St. John.
Not finding the expected passage, they altered their course
towards the south and cruised along the coast as far as Florida,
when they returned to England, bringing with them two
natives from St. John Island, no doubt as a proof of their
discoveries. It is interesting to note that up to this time the

voyages of Columbus had not extended beyond the Islands in the Gulf of Mexico, he not yet having reached the mainland. The claim of England on Nova Scotia, and in fact, on the whole of North America, was, therefore, based on this discovery. It became known to the French shortly afterwards, who called it "Acadie," supposed to be derived from the Indian word " Cadie," a place of abundance. The aborigines of Acadia, now known as the Micmac Indians, belonged to the family of the Algonquins, by far the most numerous of the eight great families of the North American Indians inhabiting the region between the Mississippi, the Atlantic, and the country of the Esquimaux. It is believed that their numbers have been very much exaggerated. Although savages in their mode of life, they were originally savages of the highest type, veritable Spartans in spirit, eloquent, brave, and honorable, with some of the noblest qualities and virtues met with in civilized races. With the advent of the white man, however, they began to deteriorate, and as a foe they were dangerous, crafty, and cruel, resorting to all kinds of strategy, devices, and treachery to outwit or surprise their enemies. These propensities were fostered and utilized by the French Acadians in their relations with the English.

We find that the rich fisheries on the coasts were known to the Europeans in 1504, and in that year many adventurers visited these shores to fish or barter with the Indians, so that the coasts were already familiar to these traders and fishermen before there were any white settlers. For several years no steps seem to have been taken in England to follow up the advantages of Cabot's discovery. The French attempted to effect a settlement in 1518 but failed, as also did the attempt by an Englishman named Howe in 1536. In 1583 Sir Humphrey Gilbert, a half brother of Sir Walter Raleigh, set out for the new world to found a colony, but after reaching St. John's, Newfoundland, of which he took formal possession in the name of England, he met with disaster on continuing his voyage, which he, therefore, abandoned, and was lost on his return to England. Whilst England based her claims on discovery, France based hers on settlement.

NORSE STONE.

The Stone bearing this Runic description was found at the head of
Yarmouth Harbor by Dr. Fletcher near the end of the last century.
Weight about 400 pounds. Photographs and rubbings have been sent
abroad to the several Historic and Antiquarian Societies. In 1884, May
2nd, Proc. Am. Phil. Soc., Mr. Henry Philips gave a full description and
translation of the characters. The following is an extract : "In 1875 a
rubbing was placed in my hand and I have become ultimately satisfied of
its bona fide nature. . . . . . Word after word appeared in disjointed
forms and each was in turn rejected until at last an intelligible word came
forth followed by another and another until a real sentence stood forth to
my astonished gaze : HARKUSSEN MEN VARN, Hako's son addressed the
men. In the expedition of Thorfinn Karlfsefne in 1007, the name of Haki
occurs among those who accompanied him." This stone has been in the
possession of Mr. S. M. Ryerson for the last thirty years, and may be seen
at any time on the lawn at his residence, Yarmouth, Nova Scotia.

It appears uncertain when the first settlement took place, but it is stated by French writers that one was made in Cape Breton by France in 1541. However, De Monts, a Frenchman, having been appointed Lieut.-General of the Territory of Acadie took out an expedition in 1604, when he founded Port Royal (afterwards changed to Annapolis by the British). In 1613 Port Royal was attacked and destroyed by Captain Samuel Argel, on the claim made by England to the whole continent of North America, founded on the discoveries of the Cabots.

This claim the English continued to assert, and the French who remained there were merely regarded as interlopers, whose presence, like that of the Indians, was simply tolerated for the time. In 1621 Sir William Alexander received from King James 1. a grant of the whole of the Provinces of Nova Scotia, New Brunswick and the Gaspè Peninsula, which territory was to be known by the name of Nova Scotia, and to be held at a quit rent of one penny Scots per year, to be paid on the soil of Nova Scotia, if demanded. Sir William Alexander accordingly visited Acadia, but the French being again in possession of Port Royal, no permanent settlement was made by his people, and nothing was done for several years. James I. dying, Charles I. confirmed his father's grant to Alexander, and for the purpose of facilitating the settlement of the Colony and providing funds for its subsistence, an order of Baronets of Nova Scotia was created, to consist of 150 gentlemen who were willing to contribute to the founding of the Colony, each of whom was to receive a tract of land six miles by three in Nova Scotia, which Sir William Alexander re-leased to them in consideration of their aid in the work of colonisation. This scheme, if vigorously carried out, would probably have ended in establishing a strong Colony, but meantime a war suddenly broke out between England and France. Sir David Kirk captured Port Royal, and when the war ceased Acadia, with the exception of Cape Sable and Fort St. Louis, was in the hands of the British. In 1632 a treaty was signed, ceding Acadia to France. In 1654 a force under Major Sedgewick attacked the French, and Nova Scotia for the third time became English.

CAPE BLOMIDON.

However, by the treaty of Breda in 1667, France again obtained possession of Acadia. Hostilities having broken out in 1689, the New England people, mindful of what they had formerly suffered from the French, fitted out an expedition under Sir William Phips, which again took Port Royal. It was once more given up in 1696 by the treaty of Ryswick and restored to France. However, in 1710 it was finally captured by General Nicholson, and the name changed to Annapolis Royal.

By the treaty of Utrecht in 1713 Acadia was formally given up to England as well as the whole of Newfoundland, but France retained the Island of Cape Breton and the other islands in the Gulf of St. Lawrence, including the Island of St. John, now known as Prince Edward Island, and the right to erect fortifications for defensive purposes. The value and extent of the fisheries on the shores of Cape Breton, already referred to, had been very early recognised by the French, and we find in 1710, according to their returns, that upwards of 23,000 men and 2,100 vessels of various sizes were engaged in that industry.

Their fleet usually arrived in April and fished till August, when they commenced to return. A very large revenue was derived from this trade, which gave France the monopoly of the chief European fish markets, and the men it provided for drafting into its Navy impressed upon France the importance of retaining these possessions at all hazards. The building of the town and fortress of Louisburg were therefore commenced for the protection of these fisheries, the shelter of its fleets, and for a base of operations in time of war. Its magnificent harbour was capable of accommodating its large and numerous warships. It mounted 400 cannon, and its garrison numbered at times 10,000 men. After 30 years of labour and vast expense it became a mighty stronghold and a standing menace to New England, harbouring a swarm of privateers which preyed upon English commerce. Its existence became intolerable to the British Colonists, and they determined to make an attempt to capture it. Although most of the troops, which were raised in New England for the purpose, were inexperienced, success

PART OF OLD FORT, ANNAPOLIS ROYAL, SHOWING BARRACKS.

crowned their efforts; for aided by the ships of war under Commodore Warren, after a siege of 49 days, the garrison capitulated. The benefits of this remarkable exploit were not very lasting, for Louisburg was soon restored to France by treaty. In 1758, however, it was again necessary to reduce the place, and an attack was delivered by General Wolfe, shortly after which the garrison surrendered; the capitulation including the whole Island of Cape Breton and the Island of St. John. It was decided to demolish the fortifications of Louisburg; the work took two years to accomplish, and very little now remains to mark the site of this once famous fortress.

Meantime owing to the representations of the New England Colonists for the necessity of a fortified post on the shores of Nova Scotia to counteract the influence exercised by Louisburg, then all powerful, the English Government founded Halifax in 1749. This work was entrusted to the Hon. Edward Cornwallis, who landed there on the 21st June with about 4,000 settlers. A town site was very soon selected, cleared and laid out and named after Lord Halifax, then at the head of the Board of Trade in England. During these building operations great annoyance and trouble was experienced from the hostile actions of the Indians, who, instigated by the French settlers, practiced all sorts of barbarities on the new comers, whom they treacherously cut off in the woods surrounding the town, whenever opportunity offered. These Acadians after remaining for 40 years on their lands without carrying out their undertaking to become British subjects, on which conditions they were permitted to remain, had allowed themselves to be made the tools of French intrigue. They were now called upon to take the oath of allegiance or to leave the country. They unanimously refused to do either, claiming that they were neutrals. England now being engaged in a life and death struggle with France for the supremacy of North America, the Acadians became her enemies within the gate. If such a state of things were to be tolerated, English settlement in an English Province would be impossible. Diplomacy and argument proving equally unavailing upon this misguided people, nothing remained but expulsion, which accordingly took place with as little hardship as possible under the circumstances. Their masters were so constantly being changed, that they may have awaited the next swing of the pendulum to restore them to their original condition. It is conjectured that of the total number of those forcibly removed at least two thirds afterwards returned to Nova Scotia.

In view of the Independence of the United States, many thousands of Loyalists came to Nova Scotia in 1783, and settled in different parts of the Province. They proved a very acceptable acquisition, being principally persons of position and education, who preferred to reside within the sphere of the flag, under which they were born. They soon became the leading men, and to their courage, loyalty and devotion Nova Scotia owes many of the beneficent laws and institutions of which she is so justly proud.

In 1867 Nova Scotia joined the Confederation of the Canadian Provinces, and thus became part of the Dominion of Canada.

NOVA SCOTIA.

Nova Scotia forming the most easterly Province of the Dominion of Canada, juts out into the Atlantic, and is connected to the mainland by a narrow isthmus about 14 miles across. It lies between 43 and 47 North latitude and 60 and 67 West longitude. Varying in breadth from 50 to 100 miles, it is 350 miles long, and contains about 21,000 (20,907) square miles. In this small area is crowded a wealth and variety of natural resources which no region of similar extent can equal. Its coast line has been so cut into by the sea, that it measures many thousands of miles, and is broken by a myriad of beautiful bays and natural harbours, which provide that no part of the Province is more than 30 miles from navigable waters. The number of harbours capable of accommodating ships of the largest size, being more than double those on the entire seaboard of the United States, from Maine to Mexico, and include to the east of Halifax 26 ports, 12 of which can accommodate ships of the Navy, the remainder with capacity to take average sized merchantmen, whilst to the west are 15 ports and harbours of magnificent proportions and beauty.

Unsurpassed in North America, if not in the world, is Halifax Harbour; accessible at all seasons, safe, deep and easily approached, it is said to be sufficiently extensive to contain the united Navies of Britain, France, America, and Germany with room to spare, or to permit the entire British Navy to manœuvre upon its waters with ease and safety. Within the interior of the Province there are innumerable rivers and streams and more than 400 lakes, the largest and most famous being the Bras d'or in Cape Breton. This is about 45 miles long and varies to about 15 miles in width, combining the charm of a fresh water lake, with all the advantages of a salt water one. It has been called the Loch Lomond of North America, a gem of nature so beautiful and so picturesque that it challenges the admiration of even the greatest travellers. Generally throughout the Province the scenery is varied and beautiful, with the alternating views of lakes, meadows, and woods. Special mention should be made of Grand Prè, rendered famous by Longfellow's " Evangeline." It is now called the Land of Evangeline, and is yearly visited by many thousands of tourists, no doubt attracted as much by its natural beauties as by the romantic and tragedic interest with which Longfellow has surrounded it. I think the inhabitants owe to the great poet a deep debt of gratitude for the indirect benefits that have accrued to them in making their district a favourite summer resort.

CLIMATE.

The climatic conditions are very favourable. The heat of summer is tempered by the sea breezes and the winter cold is also moderated by its proximity. Extremes of temperature are not great. It is exceedingly healthy, and the air is vitalizing, exhilarating and recuperative, and in winter, although the mercury falls lower than it does here, it causes no discomfort, owing to the dryness of the atmosphere and its bracing qualities. The spring seasons are, however, somewhat backward, but the weather in the autumn is usually very fine and that period called Indian summer requires to be experienced to be appreciated. Then the sun appears with its mellow

THE LAND OF EVANGELINE.

light and temperate warmth and bathes all things in a golden
glory. The air is still and nature seems to rest after her
exertions in bringing vegetation to maturity. The leaves of
the forest, touched by the magic fingers of the frost, put on
fresh shades, and changing in intensity, become a mass of
bewildering colour, ranging from light yellows, through the
vermillions, to deepest purples, completing a scene of wondrous
beauty.

EDUCATION.

Nova Scotia is justly proud of her free educational system
established by the Government in 1865. It works well and is
extremely thorough, and I do not think there is any other
country which pays greater attention to this most important
subject. The system is so arranged that the younger generation
can pass by progressive steps from the early stages of tuition
through the various grades until the universities are reached.
The schools are non-sectarian. The teachers are trained at the
Normal School at Truro, where thorough provision is made for
a most efficient and comprehensive preparation. The number
of schools open in 1904 was 2,331, at which 96,886 pupils
attended, being a proportion of about one in five of the total
population. There are five colleges, and each of the eighteen
counties has a high school or academy, which is the connecting
link between the public schools and the universities.

The movement for the confederation of the colleges in the
maritime provinces has been revived, as it is pointed out that
many of the departments are duplicated, whilst others might
with advantage be extended, particularly those relating to
science. The object is to establish a central university, offering
as many inducements in the higher branches as can be obtained
in the United States or at the McGill University. The benefits
of such a course of scientific training would nowhere be more
appreciated than in the maritime provinces, owing to the great
importance of their natural resources. ·

N.B.—The Nova Scotia Legislature realising the enormous possibilities
that lie in the efficient development of the great natural resources of the
province, has recently made provision for the immediate establishment and
equipment in Halifax of an Institute of Technology of university grade.
This Institution will have departments providing instruction and professional
training in metallurgical, civil, mining, mechanical, chemical and electrical
engineering.

PROVINCIAL NORMAL SCHOOL, TRURO, N.S.

WOLFVILLE, N.S., AND GRAND PRÉ DIKE.

AGRICULTURE.

The province is eminently adapted by nature for agriculture and fruit growing. The soil is rich and easy to work, whilst the climate, neither too hot in summer nor too cold in winter, lends itself to the rapid development of vegetation of all kinds. Farming is largely carried on, by itself, and in conjunction with fruit growing, dairying, and stock raising. The district known as the Annapolis Valley is probably the most favoured part of the province, and it is here that the fruit growing industry prevails. As an apple producing country Nova Scotia is unexcelled, and they are famed both in Great Britian and the United States for their excellent flavour and fine keeping qualities. The possibilities with regard to this important industry are very great, and the Nova Scotia Government recognising this, recently established experimental orchards in every county of the Province, and it has been found that apples can be successfully cultivated practically in any district. This has resulted in the "fruit belt" being largely increased, and it is estimated that there are 2,000,000 apple trees in Nova Scotia, although not 10% of the land suitable for apple culture has yet been planted.* Dairying is now having much more attention paid to it than formerly, but there is still room for the introduction of improvements in this respect, which the local government is encouraging by money grants towards the establishment of cheese factories and creameries, which are now becoming more numerous. In 1905 the value from dairy products was about three million dollars. There is a good local demand for these products at remunerative prices, whilst a fair amount is exported, and with further energy there is no reason why this export trade should not assume very large and profitable proportions. With regard to stock raising, this was chiefly carried on to supply the local markets. Latterly, however, farmers have found out that this business can be extended with advantage, and they are now supplying the home demand, as well as looking for markets abroad. The class of

*N.B.—Statements by responsible fruit farmers in Nova Scotia show that orchards in full bearing return on an average a net profit of over 25% per annum, valuing an acre of orchard at $610.

APPLE ORCHARDS.

APPLE PACKING.

stock has been much improved, and at the last Provincial
Exhibition held at Halifax (Nova Scotia), the animals shown
compare favourably with those to be seen in Scotland. The
marsh lands are a prominent feature in the country, and derive
their fertility from deposits of mud left by the tide, which is
sometimes allowed to overflow. In many cases these lands
have been reclaimed by a system of dyking, forming very rich
meadows, the crops of hay from which are exceptionally good,
in some cases reaching four tons to the acre. In the south
western district of the province there are lands suitable for
sheep raising, as they can be pastured along the shores and on
the islands during most of the winter, and considerable attention
is now being given to this industry by capitalists. There is
also a growing demand for bacon and pork at fair prices, and
the attention of farmers has been directed to this branch of
stock breeding, with a view of securing a larger share of the
increasing export trade, and the amount of live stock sold in
1905 is returned as over one million dollars. Poultry farming
has already made some considerable headway, and large
numbers are raised, and as it requires but a small capital, it
should be carried on with satisfactory results, if not by itself,
then in conjunction with fruit growing or mixed farming. In
Cape Breton the prospects for market-gardening, dairying, and
poultry farming are very bright, especially near the mining
towns, where the exceptional increase in the population has
caused a corresponding demand for these commodities. With
the cold storage system now in operation, I think that more
advantage should be taken of it in exporting Nova Scotia
produce of this description. Arrangements have already been
made with a Canadian Produce Co. (of Toronto), to take all the
poultry of some Nova Scotia farms, and ship it by this system
from Halifax to Great Britain.

I have already alluded to the apples of Nova Scotia, but in
addition to this fruit, plums, raspberries, blackberries, peaches,
grapes, cranberries, strawberries, &c., are grown with success.
Cranberries especially are an increasing crop. They grow on
boggy land where nothing else will succeed, and require very
little attention after the beds are made. They come into

bearing in about four years, and are said to last forty years. The average crop is some forty barrels per acre, but much larger returns have been received from certain districts. A considerable quantity is shipped to London, where they find a ready market at good prices.

There is a Horticultural and Agricultural College at Truro where students are received and instructed in the best and most scientific manner of fruit growing and agriculture, &c., in all its branches. The only expense incurred is a merely nominal one for board. An experimental farm is attached to the College, · so that practical as well as theoretical knowledge is imparted.

The foregoing slight sketch of the agricultural resources of Nova Scotia, will, I think, prove that there are plenty of openings for men with a small capital who are not afraid of hard work. The suitable climate, beauty of the country, richness of the soil, variety of crops that may be raised, and the accessibility to the markets of the world, constitute advantages, which I venture to affirm, are not exceeded by any of our colonies.

FISHERIES.

One of the chief industries is that of the fisheries. Probably no other country is so favoured as Nova Scotia in this respect, the yield appearing to be inexhaustible. The varieties include the cod, mackerel, herring, haddock, salmon, hake, halibut, sea trout, &c., as well as shell fish in profusion. In 1903 the value of Nova Scotia fish is returned as nearly eight million dollars (7,841,602 dollars), of which about five millions was exported. From 1869 to 1899 the value of the total yield of the Dominion was nearly six hundred million dollars (577,312,478 dollars), of which Nova Scotia contributed over two hundred and a quarter million dollars (234,499,716 dollars), or about forty per cent. The total number of boats and vessels engaged in the trade in 1903 was over 6,000, and the number of men nearly 17,000. In 1905 the value of the fisheries was 9,000,000 dollars.

GOLD.

Gold was first discovered about 1860 and is found scattered all over Nova Scotia, the auriferous area being about 3,000

SCENE ON ANNAPOLIS RIVER.

FISH CURING IN NOVA SCOTIA.

square miles, extending principally along the southern shore. The veins are of varying richness and width and carry gold in amounts ranging from a trace up to several ounces. The great future in the gold mining of Nova Scotia, however, lies in the so-called "low grade" ores, which are met with in wide belts in many districts Trials on a working scale have been made with these ores, and the field appears even more promising here than in many other gold mining countries, and nowhere can labour and the usual supplies be procured more cheaply than in Nova Scotia. As this industry has not been systematically developed until within a short time, the returns appear to be comparatively small, but there is no doubt that a great future lies before it, and the Provincial Government in April, 1903, passed an Act to encourage deep level Gold Mining, and appropriated a sum to assist in the sinking deep shafts in the Goldfields of the Province.

In addition to gold—copper, lead and zinc are found in considerable quantities ; copper in particular having been discovered in many districts. A writer in the *Mining Journal* states " That 'ere ten years have passed, Nova Scotia will have "arrived at a point in the development of its copper deposits "that will make it a factor in the production of this important " metal."

Among the miscellaneous minerals are the following : Gypsum, found in large deposits, some beds being 100 feet thick ; manganese ; antimony, which carries considerable traces of gold ; celestine ; sulphur (molybdenum) ; cobalt ; nickel and arsenic, and others.

COAL.

Coal was first discovered at Stellarton, in Picton County, in 1791. It was not, however, until after 1825 that it was systematically worked and shipped by the General Mining Association, which had acquired the lease of all the mines originally granted to the Duke of York by the government. In 1865 it was decided to abolish this monopoly, and since then the industry has been dependent upon individual enterprise for

SCENES IN THE TOWN OF SYDNEY, CAPE BRETON, N.S.

its development and progress. The coalfields of Nova Scotia, the best known of which are situate in Cape Breton, Cumberland, and Picton, embrace an area of about 4,000 square miles, and the total deposits available are estimated at 40 billion tons. The Manager of the Dominion Coal Company is reported to have said that they have sufficient resources to produce 3,000,000 tons a year for 10,000 years, and the same gentleman also mentions that the Shaft of Dominion No. 2 was perhaps the largest in the World, and would alone give employment to 2,000 men. During parts of 1904 the pay roll of this Company was for 8,000 men.* The development has only assumed important proportions during the last few years. In 1900 the total output was 3,358,000 tons, whilst that of the whole of Canada amounted to 5,333,000 tons, so that Nova Scotia contributed 3/5ths, or 60%, while the returns for 1903 show that Nova Scotia contributed 5,653,338 tons out of a total output for the Dominion of 7,643,999 tons. In 1905 the value of the coal raised was estimated at 11,250,000 dollars, and in 1906, 12,575,000 dollars. During the five years ending in 1905, more coal was raised than during the previous ten years, when over 20,000,000 tons were mined, and it is computed on this basis, that by the end of 1910, close on 50 million tons will be raised.

The coal is of the bituminous variety, and may be divided into coking, free burning, and cannel coal. The official reports on the Sydney Harbour Seam show that it is practically equal to Welsh steam coal, and during some trials made by H.M.S. "Gannet," a saving of $12\frac{1}{2}\%$ was obtained over the use of Welsh coal alone, by mixing the latter with twice its weight of Sydney coal.

Now that the supply of coal is increasing so rapidly in the province, it has been advisable to look for more extended markets, with the result that considerable shipments have been made to Norway and Sweden, the Mediterranean Ports, and France. The total exports to September 30th, 1904, amounted to about $3\frac{1}{4}$ million tons. The difficulty in connection with this trade is the question of securing return freight, and it has been

---

*N.B.—11,000 men are now (1907) employed in the Dominion Coal Company's Mine.

met in many instances by taking back suitable iron ores for mixing with those at present in use in Nova Scotia, for the manufacture of iron and steel of the grades required, which is another most important industry of the province.

IRON AND STEEL.

With regard to the quality and extent of the iron ores found in Nova Scotia, Sir William Fairburn, in writing on the subject states, " In Nova Scotia some of the richest ores yet discovered occur in boundless abundance." Also Sir William Dawson, in referring generally to the distance of the iron ore from the fuel required in so great a quantity, whenever smelting processes are undertaken on a large scale in Canada, says: " It " should, however, be borne in mind that the great iron deposits " of Nova Scotia, equal in extent and value to any others in the " Dominion, are not so situated, but lie in close proximity to " some of the greatest coalfields in the world. Even in Great " Britain itself, the two great staples of mineral wealth are not " in more favourable contiguity, and the iron ores of Great " Britain are neither so rich nor so accessible as those of Nova " Scotia."

With these vast and rich deposits of iron ore at hand, and with coal and limestone in close proximity, great facilities exist for the manufacture of iron and steel in the province. Several plants of moderate capacity had been in existence for some time, but their locations were, as a rule, against their favourable working. In 1899 at Sydney, C.B., the Dominion Iron and Steel Co. erected large works for the manufacture of iron and steel, occupying some 250 acres, with a front of 3,000 feet to deep water. They comprise four blast furnaces, ten open hearth furnaces, blooming mill, rail mill, 400 by-product coke ovens, coal-washing and sulphuric acid plants, machine shop, and foundry. The store yards have a capacity of 645,000 tons and are close to the unloading piers, on which are two towers, each capable of dealing with some 1,500 tons in the twenty-four hours. The iron ore at present used is conveyed from Wabana, Bell Island, in vessels of over 6,000 tons, which are loaded and discharged in about sixteen hours. The limestone

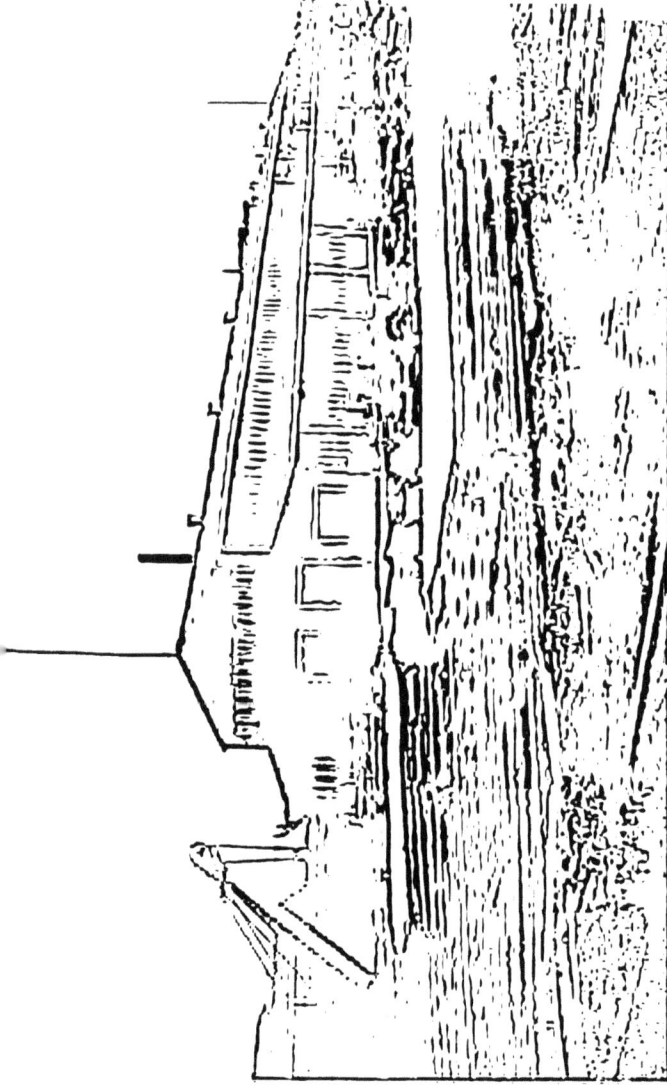

MACHINERY HALL, SYDNEY, C.B.

FURNACES, SYDNEY, C.B.

is brought by rail from a quarry about forty miles away, and there is also railway connection with the coal mines. Owing, therefore, to the raw materials being obtained within a very small radius, with coal on the spot, the cost of assembling them is very small, and on this subject the General Manager of the Company says: "That Sydney, C.B., can furnish steel to the world's markets at $6 per ton less than Pittsburg (which, for purposes of comparison, is selected as the cheapest producer), for the following reasons:—the cost of assembling the raw materials at Pittsburg, is at the lowest estimate $3.25 per ton, to which must be added the cost of conveying the manufactured iron to the seaboard, viz., $2 per ton, whilst the cost of assemblage at Sydney, which is on the seaboard, and 1,000 miles nearer to the great markets, is given at 79½ cents per ton, the difference in favour of Sydney being calculated at $6 per ton as stated." All the works are now completed, and employ a large number of hands. The machinery is of the very latest type, electrical power is used wherever practicable, and the whole plant is thoroughly up-to-date. The output for 1903 was coke 310,641 tons, pig iron 155,139 tons, steel billets and slabs 117,986 tons.

At a freight of $2 per ton, pig iron has been shipped to Glasgow, cheaper than Cleveland is shipping to that market. Steel billets are now being made and shipped to the United States, the quality of which is said be excellent. At Sydney mines the Nova Scotia Steel and Coal Co. have built extensive works alongside their coal supply. The open hearth plant was put into operation in July, 1905. In 1905, 28,723 tons of pig iron were sold, and 28,225 tons of finished material shipped from the Steel Department.

It is proposed to establish a manufactory for the making of general railway supplies to cost $400,000, and the Sydney council have voted $20,000 on completion; $10,000 for three years; exemption from taxes for ten years; and water at six cents. for 1,000 gallons. This shows the encouragement given by the city to new industries. As a result of the recent rapid development of these important industries, I may point out that the population of Sydney, which a few years ago was between 2,000 and 3,000, had increased in 1901 to nearly 18,000, and it is anticipated that the era of prosperity upon which this district has entered, will disclose possibilities, the far reaching effects of which upon the province generally, it is at present difficult to guage.

STEEL SHIPBUILDING.

With such magnificent coal resources in our possession and the establishment of the steel and iron works already referred to, producing the cheapest material in the world, the attention of capitalists was bound to be drawn to the advantages which exist for the building of steel ships. Thirty years ago, when most of the carrying trade of the world was done in wooden ships, Nova Scotia owned more tonnage in proportion to population than any other country, and her flag was to be met with in every important port in the world. With suitable timber at hand to be had for the cutting, almost every little fishing village boasted its ship-building yard, and the busy hammers of the workers were to be heard all day long. As, however, the substitution of iron ships increased, this industry gradually decayed, and although there are still a large number of small wooden vessels being built, it does not attain to anything like its former proportions. There is no doubt that, if Nova Scotia is again to compete in the carrying trade of the world with a view of regaining her old proud position, it must be achieved by iron or steel vessels. I may point out that New Glasgow has already had the distinction of constructing some steel vessels (due, no doubt, to the proximity of the old Nova Scotia Steel Company's works), but they have been of small tonnage owing to the character of the East River. It is to be hoped, however, that in this respect New Glasgow will follow in the footsteps of her old namesake. The Nova Scotia Government has already offered a large bonus to the first company establishing a ship yard capable of building steamers

N.B.—In 1905-6, the production of iron and steel was—

Pig iron	$3,500,000	$1,500,000.
Steel	3,800,000	6,500,000.
Steel rails	1,500,000	4,250,000.

of 5,000 tons, with exemption from taxation for 20 years on all ships built and registered in Nova Scotia. A member of one of the principal shipbuilding firms in this country, who has lately visited the province, after emphasizing the difficulties in the way of establishing a steel shipbuilding plant in Nova Scotia, is reported to have said: "I am persuaded that the cheapest steel "ships in the world can be produced in Nova Scotia." Whilst reminding us that the cost of skilled labour will probably remain higher than in Europe, he confidently anticipates a very superior class springing up which will do credit to the country. The higher cost of labour, moreover, will be more than counter-balanced by the cheapness of steel production and in other directions. In view of the disabilities under which this industry would labour in its initial stages, it is thought necessary that sufficient encouragement should be forthcoming from the Dominion Government in the shape of a bonus, in order to equalise the situation with that prevailing in Great Britain, Germany and the United States. It may safely be said that there is no industry so well calculated to meet the aspirations of Nova Scotians as that of steel shipbuilding, their strong maritime characters, knowledge of construction, and practical acquaintance with its details, are such important factors in the successful prosecution of an enterprise of this nature, that we may look forward to the result with every confidence.

FAST SERVICE.

The question of a fast service between the United Kingdom and Canada, has been attracting so much attention lately, and is of such importance to the advancement and development of the Dominion as a whole, and to Nova Scotia in particular, that some reference to the matter may not be without interest on this occasion.

It appears to be generally agreed that such a line, to give due prominence to the Canadian route for passenger traffic over the New York route, should consist of the fastest boats possible to build; and that the shortest available sea passage be adopted. Several schemes for this service have been put forward, and much discussion has taken place in the press. It seems certain that if sufficiently large subsidies are given by the Imperial and Dominion governments, the fastest passenger service in the world can be inaugurated between this country and Canada. This would attract and divert the present large volume of passenger traffic now going direct to New York, to some point in Nova Scotia, where, after a sea voyage occupying, according to different calculations, between three and four days, express trains would be in readiness to convey mails and passengers to their destinations to all points on the North American continent, many hours, and in some cases days, in advance of the time now occupied. We have it on the authority of our noble chairman, who, if the Nova Scotia papers can be relied on, is reported to have said at Montreal, last September, "that he " was convinced that this project would tend to make Canada "boom more than any other enterprise. He believed the "passage across the ocean could be made in four days, and he "hoped the people would realise that an additional grant would "be a trifle to the benefits that would certainly come to Canada "when such a service was an accomplished fact."

The Halifax Board of Trade in their annual report, January, 1901, when referring to this subject, states that "An "express service between Canada and the United Kingdom is "bound to come, and the only question is, how far the Govern- "ment is justified in going towards the support of such an "enterprise. In our opinion, no service should be subsidized "at all, which is not first-class in every particular, and equal "to any now crossing the Atlantic. The special advantage we "possess is our geographical position, which ensures a short sea "voyage, and we would be throwing that advantage away, if "other than the fastest boats were employed. We have always "contended that such a service, to be successful, must run "continuously all the year round from one port, and we believe "Halifax to be the best port available for that purpose in "Canada. The prestige of such a service would be of immense "benefit to Canada and the Empire, and both the Federal and "Imperial Governments should encourage the undertaking by "generous financial assistance."

A member of a well-known shipbuilding firm in this country, when in Canada recently, is reported to have said, " I am of opinion that a fast Canadian service would well repay " its establishment. I am in favour of 25 knot boats being put " on in order to make this route the quickest between Europe " and America. The success of the enterprise depends upon " the mails being landed at some point in Nova Scotia. I am " in favour of a fast service—in fact, a very fast service—so fast, " that no other steamship service on the Atlantic could compete " with it. Canada can have such a service if the country thinks " desirable. I am a little doubtful, however, whether Canada " is quite aware of the importance of this subject, or of the " greatness of this country's opportunity. In this connection, I " may say that New York is not shutting its eyes to the facts, " but on the contrary, capitalists from that city have already got " their attention firmly fixed on the Canadian route. If you " will carefully examine the subject, you will find that Canada " has been practically standing still in the matter of an Atlantic " mail and passenger service for the past forty years. The " Cunard line, as far back as 1860, used to make the passage " between Liverpool and Halifax in about eleven days. There " is hardly any substantial improvement in that time to-day in " the Canadian service, whilst New York, in the meantime, has " halved the time.* The fast mail delivery must form an " attractive feature. So speedy and satisfactory. must this " service be, that the European mails for the East shall go by

---

*It is extremely interesting to note that, since the above was written, the time occupied in crossing to Canada has been appreciably shortened. The Allan Line with their new Turbine boats have covered the distance from Rimouski to Moville (Ireland) in 5 days 17 hours, while the C.P.R. in their new ''Empress'' boats took 5 days 6 hours from Rimouski to Liverpool. This shows improvement, but the real fast service as fore-shadowed in the foregoing notes is yet to come. Indications, however, all point to such a service being established at no distant date, and at the time of writing (1907) a scheme is being formulated whereby it is stated that the sea passage will be accomplished in 3½ days, the terminal port in Canada to be Halifax, N.S. It is also interesting to note the advantages in distance Halifax possesses as a port, which will be best gathered from the following table :—

Halifax is 655 miles nearer Liverpool than New York.
,, 1,050 ,, ,, Gibraltar ,, ,,
,, 750 ,, ,, Cape Town ,, ,,
,, 200 ,, ,, ,, ,, Liverpool, G.B.
It is nearer every South American Port South of Pernambuco than any United States shipping Port.

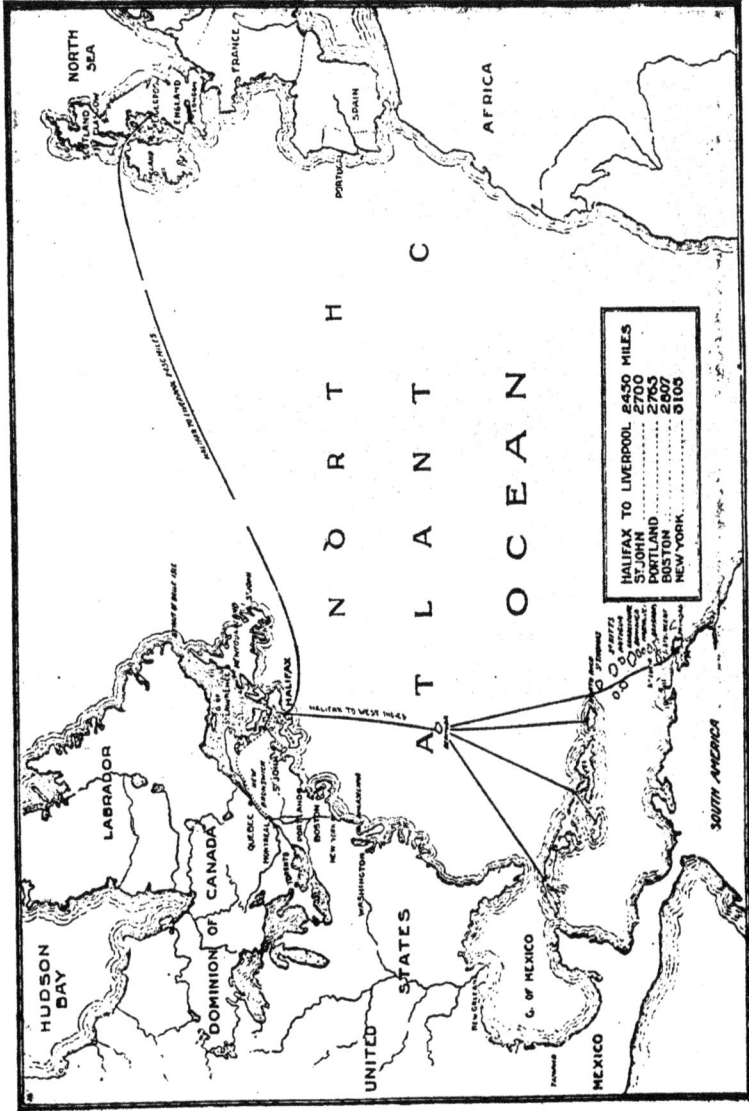

MAP OF NORTH ATLANTIC OCEAN.

"the Canadian route, and across the continent by the Canadian
"Pacific Railway, instead of going as at present, via New York
"and San Francisco. It will be a very fine thing for Canada
"to have the cream of the world's travel streaming through the
"country. It cannot fail to be a magnificent advertisement for
"the country, and will result, I am convinced, in very much
"additional foreign capital coming into Canada for development
"purposes."

These are but a sample of the opinions expressed by
capable men, of the great advantages which would accrue to
Canada, on the establishment of this fast service.

Halifax, with its fine harbour, extensive docks, and open
water all the year round, has strong claims to be the terminal
in Nova Scotia, and it does not appear that any other port, at
present, has equal facilities for dealing with the class of boats
which will run when this line is inaugurated. Sydney, Cape
Breton, is some hours steaming nearer England, and would
make an excellent terminus during summer months. But
whatever terminal is selected, I venture to affirm that the
establishment of the Canadian Fast Service will be an epoch in
the annals of the Great Dominion.

RAILWAYS.

Nova Scotia is well served by railways, 1,174 miles being
in operation in 1905; thus ensuring ready access to markets.
There is also very efficient communication by water, so that the
inhabitants of Nova Scotia enjoy, perhaps, more facilities in
the way of transport than the other Maritime Provinces, as
there are not many parts of the country more than a few miles
distant from a railway.

The Government are, however, still pushing on with the
construction of other lines, recognizing the importance of this
means of communication in the development of the many
natural resources of the Province.

SPORT.

Nova Scotia is a sportsman's paradise, there being
excellent shooting, hunting and fishing in nearly every county.

FISHING SCENES.

There are bears, foxes, moose cariboo, otter, mink, sable musquash, hares and racoons; whilst of feathered game, wood-cock, snipe, plover, partridges, geese, ducks, brant, curlew, &c., abound. The game laws are simple, and are made only to protect game when out of season. There are no private game preserves in the province, and the whole of the interior is a grand hunting ground open to everybody. The lake region is the natural haunt for numbers of large and small animals. One feature of autumn shooting is that of wild duck and geese. They swarm over the lakes in millions, and in October one could kill them as fast as a gun could be discharged But the great ambition of the sportsman visiting Nova Scotia is to kill a moose. These animals are frequently 8 feet high, weighing 1,500 lbs., with horns measuring from 5 to 6 feet across, and when wounded they are inclined to show fight The most fascinating form of moose hunting is known as "calling." The hunter secretes himself behind bushes or rocks on some favourable night in September or October, when the moon is full and no wind disturbs the foliage. He imitates the call of the moose through a birch bark trumpet. Perhaps the call is answered and repeated, until, amidst the crashing of branches and breaking of trees, the victim appears upon the scene, only to see a flash and hear a report, when he rolls over if fairly hit, or if merely wounded rushes upon his assailants to fight it out.

HALIFAX.

Halifax, the capital, has a population of 45,000 It has been called the Gibraltar of North America, and while it has no similarity in appearance to that stronghold, it would just as stubbornly refuse capture. The citadel, occupying the summit of a hill 250 feet high at the back of the town, commands the surrounding country and harbour, the entrances to which are protected by numerous forts and batteries.

The original fortress was built by the Duke of Kent, father of our late lamented Queen, when Commander-in-Chief there, and he utilized the labour of some 500 maroons, who, having burned plantations and murdered slaves in Jamaica, were exiled to Nova Scotia. The explanation of their arrival contains a

PUBLIC HALL, HALIFAX, N.S.

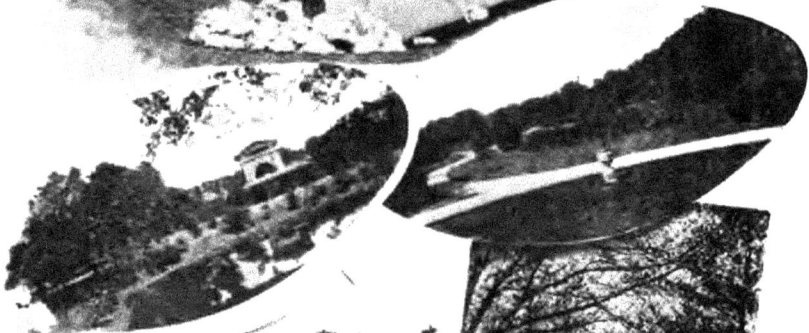

PUBLIC GARDENS, HALIFAX, N.S.

reflection on the climate of the province. It was thought in Jamaica that the winters in Nova Scotia were very severe, and that by sending them from a hot district they would soon be finished off. It speaks volumes for the climate, however, for they not only survived, but re-invigorated by the beneficial climatic conditions, their old propensities developed to such an extent, that they had to be sent to the West Coast of Africa, where they promptly commenced to depopulate a wide stretch of country in the interests of the Arab slave traders.

Halifax is the seat of the Local Government and a University and Cathedral City, with a Lieut.-Governor, an Archbishop and a Bishop. In addition to this, it is the head-quarters of the Navy in British North America, and is besides a garrison town, which adds a brilliancy to the social life not met with elsewhere in Canada.* There are few places which combine so many advantages for pleasure, quiet, or health, according to inclination. The people are most hospitable, and the prevailing object appears to give the visitor as favourable an impression of the place as possible, which is invariably attained. The opportunities for open air relaxations are many. In summer all Halifax rides and drives, sails and rows. In winter it devotes itself to those sports which put roses in the cheeks of her women and vigour and health in her men.

The Public Gardens are a great credit to Halifax, as well as the Park at Point Pleasant. The gardens, containing 14 acres, are admitted to be as fine as any in America, and form a favourite resort for the people. A great dry dock is the natural accompaniment to a great seaport, and Halifax prides itself upon its possession. The dock is 613 feet long, 102 feet wide at the top and 70 feet at the bottom, and it is rarely unoccupied.

DESCRIPTION OF INHABITANTS.

The inhabitants of Nova Scotia are known locally as "Blue Noses," a name which originated with the loyalists who

*Since the above address was delivered the British Regiments have been withdrawn from Halifax, which is now garrisoned by Canadian Troops; and it is no longer the Head-quarters of the British North American Squadron.

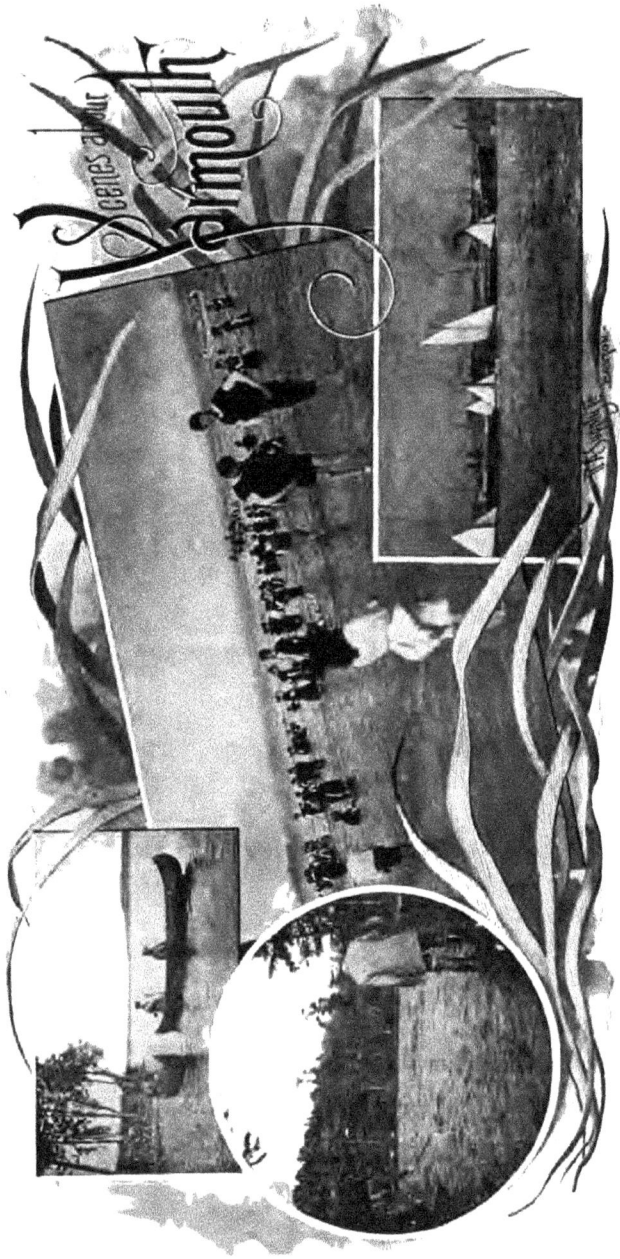

Scenes about Yarmouth

left the revolting colonies in 1776 to settle in Nova Scotia, calling themselves "True Blues." This was changed by the rebels to "Blue Noses," as a term of contempt, but it was always accepted by every Nova Scotian as a most flattering acknowledgment ot their unwavering loyalty. The experiences of the early settler, and the many parts he was called upon to play, not only in surmounting the hardships connected with colonization, but in preparing for and repulsing sudden hostile attacks, made him and his children self reliant to a degree, and this trait has been developed in subsequent generations.

To this no doubt is due the remarkable versatility of the Blue Nose, who is able to turn his hand to any trade or calling. He may be called a "Jack of all trades," but I think it must be admitted, that to such a type of the "handy man," is attributable the founding, retention and success, not only of our own province, but of the colonies generally. It is a fact not generally known, but worthy of note, how many of Nova Scotia's sons have become famous outside their province, and rendered valuable services to the Empire. In 1753 we find at the attack on Fort Beau Sèjour. N.S , a crippled boy rounding up a herd of cattle which had strayed towards the French. Thanks to the chivalry of the French Commander, who, admiring his pluck, refused to open fire; the boy accomplished his task, returning with the cattle. After being in a Halifax office he drifted to London, where he subsequently rose to the dignity of Lord Mayor as Sir Brook Watson, so far as I know, the only colonist who has attained to this proud distinction.

Of Britain's military heroes, Nova Scotia has given General Fenwick Williams, the hero of Kars, and Sir Provo Wallis, still fresh in the memory of the English. Inglis, of Lucknow, though not born there, was the son of a Nova Scotian Bishop, and more recently, Stairs, who, connected with Stanley's expedition, lost his life in Africa, where alas, so many of his brother Canadians have died in the cause of the Empire Then we have Samuel Cunard, whose steamships placed Great Britain .in the van, being also a native of Halifax. Our patriot Joseph Howe, who, in a speech at Southampton nearly fifty years ago, advocated that United Empire, which recent circumstances

have done so much to bring about. The humorous writings of
Haliburton, the Nova Scotian judge, are not forgotten here.
Nor will we omit the name of Sir Charles Archibald, the only
Canadian who ever attained the position of judge of the High
Court of this country, for he was a native of Truro, Nova
Scotia. Closer at home, but still with records of Imperial
services, we have Sir Charles Tupper, one of the founders of
Confederation, late High Commissioner for Canada, and after-
wards Premier of the Dominion. We also claim another
Dominion Premier in the person of Sir John Thompson, whose
death occurred under such distressing circumstances at Windsor
Castle. Other distinguished names such as Dawson, Grant,
O'Brien, Fielding, and Borden, belong to Nova Scotia, in
addition to Murray and Longley, at present watching over the
destinies of our province. It is a good list, and one of which
we may well be proud, and my only regret is that we are not
able to claim our noble chairman here to-night as a " Blue
Nose."

Information for Intending Emigrants.

Nova Scotia, distant from Liverpool 2,450 miles, is reached by steamers in from 7 to 10 days. Passages can be booked by the "Allan," "Dominion," and the "Canadian Pacific" S.S. Lines, either direct or *via* the St. Lawrence route, while the Furness Line have a service from London to Halifax at frequent intervals. Full details as to dates of sailing, luggage, &c., will be supplied upon application to any of the above Companies. (See Advertisements.) There are no assisted passages, and the fares range from £5 10/- third class. At Halifax, emigrants are met by a Government Agent, and accommodation is provided for them if required on landing, pending their departure for outlying districts to which they may be proceeding. Letters of introduction to the Agent will be supplied upon application to the Agent-General, and intending settlers are advised in all cases to avail themselves of this opportunity, and to avoid being guided by any irresponsible persons.

TRANSPORT.

The Province is well served by Railways and communication is rendered easy by these means, as well as by water, between the various towns, villages, and the City of Halifax. Farmers in outlying districts are thus enabled to send their produce to the central market with very little delay.

HOUSE RENT.

Rents vary according to locality, those in the country being naturally lower than for houses in or near the large towns. In Halifax flats in tenements, suitable for working men, of 4 to 6 rooms cost from $5 to $10 per month, while houses of 4 to 8 rooms cost from $7 to $15 per month. Miners cottages about $6 per month. Board can be obtained at most places at from $2·50 to $3·50 per week.

Taxes are very low, from 1 to 1½% and in some districts even less.

A Dollar $ = 4/2 Sterling.

PROVISIONS.

Generally speaking the prices of provisions (that is the necessaries of life) may be said to be cheaper than in England, while the luxuries are somewhat dearer. Clothing is about 10% or 2/- in the £ dearer than in this country.

LABOUR OPENINGS.

There are openings for nearly all classes of labour, but it must be understood that the demand varies, and it is advisable for the intending emigrant to apply in the first instance to the Agent-General (who is in touch with employers of labour in the Province), and ascertain if there are opportunities in his particular line, thus preventing possible delay or disappointment on arrival.

The following are generally in demand :—

FARM LABOURERS (Handy Men).—Wages from $16 to $20 a month, according to ability, with board. The best time for this class to start is March or April. Men desirous of acquiring a knowledge of farm work previous to starting on their own account and having no acquaintance with agriculture, would receive their board and a nominal wage according to capacity. Domestic Servants and women accustomed to farm life, and who can make butter, &c., are also needed.

COAL MINERS AND LABOURERS.—There are usually plenty of openings in the various mines for this class of men. Wages average for miners $2·89 per day of 8 hours. Surface labourers earn about $1·28 per day of 10 hours. Boys 61 to 97 cents. per day of 10 hours. The demand for this class of labour is good, and the prospects of the coal industry were never better.

Carpenters, machinists, fitters, iron workers and kindred classes of labour earn from $9 to $12 per week.

Marine engineers, at Halifax, holding first and second class certificates, have little difficulty in obtaining employment at wages from $60 to $90 per month

CROWN LANDS.

Crown lands can be obtained at the rate of £8 per 100 acres, but there are practically none now left suited for agriculture.

FARMING AND FRUIT GROWING.

Improved farms, with or without buildings on them, can be acquired at very reasonable rates throughout the Province according to size and locality, and afford openings for men with a comparatively small amount of capital. There are also openings for market gardeners in the neighbourhood of Sydney and Cape Breton, as owing to the development in the iron and coal industries, a great deal of produce has to be imported, the demand being in excess of the supply. Recent advices state a strong local demand for farm produce, and that men holding small farms of say 100 acres of mixed land, pasture, arable, and wood are making comfortable livings and putting by substantial sums yearly, and that similar opportunities are practically open to hundreds of others. A list of various farms for sale can be obtained and full particulars will be supplied upon application to the Agent-General, together with printed matter, maps, &c.

SETTLER'S EFFECTS FREE OF DUTY.

The following is an extract from the Customs tariff of Canada, specifying the articles which can be so entered by settlers free of duty :—

SETTLERS' EFFECTS, viz. :—Wearing apparel, household furniture, books, implements and tools of trade, occupation or employment, guns, musical instruments, domestic sewing machines, typewriters, live stock, bicycles, carts and other vehicles, and agricultural implements in use by the settler for at least six months before his removal to Canada; not to include machinery, or articles imported for use in any manufacturing establishment, or for sale; also books, pictures, family plate or furniture, personal effects and heirlooms left by bequest; provided that any dutiable articles entered as settlers' effects may not be so entered unless brought with the settler on his first arrival, and shall not be sold or otherwise disposed of without payment of duty, until after twelve months' actual use in Canada; provided also, that under regulations made by the Controller of Customs, live stock, when imported into Manitoba or the North-west Territories by intending settlers, shall be free until otherwise ordered by the Governor in Council.

The settler will be required to fill up a form (which will be supplied to him by the customs officer on application) giving description, value, &c., of the goods and articles he wishes to be allowed to bring in free of duty. He will also be required to take the following oath :—

I...............................do hereby solemnly make oath and say, that all the goods and articles hereinbefore mentioned are, to the best of my knowledge and belief, entitled to free entry as settler's effects, under the tariff of duties of customs now in force, and that all of them have been owned and in actual use by myself for at least six months before removal to Canada; and that none of the goods or articles shown in this entry have been imported as merchandise or for use in any manufacturing establishment, or for sale, and that I intend becoming a permanent settler within the Dominion of Canada.

BANK OF MONTREAL.

(ESTABLISHED 1817.)

CAPITAL (Paid-up), 14,400,000 dols.   RESERVE FUND, 11,000,000 dols.   Undivided Profits, 422,689 dols.

HEAD OFFICE, MONTREAL.

Board of Directors—

Rt. Hon. LORD STRATHCONA AND MOUNT ROYAL. G.C.M.G., Hon. President.
Hon. Sir GEORGE A. DRUMMOND, K.C.M G., President.   E. S. CLOUSTON, Esq., Vice-President.
A. T. PATERSON, Esq.   E. B. GREENSHIELDS, Esq.   Sir W. C. MACDONALD.
R. B. ANGUS, Esq.   JAMES ROSS, Esq.   Sir R. G. REID.
Hon. ROBERT MACKAY.

General Manager—E. S. CLOUSTON, Esq., Montreal.

BRANCHES IN THE DOMINION OF CANADA.

Alliston, Ont.	Goderich, Ont.	Montreal, St. Catherine	St. Mary's, Ont.
Almonte, Ont.	Grand Falls, N.B.	Street.	Sarnia, Ont.
Altona, Man.	Grand Mere, Que.	„ St. Henri.	Saskatoon, Sask.
Amherst, N.S.	Greenwood, B.C.	„ Seigneurs Street.	Sawyerville, Que.
Andover, N.B	Guelph, Ont.	„ Westmount.	Shediac, N.B.
Armstrong, B.C.	Halifax, N.S.	Mount Forest, Ont.	Stratford, Ont.
Aurora, Ont.	*Sub-Branch*	Nelson, B.C.	Sudbury, Ont.
Bathurst, N.B.	„ North End.	New Denver, B.C.	Summerland, B.C.
Belleville, Ont.	Hamilton, Ont	Newmarket, Ont.	Sydney, N.S.
Bowmanville, Ont.	*Sub-Branch*	New Westminster, B.C.	Toronto, Ont.
Brandon, Man.	„ Sherman Avenue	Nicola, B.O.	*Sub-Branches*
Brantford, Ont.	Hartland, N.B.	Oakville, Man.	„ Carlton Street.
Bridgewater, N.S.	Holstein, Ont.	Ottawa, Ont.	„ Queen Street.
Brockville, Ont.	Indian Head, Sask.	*Sub-Branches*	„ Richmond Street.
Buckingham, Que.	Kelowna, B.C.	„ Bank Street.	„ Wellington Street
Calgary, Alta	King City, Ont.	„ Hull, Que.	„ Yonge Street.
Canso, N.S.	Kingston, Ont.	Paris, Ont.	Trenton, Ont.
Chatham, N.B.	Lake Megantic, Que.	Perth, Ont.	Tweed, Ont.
Chatham, Ont.	Lethbridge, Alta.	Peterborough, Ont.	Vancouver, B.C.
Chilliwack, B.C.	Levis, Que.	Picton, Oct.	*Sub-Branch*
Collingwood, Ont.	Lindsay, Ont.	Portage La Prairie, Man.	„ Westminster Ave.
Cookshire, Que.	London, Ont.	Port Arthur, Ont.	Vernon, B.C.
Cornwall, Ont.	Lunenburg, N.S.	Port Hood, N.S.	Victoria, B.C.
Danville, Que.	Mahone Bay, N.S.	Port Hope, Ont.	Wallaceburg, Ont.
Deseronto, Ont.	Medicine Hat, Alta.	Quebec, Que.	Warsaw, Ont.
Edmonton, Alta.	Millbrook, Ont.	*Sub-Branches*	Waterford, Ont.
Edmundston, N.B.	Moncton, N.B.	„ St. Roch.	Winnipeg, Man.
Eglinton, Ont.	Montreal, Que.	„ Upper Town.	*Sub-Branches*
Enderby, B C.	*Sub-Branches*	Queensville, Ont.	„ Fort Rouge.
Fenelon Falls, Ont.	„ Hochelaga.	Raymond, Alta.	„ Logan Avenue.
Fort William, Ont.	„ Papineau Avenue.	Regina, Sask.	Wolfville. N.S.
Fraserville, Que.	„ Point St. Charles.	Rosenfeld, Man.	Woodstock, N.B.
Fredericton, N.B.	„ St. Anne de	Rossland, B.C.	Yarmouth, N.S.
Glace Bay, N.S.	Bellevue.	St. John, N.B.	

IN NEWFOUNDLAND.—Birchy Cove; St. John's.

IN UNITED STATES.—New York: 31, Pine Street.   Chicago;   Cor: La Salle and Monroe Street.
Spokane: State of Washington.   IN MEXICO.—City of Mexico.

LONDON OFFICE : 47, THREADNEEDLE STREET, E.C.

Committee—Rt. Hon. LORD STRATHCONA AND MOUNT ROYAL, G.C.M.G., THOMAS SKINNER, Esq., ALEXANDER LANG, Esq.
FREDERICK W. TAYLOR, Manager.   H. HAYLOCK, Assistant Manager.
LONDON BANKERS—Bank of England.   National Provincial Bank of England, Ltd.
London and Wetminster Bank, Ltd.   Union of London and Smiths Bank, Ltd.

THE BANK OF BRITISH NORTH AMERICA.

Established in 1836.

INCORPORATED BY ROYAL CHARTER, 1840.

PAID UP CAPITAL, £1,000,000.   RESERVE FUND, £460,000.

Head Office: 5, GRACECHURCH STREET, LONDON, E.C.

Secretary : A. G. WALLIS.
London Office Manager : W. S. GOLDBY.

The Bank grants Drafts and Telegraphic Transfers upon its Branches in the Provinces of Quebec, Ontario, Manitoba, Nova Scotia, New Brunswick, Saskatchewan, Alberta, and British Columbia, also upon its Agents in New York, San Francisco, and Chicago. Similar remittances can be arranged through the Bank's Correspondents in the chief cities and towns in the United Kingdom.

Drafts and Telegraphic Transfers are also issued upon Dawson City (Klondyke).

BILLS, COUPONS, &C., PURCHASED OR FORWARDED FOR COLLECTION.

Purchase and sale of stocks, collection of dividends, and banking business generally undertaken throughout the Dominion of Canada and the United States.

DEPOSITS ARE RECEIVED IN THE LONDON OFFICE

CANADIAN PACIFIC
ATLANTIC STEAMSHIP LINES.

THE ROYAL MAIL STEAMERS
"Empress of Ireland"
AND
"Empress of Britain"
Are the LARGEST, FASTEST and MOST COMFORTABLE to Canada.

R.M.S. "Empress of Britain" (14,500 tons, 18,000 h.p. twin screw) which
with her sister ship the "Empress of Ireland," holds the record for quickest
passage in both directions from Liverpool to Rimouski.

Luxurious Accommodation for Tourist and Emigrant.

Weekly Sailings from Liverpool.

Frequent Sailings from Bristol, London and Antwerp.

Cargo Booked on Through Bills of Lading.

Parcels Despatched Swiftly by Dominion Express to all Points in
Canada and the United States.

Apply personally, or by letter, for gratuitous and post free accurate Maps and Hand-
somely Illustrated Guide Books to the **PASSENGER DEPARTMENT, CANADIAN
PACIFIC RAILWAY, 62-65, Charing Cross, London, S.W.**; 67 & 68, King William
Street, London, E.C.; 24, James Street, Liverpool; 67, St. Vincent Street, Glasgow

PRODUCTS OF NOVA SCOTIA.

	1905.	1906.
Coal ...	$11,250,000	$12,575,000
Coke ...	650,000	1,350,000
Gold	320,000	260,000
Iron Ore ...	80,000	—
Other Minerals ...	620,000	640,000
Pig Iron	3,500,000	1,500,000
Steel ...	3,800,000	6,500,000
Steel Rails	1,500,000	4,250,000
Fisheries	9,000,000	9,000,000
Manufactures, Ships & Freights	42,350,000	38,000,000
Products of the Farm ...	18,400,000	20,500,000
Products of the Forest	3,200,000	4,750,000
	$94,670,000	$99,325,000

NOVA SCOTIA

ITS

Resource AND Opportunities

FOR

LABOURERS

AND

INVESTORS.

A Country of

Fruitful Orchards and Farms,
Rich Mines of Coal and Iron,
Valuable Fisheries and Forests,
Prosperous Manufactures and Merchandize,
Good Schools and Colleges,
Happy Homes, Contented People.

ISSUED BY AUTHORITY OF THE NOVA SCOTIA GOVERNMENT

NOVA SCOTIA

FOR

LABOURERS AND INVESTORS.

GEOGRAPHY.

Nova Scotia forms a peninsula at the south-eastern corner of Canada. It is divided by nature into the Island of Cape Breton and Nova Scotia proper. Its total length is about 360 miles, and it is from 45 to 100 miles in width. Its area is estimated at 21,730 square miles.

Its shores are indented by numerous good harbours.

The climate of Nova Scotia may be described as similar to that of the northern part of the United States, but sharing neither the excessive heat nor the extreme cold of the latter. The summer temperature rises to 93 degrees, while that of winter occasionally falls to 12 degrees below zero. The average summer temperature is 62 degrees, that of winter about 25 degrees, and the mean annual temperature about 43 degrees. The annual rainfall ranges between 40 and 45 inches, and is spread nearly equally over each month of the year. The rapidity of vegetation permits of ample time for ripening in the autumn.

Statistics show that the climate is among the healthiest in the northern hemisphere.

The population of Nova Scotia is estimated at 460,000, and is engaged largely in farming and fishing, as well as in mining, lumbering, manufacture, merchandise, &c.

HISTORY.

The claim of Great Britain upon Nova Scotia was originally based upon the discoveries of Columbus and the Cabots. The first settlement was made by the French in 1605 at Port Royal, now Annapolis The territory was later taken by the English, and for a century these two nations disputed with each other and fought for the possession of this favoured land, each having gained possession several times, until it was finally ceded to Great Britain by the Treaty of Utrecht in 1713, as was also the whole of Newfoundland. France, however, retained possession of Cape Breton, but in 1758 this island also was taken by the English ; Halifax, the capital, was founded in 1749, and the entire province became a part of the Dominion of Canada under the Confederation of 1867.

VIEW OF HALIFAX AND HARBOUR.

GOVERNMENT.

The province of Nova Scotia has, under the Confederation Act of Canada, certain powers of self-government. All matters of general importance, such as the militia, excise, customs, banking, commerce, &c, are controlled by the Federal Government at Ottawa. Other matters, such as education, administration of justice, provincial taxation, municipal matters, roads, bridges, &c., come under the Provincial Government at Halifax.

It will be seen that the principle of government by the people, and for the people is carried to its fullest extent in Nova Scotia as well as in Canada generally.

COMMUNICATION.

Railway lines are numerous, as the accompanying map will show, and these afford every accommodation for freight and passenger requirements. There are about 1,800 post offices in the province.

Telegraph lines follow the railways and the coast, while telephone lines connect the intermediate settlements. The shore ports are connected by ferries and steamers subsidised by the province. It will thus be seen that the means of communication and transportation are ample.

AGRICULTURE IN NOVA SCOTIA.

Situated in the temperate zone, from three to six degrees nearer the Equator than the most southerly point in Great Britain, and almost entirely surrounded by sea, Nova Scotia possesses a temperate, humid climate, well suited to the highest forms of agriculture. The warm waters of the Gulf stream approach very nearly to the south-western end of the province, and although their influence is partially offset by the Arctic currents, which delay spring vegetation about two or three weeks, yet the climate is moderate and free from extremes, the average temperature in summer being 62 degrees F., and in winter 25 degrees F. The rainfall averages about 42 inches per annum, providing ample moisture for the growth of good crops and for the maintenance of excellent pastures.

The surface of Nova Scotia is of a varied character, being intersected by ranges of hills and mountains, between which lie fertile valleys. The Atlantic sea-board on the south consists of rocky highlands covered with a network of lakes, which find their outlet to the sea by numerous rivers, along the banks of which are many fertile, though somewhat isolated, areas of cultivated land. The larger agricultural areas are found north of these highlands, and consist, in some cases, of large valleys—such as the famed Annapolis Valley, described elsewhere—and in other cases of extensive level areas gradually sloping towards the northern shores.

Out of a total land area of 14,483,000 acres exclusive of the water surface, about 37.68 per cent. is occupied by farmers. Of the

remainder, comprising rocky, swampy, marsh, wooded land or
pasture, about 50 per cent. can be cultivated or grazed. Farms
average from 50 to 150 acres, although there are a few of much
larger area. Nearly all the farms are owned outright by the farmers
that till them, only 2 per cent. being leased or rented. There are
practically no Crown lands available for agricultural purposes, but
there are many sections where both cultivated and unbroken land
can be readily purchased at reasonable figures.

DYKED LANDS.

Nova Scotia is particularly noted for its large tracts of
inexhaustibly fertile dyked marsh lands that in earlier or later times
have been reclaimed from the sea. This marsh land lines the head
waters of the Bay of Fundy and spreads far inland up its river
tributaries. The Annapolis, King's, Hants, Colchester, and
Cumberland County Marshes have been providing annually from
two to four tons of hay per acre, and have, in addition, afforded fall
pasturage for large herds of cattle. Every incoming tide still continues to leave deposits of rich soil, formed from the wearing action
of the currents upon the sides and bottom of the bay. Farmers
whose lands closely adjoin haul this " marsh mud," as it is called, on
their upland fields, and thus add greatly to their fertility.

The interval lands along the rivers and smaller streams are
invariably rich and productive. The upland is of varying degrees
of fertility.

As might naturally be expected in a country where apples and
other fruit grow to such perfection, all farm crops natural to temperate climates will flourish under proper care and cultivation. Of
the grains, oats are the most largely grown, but wheat and barley
grow almost equally well. However, in this humid, temperate
climate hay and roots of all kinds, including potatoes, flourish in
the highest degree. With lands that will raise from two to three
tons of hay per acre, 1,000 bushels and more of turnips per acre and
good crops of oats and other grains, the Nova Scotia farmer, under
such excellent climatic conditions as prevail, can raise stock most
successfully and at a handsome profit.

As to markets, a most important consideration to the practical
farmer, there are few countries more fortunately situated. The
extensive development of coal, iron, lumbering, fishing, and other
industries provides a large population of people who pay the highest
market prices for all kinds of farm produce. At the present time
this local market is not nearly supplied by local farmers, much
produce having to be shipped in from others parts of the Dominion
of Canada. Potatoes and other produce are extensively shipped to
the West India Islands, which are conveniently reached by lines
of steamers. Lambs are largely shipped to the United States.
With its splendid system of railways and shipping ports the possibilities of developing other markets are unlimited. Apples are, as
yet, the only farm produce shipped to Great Britain, but, with the

improvement of transportation facilities (which is now assured), there is no reason why dairy produce, beef, mutton, and various fruits cannot be shipped equally successfully.

The development of agriculture in Nova Scotia has been somewhat peculiar. Owing to the varied resources of the province, a large proportion of those living upon farming lands have, up to the present time, not confined themselves to pure agriculture, but have entered into mining, lumbering, fishing, and other enterprises. On this account many of the farms have been neglected and now need better cultivation. Moreover, especially in the days when farming implements were crude and farm work was little better than drudgery, the nearby cities of the United States held out attractions to many of the young men, who were thus induced to leave their farms. Following this there has been the natural exodus to the Western lands. The general result of these factors has been to produce a restlessness among the young men which has not been conducive to thorough farming methods, and, as a result, there are to-day many splendid farms, somewhat under-cultivated, it is true, but capable of providing a good livelihood, which can be purchased at very reasonable figures.

A NOVA SCOTIA TURNIP FIELD.

KINDS OF FARMING.

Setting aside the fruit sections, Nova Scotia is particularly adapted to general farming. While there are successful farmers who make a specialty of dairying, hog raising, beef, sheep, and horses, yet even these usually keep other kinds of stock and raise potatoes, turnips, and vegetables of all kinds for markets. The day for exclusive grain growing has passed by.

DAIRYING.

The greatest successes have been achieved in dairy farming, especially when that has been combined with hog raising. At the present time large quantities of butter have to be imported to supply the local demand. The outlook is particularly good for those who will take the pains to produce first-class butter, which meets a ready sale at from 25 cents to 35 cents per pound the year round. Cream and milk are shipped to Halifax, Sydney, and other places from points over 200 miles away. The only drawback to successfully carrying on the dairy industry is the scarcity of labour, but those farmers who have families to assist in the work have no difficulty in making dairy farming successful.

BEEF.

Beef raising is pursued along the dyked marsh land areas and in some of the more isolated sections described above. While, for the most part, pure beef farming does not hold out such attractive promises as dairying, yet, when combined with partial dairying or sheep raising, or carried on in combination with fruit growing, good results have been realized.

SHEEP.

Sheep raising pays exceedingly well. The quality of lamb and mutton is much superior to that raised further inland, and as a result these products realize splendid prices and are largely sought by buyers from the United States and elsewhere. Although considerable numbers are shipped away, yet, as often transpires, the local market is far from supplied, especially in the early spring and during the winter months. The quality of wool produced under Nova Scotian conditions is also much above the average. For the most part sheep are kept in small flocks, varying from 15 to 75. Kept in these numbers they have paid well. The average Nova Scotia farmer does not possess the expert knowledge necessary to the keeping of larger flocks. The following statement, made by a sheep raiser, will give a general idea of the profits that have been made. His flock consists of 50 grade ewes and one pure-bred ram. He values these at $275.

INCOME.

70 lambs at $3	...	...	...	...	...	$210.00
300 lbs. wool at 27c.	...	...	...	...	...	81.00
						$291.00

EXPENSES.

500 bushels turnips at 10c.	...	...	...	...	...	$50.00
50 ,, oats at 40c. ...	...	...	.	...	...	20.00
1 ton wheat bran	...	...	...	...	...	20.00
7 tons clover hay at $8 ...	...	...	...	...	...	56.00
Extras ...	...	...	...	...	...	2.00
						148.00
Net profit ...	...	...	...			$143 00

In this statement no allowance is made for labour or pasture, the owner considering that the value of the sheep as weed destroyers, together with the manure produced, would offset these items. Granting, however, a liberal allowance for these items, there is still a good profit

A FARMERS' PICNIC AT THE COLLEGE OF AGRICULTURE, TRURO, N.S.

The one great drawback, as the reader will see from the above figures, to the development of the sheep industry is the fact that the shepherd must count on winter feeding his sheep for nearly six months each year. As, however, hay and turnips can be raised cheaply, it is quite possible to do this and make, as in the case quoted, a handsome profit. Some sheep farmers in Nova Scotia to whom the above figures were submitted, considered the cost of winter feeding much higher than it need be. In addition, the price allowed for the lambs ($3) is below the amount received by many sheep farmers, who have had no difficulty in getting from $4 to $5 each from American buyers. It is, therefore, quite evident that the farmer who understands the art of sheep raising has a good opening in Nova Scotia, and if he will combine this with some dairying, beef raising, or horse breeding, he has a sure proposition.

HORSES.

Horse breeding also holds out excellent prospects. At the present time there is a dearth of good horse flesh, and the highest prices are obtained for anything like a first-class horse. The Government makes a practice of importing at frequent intervals high-class stallions of the various breeds, whose services are available at a nominal figure. The only drawback at the present time is a scarcity of first-class brood mares. However, a good judge of horses who knows better than to part with his best fillies and brood mares, can soon overcome this difficulty, and with the excellent markets at his door can do well with almost any breed of horses.

POULTRY.

With such good markets within convenient access, it goes without saying that poultry raising, rightly pursued, holds out good inducements. While we would not recommend anyone, unless he has a peculiar genius for such work, to follow this exclusively : the markets prices of the various poultry feeds are too high for such a practice. But, on the other hand, there are few who cannot make a nice little revenue from a fair-sized flock of poultry on the farm. There need be no fear about a good market either for eggs or for dressed birds, and there is no reason why farmers, especially if helped by their wives and families, cannot do, as many are doing here now, make the hens pay the ordinary household necessities, such as tea, sugar, flour, &c., &c.

HOGS.

Hog-raising, especially in conjunction with dairying, is also a profitable industry. For the most part it is carried on, like poultry raising, on a small scale on each farm. A large proportion of the miners and fishermen feed from one to three or four pigs every summer and, on a limited scale, the raising of spring litters of pigs to sell when weaned yields a large profit.

GARDEN PRODUCE.

There is also the usual proportion of men who make a specialty of raising market truck of all kinds. Strawberries and all small fruits grow well and meet a ready market, and· especially in the vicinity of the towns and cities, there is yet plenty of scope for the raising and selling of all kinds of garden produce.

GOVERNMENT AID TO AGRICULTURE.

The Local, as well as the Federal Government, are exceedingly generous in the appropriation of funds to aid agricultural industry in Nova Scotia. Under the direction of the Federal Government there is a large Experimental Station at Nappan, in Cumberland County, where tests are annually carried on with the various varieties of grain and other crops. At Truro, in the centre of the province, the local government has established a well-equipped Agricultural College, of which some views are given in this pamphlet. At this institution instruction is given, free of cost, to farmers and farmers' sons in the various lines of agriculture. Courses are arranged, varying in length from a two weeks' practical course to a two years' course, in which there is combined with a detailed practical course, general educational instruction along business and other lines. There is connected with this college a farm on which are kept

HAYMAKING IN NOVA SCOTIA.

splendid stables of pure-bred cattle, horses, and other farm stock. There are also scattered throughout the province 166 Agricultural Societies, which are aided by a Government appropriation to purchase pure-bred bulls and other stock. At the present time, for every $1 subscribed by members of these societies the Government adds a little over 80 cents, so that if a group of farmers had $100 to invest in a pure-bred bull, they would have this amount supplemented by $80, and thus be enabled to purchase a much better sire. The Government, also, at frequent intervals, makes appropriations for the purchase of pure-bred stock of various kinds which is ordinarily sold at very reasonable figures under bond to remain in the province. Agricultural meetings, addressed by experts, are regularly held in the various parts of the province. There is a well-organised Farmers' Association, with a membership representative of every part of the province, and a Fruit-Growers' Association, whose membership is practically confined to the fruit-growers of the province. A large Provincial Exhibition is held in Halifax, the capital, every year; an Annual Fat Stock Show at Amherst, in Cumberland County; and, at frequent intervals, County Exhibitions, in a number of the counties. From this it will be seen that a strong agricultural sentiment exists throughout the country, indicating the high estimate in which the farmers, the mainstay of the province, are held.

WOOD.

A large proportion of the Nova Scotian farms, unlike those in Great Britain, contain a wood lot, and this is a great advantage to the farmer. He has arable land enough for raising all necessary crops, including hay; pasture land enough for the requirements of the stock he keeps, and wood land enough to supply all necessary fuel, fencing, and building timber. Although the coal supply in the province is abundant, farmers usually provide their fuel from their own wood lot, and do so in winter months when farm work does not press upon their time. When the farm embraces 200 acres or more, the wood portion is often large. These wood areas, combined with the larger forests owned by the lumber manufacturers, perform the very important function of preventing high winds, of promoting more frequent rainfall, and of holding, by their shade, the moisture in the soil, discharging it gradually to the streams, and thus preventing droughts as well as early frosts. Should the farmer wish to sell cord wood, he can easily do so to near-by towns and villages at good prices.

FRUIT CULTURE.

The fame of Nova Scotia as a fruit-growing country, and the profits attending the raising of fruit, are of special interest to those thinking of moving westward from the Old Country.

The principal fruit section of the province is the Annapolis Valley so-called, extending from the town of Windsor to Digby and comprising part of Hants, King's, Annapolis, and Digby Counties. This section is about one hundred miles long by from five to fifteen miles in width, and the greater part of it is admirably suited to fruit, though as yet only a relatively small part of it is planted to orchards.

This valley extends east and west and is sheltered from the north and south by high ranges of hills, which cut off not only the heavier winds, but also the fogs from the Bay of Fundy, and the bright sunny weather of summer and autumn combine with other conditions to bring fruit, particularly apples, to a state of perfection rarely excelled.

Aside from this valley there are several smaller sections in outlying counties, comprising parts of the counties of Queen's,

Lunenburg, Shelburne, Yarmouth, Pictou, and Antigonish, where fruit-growing, though as yet in its infancy, can undoubtedly be carried on with great success.

To test the culture of fruit in these newer sections, and to encourage the adoption of better methods there, the Government has established over thirty experimental orchards. These are doing excellent work, and together with private enterprises of the same kind, are proving that these new sections can in many cases produce equally as good fruit as the famed Annapolis Valley.

SPECIAL ADVANTAGES.

In the opinion of experts the province of Nova Scotia has a number of distinct advantages over almost any other portion of the

A NOVA SCOTIA ORCHARD IN BLOOM.

American Continent as a fruit-producing country. Briefly stated, some of the more important of these are as follows :—

First.—Nova Scotia is nearer the British and other European markets than any other part of the Continent. Some of the very best fruit-growing sections of Canada are near the Pacific Coast, but the eight or ten days required to bring their fruits to Atlantic ports, not to mention the extra freight charge, must certainly act as a serious handicap in catering to European markets, which for the present, and for the future, probably, will be the chief markets for Canadian fruits.

Second.—Nova Scotia can produce fruit of the very highest quality. It is a generally recognised principle among horticulturists that the further north fruits can be brought to full maturity the higher the quality is likely to be, and this certainly holds good for Nova Scotia.

Third.—Among varieties most largely grown in this province, the greater proportion are well-known English and European sorts, like Ribston Pippin, Gravenstein, Blenheim Pippin, &c., which are sure of a market among English consumers.

Fourth.—The life of an apple-tree in Nova Scotia is from 60 to 100 years, a very great advantage over more trying climates, where from 20 to 30 years are all that can be counted on. Thus in Nova Scotia, when a man has once established an orchard he is reasonably sure of an income from it, not only throughout his own lifetime, but for the next generation. Indeed, while 100 years may be

APPLE PICKING IN THE ANNAPOLIS VALLEY.

given as the probable limit of an orchard's life, there are many trees in the Annapolis Valley known to have been planted by the French, and which cannot therefore be less than 150 to 200 years old, but are still vigorous.

Fifth.—In small fruits, and in the more perishable of the tree fruits, particularly plums, Nova Scotia growers have a great advantage, inasmuch as the lateness of its season brings on these products at a time when the United States markets are practically bare of such delicacies. Boston, in particular, affords an excellent outlet for them, and this trade is growing year by year.

With these advantages and others that might be mentioned, it is not surprising that the fruit industry is growing rapidly. It has developed within the memory of men now living from practically no export to an annual shipment abroad of 500,000 barrels, and the prices instead of declining with the increased output, have steadily advanced. In the opinion of well-informed growers, there are apple-trees enough now planted in Nova Scotia to yield in a very few years one million barrels of fruit, yet orchardists continue to plant and revenues continue to be satisfactory.

EXPERT OPINION.

Prof. F. C. Sears, of the Horticulture Department of the Nova Scotia College of Agriculture, in commenting upon fruit-raising in the province, says :—

"I believe the fruit industry of this province offers excellent investments for at least two classes of men coming here as settlers : first, to those who come with a capital of, say, £2,000 to £3,000 sterling, and can therefore buy orchards already in bearing, which will yield a good interest on their investment from the start ; and, second, to those of lesser means who come with a few hundred pounds and buy up unimproved lands and develop these. Lands suitable for this latter purpose can be had for from $10 per acre upwards. By setting such land with apple-trees and then interplanting with plums and small fruits a plantation is quickly established on a paying basis. Strawberries will give returns the next year after planting, and two crops are usually taken from each area planted. Raspberries, blackberries, currants, and gooseberries require a year or two longer to yield profitable returns, but the plantation once established is good for several years, while the apple-trees, which are to constitute the permanent value of the plantation, should give working expenses in ten years or less, and by fifteen years should yield a good profit."

Professor Sears further says : "I certainly believe that fruit-growing in Nova Scotia offers to investors an excellent living and a good rate of interest on their investment, and many of our growers have certainly made profits on their orchards which would discount many other investments considered as 'gilt edged.'

"Furthermore, with the future improvements in the way of better facilities for handling, cheaper freights, &c., which are sure to come with increased production, and with the quality of fruit which Nova Scotia is capable of producing, a good profit could be still made, even if prices should decrease instead of holding their own or advancing, as we have every reason to hope and expect."

The following statements by responsible and reliable fruit farmers will afford some idea of the money that may be made out of orchards properly cared for :—

"Blink Bonnie" Orchard, Wolfville, N.S. 7½ acres. Value, $6,700 or £1,350. C. M. Vaughan, proprietor. Yearly average yield, 900 barrels. Receipt above cost of barrels, picking, packing, freights, &c., and similarly for other products grown in the orchard as follows, viz. :

900 barrels apples, at $2 per barrel net	$1,800.00
40 ,, pears, at $2.50 ,, ...	100.00
40 bushels plums at $1.50 per bushel net	60.00
100 ,, potatoes at $50 ,, ...	50 00
Vegetables, &c.	25.00
	$2,035.00

Deduct other expenses, viz. :—
Fertilisers, $100 ; ploughing, cultivating, &c., $75=$175
Interest on $6,750, $337.50 ; taxes $45, spraying $60= $442.50   617.50

Total average yearly profit	$1,417.50
,, ,, per acre	189.00
Percentage of profit on investment	21 %

I certify that the above is a correct statement based on my own experience.

C. M. VAUGHAN.

WOLFVILLE ORCHARDS.

Starr's Point, King's County, N.S.,

Feb. 18th, 1907.

The following is a statement of gross income of two areas of orchard near here, viz. :—

Apples grown on 12 acres orchard, 600 trees, 30 years old :—

1899	1,725 barrels, sold for	$3,968.00
1900	995 ,, ,, ...	1,403.00
1901	2,000 ,, ,, ...	5,330.00
1902	1,220 ,, ,, ...	2,581.00
1903	2,037 ,, ,, ...	4,554.00

The expenses of cultivating, pruning, spraying, picking, packing, and shipping are found to vary from 60 cents to $1 per barrel, according to amount of labour and the crop obtained.

J. RUFUS STARR.

Mr. B. L. BISHOP, of Greenwich, furnishes the following statement, viz. :—

Two acres of orchard at Greenwich, containing about 100 very large and closely-set apple-trees.

Annual Outlay.

Interest on value, $500 per acre, or total value of $1,000	$60.00
Taxes $5, cultivation $8, fertilizers $20	33.00
Pruning, &c. $20, spraying $24, picking $30	74.00
Packages $57.50, packing and preparing for market $20	77.50
Cr.	$244.50

Average for three years, viz. :—

250 barrels per year at $2.50 per barrel	$625.00	
Annual profit on two acres		$380.50

Mr. C. C. Brown, a native of England, and now engaged in fruit-raising at Greenwich, N.S., where he has lived for several years, after pointing out the mistakes a newcomer to Nova Scotia should guard against, says :—

" I do not think that a man can engage in a business more likely to lead to a competency than fruit-growing in Nova Scotia, provided he brings to bear capital, industry, intelligence, and a liking for the work ; and provided he avoids the initial mistakes so many fall into in a new country.

"Apples, pears, plums of all varieties, and the hardier peaches bear and ripen well. In planting out or grafting in pears it would be well to choose the later kinds, which will do for shipping to England when they bring good prices.

"When it comes to purchasing, I would advise anyone to buy a farm with an apple orchard already in good bearing on it. It is better to purchase a place that from the start will almost provide a living, and then, at small cost, additional small orchard can be planted, which, if well looked after, will greatly add to its value."

STATEMENT OF YIELD AND RETURN FROM SOME NOVA SCOTIA ORCHARDS THE LAST FIVE YEARS,

Compiled by Ralph S. Eaton, Esq., Proprietor of "Hillcrest Orchards," N.S.

Owner of Orchard.	Average number of barrels shipping apples per acre, bearing orchard has produced.	Number of acres of orchard considered in the estimate.	Average price per barrel these apples returned.	Gross return per acre orchard has given.	Net return per acre.	* Per cent. of net earnings at $610 per acre.
F. C. Johnson, Port Williams	121	9	$2.50	$302.60	$219.00	35.90
J. Elliott Smith, Wolfville	165	4½	2.12	349.80	245.00	40.16
Arthur C. Starr, Starr's Point	100	14	2.13	213.00	140.00	22.95
G. C. Miller, Middleton	111	2	1.98	219.78	143.00	23.44
Geo. H. Starr, Port Williams	117	9	2.25	263.25	130.00	29.50
Chas. E. Sheffield, Upper Canard	100	4	2.25	225.00	152.00	24.92
F. H. Johnson, Bridgetown	100	6	2.25	225.00	152.00	24.92
Ralph J. Messenger, Bridgetown	100	3½	1.90	190.00	117.00	19.18
E. J. Elliott, Clarence	128	2½	2.38	304.64	218.00	34.42
Average ...				$174.00		28.52%

* "Hillcrest Orchards" is capitalised at $610 per acre.

J. W. BIGELOW, Esq., for many years President of the Nova Scotia Fruit-Growers' Association, and whose opinion is entitled to consideration, in making his New Year's statement to the *Morning Chronicle*, of Halifax, says :—

The apple industry of Nova Scotia may be summed up as follows :—

1. It is the most notable product of the soil.
2. Not 10 per cent. of the land suitable for cultivating apples has yet been planted.
3. Wild land is best for apple culture, costing from $10 to $50 per acre, according to location.
4. It is proved that Nova Scotia can produce superior commercial apples, the King, Ribston, Blenheim, Fallawater, Golden Russet, Nonpareil, and Baldwins, of best flavour, colour, and keeping quality.
5. The markets of Europe will take all such apples as can be grown in Nova Scotia.

Fruit-growers of this province can join the toilers in our other industries in congratulating this highly-favoured, happy, prosperous province in its prosperity during the past year.

COST AND PROFIT OF APPLE-GROWING IN NOVA SCOTIA.

Cost.

25 acres land at $30 per acre...	750.00
1,000 apple trees, three years old, from nursery, 20 c.	200.00
Setting out 1,000 trees, preparing land, &c., 10c. each	100.00
Fertilizing, clearing, mulching, &c., first year	200.00
	$1,250.00
12 years' interest on $1,250 at 5 %	750.00
Cultivating during 12 years at $100 per year	1,200.00
Fertilising, mulching, replacing dead trees, and other expenses for 12 years	450.00
Total cost till 12 years old	$3,650.00

Income.

Yield the 10th and previous years, 500 barrels at $1 per barrel, clear of charges	$500.00
Yield from 10th to 20th year, average 500 barrels per year = 5,000 barrels at $1 clear...	5,000.00
Yield from 20th to 100th year, average 2,000 barrels a year, at $1 per barrel clear = 80 years, 2,000 barrels per year	160,000.00
	$165,500.00

All expenses not herein provided for are overpaid by value of other crops, and all possible loss from bad markets and insect pests is more than repaid by the average estimate of $1 per barrel, as $2 is about the average for the last 20 years. This estimate is based on 30 years' actual experience, and criticism is invited from any experienced fruit-grower in Nova Scotia.

J. W. BIGELOW.

Annandale Farm.

The profits of fruit-growing in Canada are very much greater than those of any other of the farm products. In the production of long keeping apples and in the profits resulting from growing and marketing the same, the Annapolis Valley of Nova Scotia cannot be excelled in all Canada.

JOHN DONALDSON,
President, N.S. Fruit-Growers' Association.

Labourers who have had experience and practice on English or Scotch farms are needed in various sections of the province. The pay of such a labourer would vary, according to his ability, from £3 to £5 a month with board and comfortable lodging.

During five months, including the winter, he would receive in many cases about half-pay, and be boarded as in summer.

Social distinctions do not shut him off from friendly intercourse with the people of the community.

Men of means can profitably invest in Nova Scotia farms provided they understand the right methods of farming. Orchard farms, especially those in the fruit belt of the province, can be purchased and cultivated with most excellent returns, as the foregoing statistics and testimonials indicate. This is evident from the fact that many farmers have begun with practically no means and have become financially independent. The man who in addition to keeping up his older orchard can from time to time add an acre of young trees, will in his middle life have a property that will render him and his family independent, beside having an occupation of intense interest and never-failing pleasure as he watches the growth of his fruits.

MINERALS.

COAL.

The coalfields of Nova Scotia are very extensive. They are being worked in Cape Breton on a large scale, as well as in Pictou and Cumberland counties, and embrace an area of about 4,000 square miles. The manager of the Dominion Coal Company is reported to have estimated that these areas have sufficient resources to produce 3,000,000 tons a year for thousands of years to come. During 1904 the pay-roll of this Company for considerable periods embraced 8,000 men.

The output of the Nova Scotia mines for 1903 was 5,653,338 tons out of a total output for the Dominion of 7,643,999 tons. In 1905 the value of the coal raised was estimated at $11,250,000. The total production of coal in 1906 was 5,566,505 long tons. The coal is of a bituminous variety, and may be divided into coking, free burning, and cannel coal. Considerable quantities are shipped into the United States, as well as into Montreal and other western cities. During the five years ending in 1905, more coal was raised than during the previous ten years. With the continuous development and operation of the Sydney Steel Works, the coal output will undoubtedly increase at a rapid rate henceforward.

IRON AND STEEL.

Sir WILLIAM FAIRBURN, in writing upon the Nova Scotia iron ores, says: " In Nova Scotia some of richest ores yet discovered occur in boundless abundance," and Sir William Dawson, in referring generally to the distance of the iron ore from the fuel required in so great a quantity whenever smelting processes are undertaken on a large scale in Canada, says : " It should, however, be borne in mind that the great iron ore deposits of Nova Scotia, equal in extent and value to any others in the Dominion, are not so situated, but lie in close proximity to some of the greatest coalfields in the world. Even in Great Britain itself the two greatest staples of mineral wealth are not in more favourable contiguity, and iron ores of Great Britain are neither so rich nor so accessible as those of Nova Scotia."

With these vast and rich deposits of iron ore at hand, and with coal and limestone in close proximity, great facilities exist for the manufacture of iron and steel in the province, and this is already demonstrated by the rapid development of the great iron and steel plant at Sydney. In 1905 the production of iron and steel was :—

Pig iron ...	...	...	...	...	...	...	$3,500,000
Steel	...	...	...	...	...	...	3,800,000
Steel rails ...	...	...	...	..	...	...	1,500,000

The result has been that the town of Sydney with a population a few years ago of between 2,000 and 3,000, has increased to about 18,000.

It is probable that steel shipbuilding will in a very short time be undertaken upon a large scale, and this will mean much for the maintenance of Nova Scotia's reputation as a shipbuilding country.

GOLD.

The goldfields of Nova Scotia have been worked with more or less attention since the year 1860. They occupy the Atlantic coast of the province. The gold generally occurs in quartz veins. Hitherto they have been worked to very small depths, and they have seldom been followed outside the pay streaks. The government is now encouraging the mining of gold at greater depths, and larger results are as a consequence expected.

The output last year was about 14,000 ounces. The cost of mining in Nova Scotia is comparatively low, owing to river communication with sources from which supplies can be purchased, and the abundant supply of fuel as wood and coal. Since 1862 there have been 1,688,649 tons of quartz crushed, which yielded $15,859,000.

Several other valuable ores are found in more cr less abundance in various parts of the province.

Sulphide of Antimony occurs at Rawdon, Hants County, and is being worked on a somewhat extensive scale.

Copper Ores are quite widespread, and during the last two or three years serious attempts have been initiated to develop some of the more promising of these properties.

Gypsum may be considered as specially characteristic of Nova Scotia, as deposits of equal extent and variety are met at few other points. It occurs as soft and hard gypsum. The amount available in the province is inexhaustible, and is principally shipped from the districts surrounding Windsor and Hants County. The export amounts to 180,000 tons a year. Manganese, building stone, and other deposits are found in various places, and there is little doubt

CHARLOTTE STREET, SYDNEY, CAPE BRETON.

that future years will witness the development of many of these deposits, and that they will be applied to important uses as the province continues to prosper. Taking the various deposits of Nova Scotia together, it is generally conceded that no country in the world, in proportion to its area, is richer in mineral wealth.

EDUCATION.

A citizen of the Old Country coming to Nova Scotia must not imagine he is about to mingle with a people only partially civilised. On the contrary, he will find an advanced civilisation. a country intersected by railroads, telegraph and telephone lines, its rivers spanned by substantial steel bridges, its ports, as before stated, connected by various steamboat lines, and its highways well maintained, thus affording superior transportation facilities to all the people.

Moreover, he will find one of the best free non-sectarian school systems in the world. This system has been in successful operation since 1865, with the result that scarcely a person among its nearly half-million people can be found who has not a common school education, while the majority may be regarded as well trained.

Provision is made for free instruction from the age of five years upwards. The course of study prescribed covers twelve years, eight years being given to the Common School grade and the remaining four years to the High School or Academy grade, fitting its students for College and University courses. Manual training and domestic science is provided for at the more important centres.

Pupils of the more advanced high school grades wishing to become teachers are given a free course in the theory and practice of teaching in the Provincial Normal School at Truro.

The number of schools in operation is 2,446, teachers employed 2,578, pupils enrolled 100,332, expended for education $1,073,719.90.

The number of high school pupils is 7,639.

High schools are located in every county throughout the province.

The newcomer from abroad, however poor he may be, has the assurance and guarantee that his children may continuously have absolutely free access to the public schools in any community in which he may make his home.

The expenses of school buildings, teachers, &c , are borne by grants from the Government, by a general county tax, and by a moderate local tax upon all the property of the school section.

Chartered colleges, with degree-conferring powers, are as follows, viz : —

King's College, Windsor ; Acadia University, Wolfville ; Dalhousie College and University, Halifax ; St. Francis Xavier University, Antigonish ; St. Anne's College, Church Point, Digby County.

These institutions of learning are quite largely attended, and their graduates are filling useful, and in many cases exalted positions in various parts of the world.

There are also several Ladies' Colleges, largely attended and well equipped.

Of special advantages to the farmer and fruit-grower is the Nova Scotia College of Agriculture, already mentioned, and situated at Truro. Its various departments provide thorough instruction and practice in field husbandry, animal husbandry, horticulture, dairying, veterinary science, farm blacksmithing, horseshoeing, &c.

In addition to the regular longer courses of study and practice, provision is also made for short, but very practical courses during winter months. Many farmers take advantage of these courses, and by the knowledge and skill thus acquired are able to adopt improved methods on their farms, and thus increase the products and profits of their occupation. To newcomers, more or less unfamiliar with conditions and methods in Nova Scotia, these short courses will be of great advantage, and can be pursued at a season of the year when the demands of the farm upon their time are at a minimum.

SPECIAL INSTITUTIONS.

Special institutions are the Halifax School for the Deaf and Dumb, the Halifax School for the Blind, the Victoria School of Art and Design, and the Medical College. These are all in successful operation and are largely attended.

A VIEW OF THE NOVA SCOTIA AGRICULTURAL COLLEGE.

INSTITUTE OF TECHNOLOGY.

The Nova Scotia Legislature realising the enormous possibilities that lie in the efficient development of the great natural resources of the province, has recently made provision for the immediate establishment and equipment in Halifax of an Institute of Technology of university grade. This Institution will have departments providing instruction and professional training in metallurgical, civil, mining, mechanical, chemical, and electrical engineering.

Original research will be carried on, and such practical tests made as will doubtless lead to important discoveries. This central

Institute, with subordinate branches later to be established in other sections of the province, will constitute a factor of great importance in promoting the prosperity of all classes of the people.

FISHERIES.

Nova Scotia has for a long time been famous for its fisheries, which are among the best in the world. The Atlantic coast of the province, the principal seat of the fishing business, is indented with innumerable harbours and coves, affording shelter for fishermen's boats, &c. The shoaler waters lying some miles off the coast are the favourite resorts of codfish, halibut, pollock, &c. These fish are caught by line from large boats and fishing schooners, and carried immediately to the shore, where they are dried. Salmon are also found along the coast. Herring and mackerel pass along the shores at certain seasons, and are caught in great numbers by nets.

The importance of this industry in Nova Scotia may be gathered from the following figures :—

PROVINCE OF NOVA SCOTIA : STATISTICS OF THE FISHERIES FOR THE YEARS 1896—1901, COMPILED FROM THE ANNUAL REPORTS OF THE DEPARTMENT OF MARINE AND FISHERIES.

Year.	Value of the Product.	Capital Invested.	No. of Fishing Vessels.	Tonnage.	No. of Boats.
	$	$			
1895	6,213,131	3,139,968	595	25,678	14,986
1896	6,070,895	3,069,753	593	25,565	14,549
1897	8,090,347	3,149,174	545	24,677	15,468
1898	7,226,034	2,972,600	537	23,718	15,358
1899	7,347,604	3,080,795	553	25,342	15,366
1900	7,809,152	3,278,623	557	26,064	14,766
1901	7,989,548	3,319,334	527	24,119	13,564

NO. OF MEN EMPLOYED IN

Year.	Vessels.	Boats.	Lobster Factory.	Total number employed.
1895	5,879	19,736	—	25,615
1896	5,801	19,174	3,839	28,814
1897	5,514	19,859	4,559	29,932
1898	5,434	20,801	5,185	31,420
1899	5,705	19,466	7,570	32,741
1900	5,816	19,396	6,447	31,659
1901	5,607	18,367	5,555	29,529

The production of lobsters in Nova Scotia is large and important. Great quantities are canned and exported. In addition to large numbers of lobsters sent in ice to the United States, there is a valuable and increasing export of fresh fish. There is room in Nova Scotia for an immense expansion of this fish trade, now that so much of it can be taken away fresh to the United States and Canadian towns, and the introduction of smoking and canning has become general.

The inland fisheries yield principally smelts, salmon, trout, and considerable quantities of eels. These are frequently frozen and sent by rail to all parts of Canada and the neighbouring cities of the American Union.

LUMBER.

The value of the lumber trade in the province amounts to about $5,000,000 annually.

It has been estimated that about one half the country's area, or 10,000 square miles of the province, is wooded. In many sections the woodland is in small patches, and used principally for domestic purposes. The greater part, however, occupies the western end of the province, and the section lying between the Atlantic tidewater and the height of land. These two sections, comparatively value-less for farming purposes, have for years yielded immense amounts of lumber.

The principal trees are spruce, fir, hemlock, pine, birch, oak, and maple. Of these, the pine, originally covering immense tracts

DIGBY PIER AND D.A.R. STEAMER "PRINCE RUPERT."

of the province with primeval growth, has become almost extinct. It has been followed by the more rapid-growing spruce. At present this tree furnishes the bulk of the export. It is worked into deal, flooring, &c., as the various markets demand. The remainder of the export consists of a little pine, beech, &c. It is probable that this tree will for ever remain the principal support of the business. The spruce grows rapidly, and after twenty-five years of growth can be utilised for deal. Next to the spruce comes the hemlock. This tree yields lumber suitable for rough boarding, floors, &c. As it is present in large amounts in a number of localities, it will form an important supply for years to come.

In many localities are large quantities of hardwood, not utilised until transportation facilities permit of its use for furniture, &c.

Included in the districts referred to, and close to the shore, and in the counties of Victoria and Inverness, are extensive growths of small spruce and fir, suitable for pulp making. These are being utilised to some extent, and the products shipped to Europe. The numerous water-powers have done much to keep up the lumber and pulp business, which, owing to favouring natural conditions, has proved steadily profitable.

There are a number of companies engaged in lumbering on the principal rivers. Usually mills are situated near tidewater, and the lumber passes directly from the mill to the vessel. The cutting, transport, and handling of the lumber gives employment to many men during the winter months.

The Government has lately paid attention to the preservation of the forests and their protection from fire, &c. It is therefore

A FARMING DISTRICT IN NOVA SCOTIA.

anticipated that the districts of the province now devoted to the lumberer may maintain a steady supply of this commodity.

MANUFACTURES.

Manufacturing has already made considerable progress in Nova Scotia, but in view of its rich and varied natural resources, only a beginning, in comparison with what is possible, has yet been made.

The immense steel works of the Dominion Iron and Steel Company, at Sydney, already referred to, have involved the largest outlay of capital of any plant, of whatever kind, in the province. These works are not only putting out large quantities of steel rails, steel rods and plates, but they have laid the foundations for various other related manufactures which are springing into existence, and destined to develop Sydney into a large city in the coming years.

At North Sydney also the Nova Scotia Steel Company are carrying on an extensive and profitable business. Their lengthened experience in this work and their uniform success in it are a guarantee that as yet they are only in the initial stages of their enterprise. In the capital city of Halifax and Dartmouth adjoining manufacturing of various kinds is carried on, including sugar refining, rope works, foundry and machine works, wood-working plants, boiler making, &c. A site has recently been purchased for manufacturing railway cars on a large scale. This plant is expected to employ a large number of men. At Truro, the large factory of Stanfield's Limited for the making of unshrinkable underwear and woollen yarns gives employment to several hundred persons. The factory where condensed milk is prepared and shipped to various parts of Canada is another of Truro's successful enterprises, not to speak of several others for want of space.

Amherst, not far from the borders of New Brunswick, is another of Nova Scotia's progressive towns. Rhodes & Curry's large establishment for the manufacture of rail cars, building materials, school furniture, &c., employs hundreds of men. Other large plants are Robb's Engine and Boiler Works, Christie's Undertaking Factory, Hewson's Woollen Mills, &c.

At Windsor, on the Avon River, are located a cotton mill, furniture factory, iron foundry, plaster mill, sash and door factory, and other smaller works. A large carriage factory, railway works, and iron foundry are located at Kentville, the headquarters of the Dominion Atlantic Railway.

Yarmouth boasts of a variety of manufactures, including stoves and other foundry work, building materials, cotton duck, woollen wears, &c. Similar enumerations could be made concerning New Glasgow, Glace Bay, Lunenburg, Liverpool, &c. There is no good reason why Nova Scotia should not manufacture all, or nearly all, her own cutlery, crockery ware, nails, hinges, farm implements, fishing gear, pickles, paper, furniture, and many other articles of common use. Here is an inviting field for skilled workmen and capitalists of the Old Country. Up to date scarcely any of these enumerated wares are made in the province, partly because the people are ignorant of the required processes and partly perhaps because they have not looked into the possibilities.

PRICES OF LAND.

The price of land varies much according to locality. Farms range in area from 50 to 300 acres, generally, however, they contain from 75 to 150 acres. In many of the districts where mixed farming is carried on exclusively, because not adapted to fruit raising, farms may be purchased on very reasonable terms, ranging from £200 upwards, and in some cases, especially where the farm is "run down" for want of proper care, and the owner is a widow, or a man too old to work, the price may be but little above the original cost of the house and other farm buildings. In such cases an Old Country farmer accustomed to pay land rent, and having only two or three hundred pounds, can make a profitable investment, owning his own land under freehold, exempt from all rents, and in a position to provide a living for himself and family.

In the Annapolis Valley land is somewhat higher, because containing greater possibilities. Should the farm contain several acres of fully developed orchard, its trees yielding good crops of fruit, especially if they be full-grown apple trees, then the price might well be double that paid outside the fruit belt, but the investment would be just as good, for the returns received for the labour and capital expended would be quite double, if not more. If the purchaser has the necessary funds to invest at the start in a farm containing a large apple orchard on the right soil, of good varieties, and under good cultivation, he will in the end probably do better than he would by buying unimproved land and waiting till his orchard can grow up to an age when it can yield a paying return; but as before intimated, planting apple trees on such unimproved land, and interplanting with plum, pear, and peach trees, as well as smaller fruits, which will bear the expenses till the apple trees come into good bearing, will pay well, and at the end of say ten years, the owner has a valuable property which will increase in value year by year. The interplanted trees can then be removed to allow the apple trees to spread their branches. They will continue to increase in size for many years, yielding larger and larger crops till they are at least thirty years old, and, as before stated, will yield good crops for forty, fifty, or sixty years longer. The late Secretary for Agriculture for the Province, Mr. B. W. Chipman, in writing concerning the price of farms in Nova Scotia, furnishes the following estimates of values in different counties :—

"At Enfield, 28 miles from Halifax, the farms are all well adapted for raising hay, grain, and all sorts of roots and vegetables. Two miles further on Elmsdale is reached, a village of about 200 inhabitants. Here the extent of farm lands widens, running west into the fertile county of Hants and east in a good farming region in Halifax county. The land is uniformly good ; farms range in size from 75 to 200 acres, with comfortable buildings, and in value from $1,000 to $4,000."*

"There are few more thrifty and attractive agricultural communities in Nova Scotia than the Middle Settlement of Musquodoboit. A railway is projected and route surveyed through this fine section of country, which when built, as it doubtless shortly will be, will enhance the profits of farming in this region. Farm lands and properties can now be bought at prices ranging from $1,000 to $8,000."

---

* £1 equals nearly $5.

" Following down the western shore of the Cobequid Bay from Truro, we find an excellent farming country for many miles, where rich dike lands and undulating upland and hills abound. First, Masstown, a good agricultural district ; next, Folly Village ; then Great Village. All this region presents a splendid field for raising stock for either beef or dairy purposes, while farming can be and is conducted with a degree of success fully equal to the efforts put forth. Masstown has a population of 150; Folly, of 400; and Great Village, 600. Below Great Village are Economy and Five Islands, both good agricultural districts, but where a good many of the inhabitants lead a mixed life of farming, and fishing in the Bay. Farm prices, $1,000 to $3,000."

" Pugwash is an important town of Cumberland County, about thirty miles from Amherst following the line of the post road, which carries one through the thriving agricultural settlements of Shinimicas and River Philip. This is all a good farming country, mostly upland, with good interval lands along the margins of the streams, comfortable farm houses and outbuildings, and prices ranging from $1,000 to $3,000."

" From New Glasgow, following the Eastern Branch of the Intercolonial Railway, and passing through a section of fine farming lands and more or less prosperous districts, the beautiful town of Antigonish is reached, a town of 3,000 population, and the shire town of the county of the same name. Antigonish is a fine agricultural county, but the inhabitants—a considerable section of it lying on the sea—combine fishing with farming as a regular employment. Antigonish has a splendid acreage of good uplands and stretches of magnificent intervals. It is one of the best grazing counties in the province, and is capable of producing large numbers of cattle and sheep. It is noted for the excellence of its dairy products, and has now in operation several cheese factories. Farms range in value from $500 to $4,000 "

Mr Chipman also refers to good agricultural districts in Cape Breton, where farms can be bought for from $1,000 to $3,000, and in the Annapolis Valley, containing good orchards, for from $3,000 to $10,000, while in the more westerly sections of the province prices range much lower.

There is, therefore, ample opportunity for English, Scotch, or Irish farmers to invest in Nova Scotia lands, both according to the means they may be able to command and according to the kind and measure of cultivation they wish to carry on.

Those wishing to inquire or negotiate concerning farms for sale in Nova Scotia may communicate personally or by letter with Agents who have ascertained what farms are for sale, and who have prepared detailed descriptions of them. These Agents will be glad to correspond direct with any person wishing to purchase, giving price, locality, and name of owner.

JOHN HOWARD, Esq., Agent-General for Nova Scotia, 57A, Pall Mall, London, S.W., will also upon application by letter or

HARVESTING IN NOVA SCOTIA.

otherwise supply a list of farms for sale and give detailed information concerning them, and will also furnish the names and addresses of Agents in Nova Scotia.

SPORT AND GAME.

Those fond of sport will find in Nova Scotia a fine field in which to test their skill.

An occasional day's hunting in the forest or near the sea shore, or fishing river pools and rapids is a most refreshing diversion from the labours of the farm.

It is an old saying that "All work and no play make Jack a dull boy," and Emmerson says—

"The incommunicable trees begin to persuade us to live with them and quit our life of solemn trifles."

Disciples of Isaac Walton will find many Nova Scotia streams inviting them to test the experiences of the "Compleat Angler." The Liverpool and Port Medway rivers are probably the best in the province for salmon and trout, and are frequented every year by many sportsmen. The upper sections of the Lahave River are also much fished.

The Margaree in Cape Breton is one of the finest rivers in America. It abounds in sea trout, and salmon occur in goodly numbers. Good trout fishing is found in scores of lakes and rivers in all parts of the province, notably in the Tasket, Sable, Sissiboo, Nictaux, and Gaspereaux rivers.

Moose are found in many parts of the province, but especially in the western counties and in the north-west sections of Cape Breton ; scores of them were shot during the autumn of 1906. Bears, foxes, mink, otters, &c., are hunted and trapped with considerable success. Deer and caribou are also occasionally found.

Wild geese and ducks frequent the streams and pools among the marshes, and the latter are found in large numbers in the lakes of the interior.

The partridge, similar to the English grouse, is also plentifully found in nearly every section of the province. The edible quality of its flesh is highly prized, and thousands are shot every year. Woodcock and plover frequent many places also.

Game laws are enforced for the protection of the wild animals, the wild fowl, and salmon, with the result that the province is still a good field for the sportsman, and no doubt will so remain.

SOCIAL CONDITIONS.

The principles of Democracy are well understood in Nova Scotia. It was in this province more particularly that the great battle for responsible government was fought out under the leadership of her great citizen and statesman, Hon. Joseph Howe.

The old " Family Compact," as it was called, consisting of a few autocrats, who assumed that they alone had the right and the ability to govern, were obliged to give way in 1848 to real representative government, responsible to the people. Under the influence of that prolonged struggle for the people's rights, citizens of Nova Scotia could not well fail to develop a strong and healthy social life. Comparatively small place is given to social distinctions. There is enough intelligence to estimate a man for what he is in character and ability rather than for the clothes he wears or the parentage of which he may boast. In the rural districts especially, as also in the towns and villages, people generally mingle together on terms of social equality. The idler and the inebriate cannot long hold social rank. He may sometimes imagine the world owes him a living, but he soon finds that he was mistaken.

Many of the leading statesmen and professional men of high standing in the province have risen from humble beginnings and from country homes, where the means of living were scanty. The people generally are fond of reading, and keep abreast of the leading social, political, and religious movements of the day, and take a deep interest in social and moral reforms.

The cost of living is somewhat higher than it was years ago, but moderate in comparison with some other portions of Canada. Groceries, such as tea, sugar, molasses, flour, dried fruits, &c., are about the same as in England. Vegetables and fresh fruit are cheaper. Clothing possibly a trifle higher. As a rule the daughters in the family assist the mother in the details of household work, especially so in all the country districts. The demand for domestic servants comes chiefly from the towns, and where they are employed they are well treated and generally well paid. The Sabbath is very well observed. It is a rest day for the people to a far greater extent than in the United States and Europe. The Lord's Day Act, passed in 1906 by the Canadian Parliament, means much for the protection of the labouring classes against unnecessary work on Sunday.

The laws of the province are framed on reasonable lines, and the guardians of the peace have little difficulty in its maintenance, owing to the law-abiding character of the population.

The eastern counties of the province proper and the island of Cape Breton were originally settled almost exclusively by the Scotch, and the central and western counties by English, Scotch, Irish, and American loyalists, while in Digby and Yarmouth counties a considerable proportion of the people are of French descent.

The various religious denominations are remarkably free from sectarian intolerance. Each sect, while loyal to its own communion, recognises the right of all others to a similar freedom. Morality stands high, and crime is exceedingly rare in comparison with other countries.

FURTHER INFORMATION FOR INTENDING EMIGRANTS.

Nova Scotia, distant from Liverpool 2,450 miles, is reached by steamers in from six to eight days. Passages can be booked by the "Allan," "Dominion," and the "Canadian Pacific" S.S. Lines, either direct or *via* the St. Lawrence route, while the Furness Line have a service from London to Halifax at frequent intervals. Full details as to dates of sailing, luggage, &c., will be supplied upon application to any of the above Companies. (See advertisements.) There are no assisted passages, and the fares range from £5 10s. third class. At Halifax, emigrants are met by a Government Agent, and accommodation is provided for them if required on landing, pending their departure for outlying districts to which they may be proceeding. Letters of introduction to the Agent will be supplied upon application to the Agent-General, John Howard, Esq., 57A, Pall Mall, London, S.W., and intending settlers are advised in all cases to avail themselves of this opportunity, and to avoid being guided by any irresponsible persons. The journey by rail or steamer to various sections of the province occupies from three to twelve hours.

HOUSE RENT.

Rents vary according to locality, those in the country being naturally lower than for houses in or near the large towns. In Halifax flats in tenements, suitable for working men, of four to six rooms cost from $5 to $10 per month, while houses of four to eight rooms cost from $7 to $15 per month. Miners' cottages about $6 per month. Board can be obtained at most places at from $2.50 to $3.50 per week.

Taxes are very low, from 1 to $1\frac{1}{2}$ % and in some districts even less.

LABOUR OPENINGS.

There are openings for nearly all classes of labour, more especially farm labourers, but it must be understood that the demand varies and is limited, and it is advisable for the intending emigrant to apply in the first instance to the Sub-Agent from Nova Scotia, or failing that, to the Agent-General (who is in touch with employers of labour in the Province), and ascertain the opportunities in his particular line and wages paid, thus preventing possible delay or disappointment on arrival.

SETTLERS' EFFECTS FREE OF DUTY.

The following is an extract from the Customs tariff of Canada, specifying the articles which can be so entered by settlers free of duty :—

SETTLERS' EFFECTS, viz. — Wearing apparel, household furniture, books, implements and tools of trade, occupation or employment, guns, musical instruments, domestic sewing machines, typewriters, live stock, bicycles, carts and other vehicles, and agricultural implements in use by the settler for at least six months before his removal to Canada ; not to include machinery, or articles imported for use in any manufacturing establishment, or for sale ; also books, pictures, family plate or furniture, personal effects and heirlooms left by bequest ; provided that any dutiable articles entered as settlers' effects may not be so entered unless brought with the settler on his first arrival, and shall not be sold or otherwise disposed of without payment of duty, until after twelve months' actual use in Canada ; provided also, that under regulations made by the Controller of Customs, live stock, when imported into Manitoba or the North-west Territories by intending settlers, shall be free until otherwise ordered by the Governor in Council.

The settler will be required to fill up a form (which will be supplied to him by the customs officer on application) giving description, value, &c., of the goods and articles he wishes to be allowed to bring in free of duty. He will also be required to take the following oath :—

I .... do hereby solemnly make oath and say, that all the goods and articles hereinbefore mentioned are, to the best of my knowledge and belief, entitled to free entry as settler's effects, under the tariff of duties of customs now in force, and that all of them have been owned and in actual use by myself for at least six months before removal to Canada ; and that none of the goods or articles shown in this entry have been imported as merchandise or for use in any manufacturing establishment, or for sale, and that I intend becoming a permanent settler within the Dominion of Canada.

BAGGAGE.

Colonists are allowed free on ocean steamers 20 cubic feet for second cabin, and 10 cubic feet third class. On railways in Canada and the United States 150 lbs. for each adult ticket, and 75 lbs. for each child's half-fare ticket are carried free, but this applies only to personal effects. No single piece of baggage of over 250 lbs. will be carried on passenger trains, but must be sent by goods train, as also must be done at additional charge for all goods not wearing apparel or personal effects.

If free baggage allowance as above is exceeded, the extra charge on steamship will be 9d. per cubic foot, second cabin ; and 6d. per cubic foot, third class. On the railways the extra charge will be 12 per cent. of the colonist fare for each 100 lbs. or part thereof.

PRINCE EDWARD ISLAND

CANADA'S GARDEN PROVINCE

ITS CLIMATE AND RESOURCES.

OPPORTUNITIES FOR SUCCESSFUL FARMING.

PRINCE EDWARD ISLAND

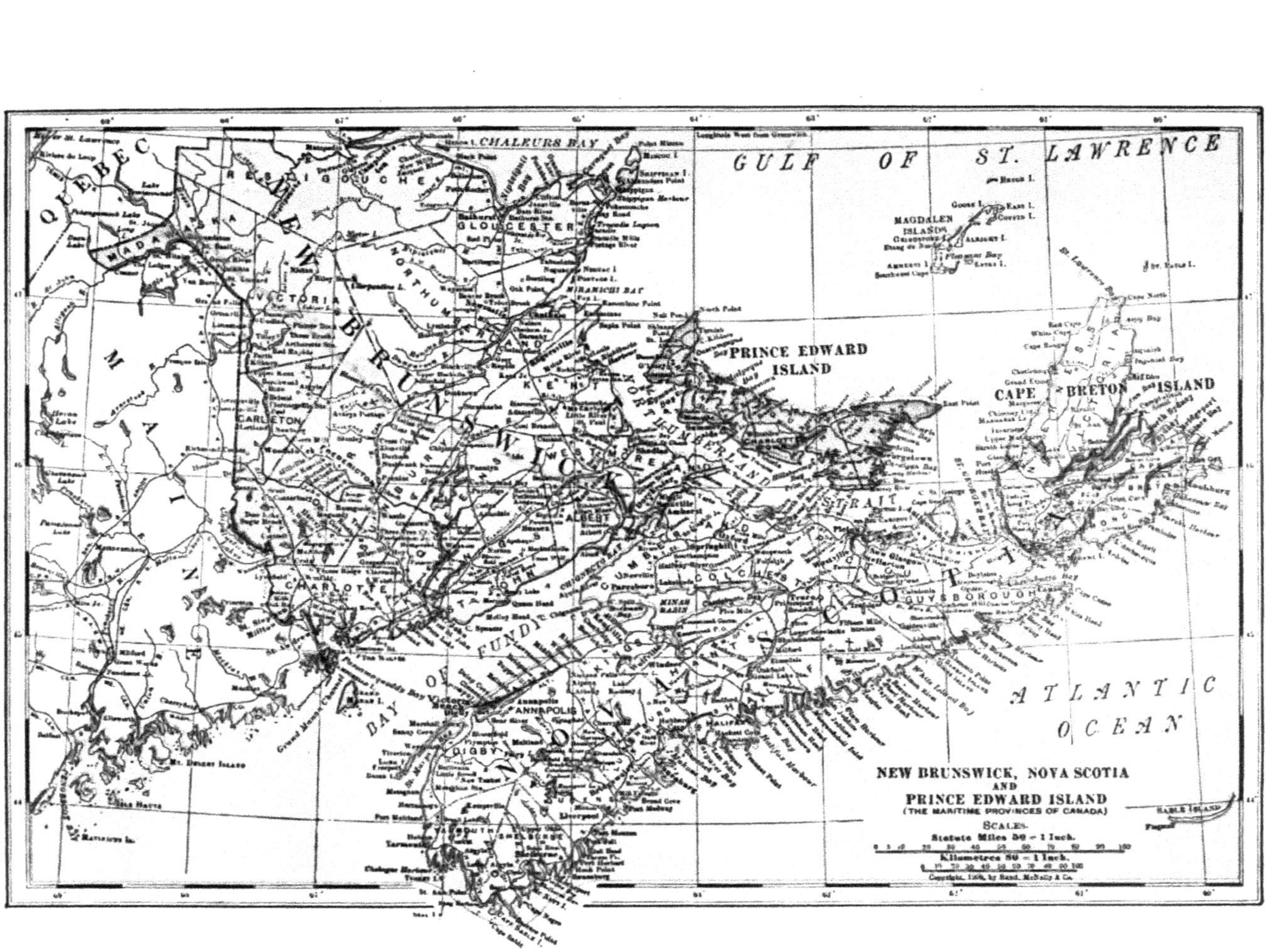

NEW BRUNSWICK, NOVA SCOTIA
AND
PRINCE EDWARD ISLAND
(THE MARITIME PROVINCES OF CANADA)
SCALES.
Statute Miles 50 = 1 Inch.
Kilometres 80 = 1 Inch.
GULF OF ST. LAWRENCE
ATLANTIC OCEAN
CHALEURS BAY
QUEBEC
MAINE
NEW BRUNSWICK
NOVA SCOTIA
PRINCE EDWARD ISLAND
CAPE BRETON ISLAND
MAGDALEN ISLANDS
BAY OF FUNDY
NORTHUMBERLAND STRAIT
RESTIGOUCHE
GLOUCESTER
NORTHUMBERLAND
VICTORIA
CARLETON
KENT
WESTMORELAND
ALBERT
CHARLOTTE
GUYSBOROUGH
ANNAPOLIS
DIGBY
SABLE ISLAND

Prince Edward Island

CANADA'S GARDEN PROVINCE

ITS CLIMATE AND
RESOURCES

OPPORTUNITIES FOR
SUCCESSFUL FARMING

ISSUED UNDER THE DIRECTION OF HON. FRANK OLIVER, MINISTER OF THE INTERIOR

PRINCE EDWARD ISLAND

CANADA'S GARDEN PROVINCE

SITUATION.

A glance at the map shows Prince Edward Island to lie in the Gulf of St. Lawrence, off the coasts of Nova Scotia and New Brunswick. Charlottetown, the capital, is situate on the south side of the Island, some three miles from the mouth of a magnificent harbour, in Latitude 46 degrees 14 minutes North, or nine degrees South of Edinburgh, and in Longitude 63 degrees 10 minutes West, or sixty degrees West of Edinburgh.

The Island is separated from its sister provinces by the Strait of Northumberland, which, at its narrowest point, has a width of about eight miles.

Allan Line Steamers sail at frequent intervals throughout the summer, as well as throughout the early spring from Glasgow and Liverpool to Halifax, Nova Scotia, direct, taking passengers, stock and cargo. The dates of sailing, usually once a month, are advertised beforehand.

From Liverpool, Prince Edward Island is distant about 2,500 miles. Halifax, Nova Scotia, which would be the most convenient port of arrival for the emigrant from Great Britain, is distant 113 miles by rail and fifty by water, the latter part of the journey being made in remarkably fine, well-equipped steamships. Leaving Halifax at noon, the traveller arrives in Charlottetown between eight and nine p.m., the same evening, by boat from Pictou.

Passengers from Great Britain, who prefer to come by Rimouski, Montreal or St. John, New Brunswick, come down from their ports of landing, by the Intercolonial Railway, owned and operated by the Dominion Government; or, by the Canadian Pacific Railway, and cross from Point du Chene to Summerside, a flourishing city on the western part of the Island. If they prefer, they may come to the Island from Montreal, by the steamers regularly plying between Canada's Commercial Capital and the Island ports.

In winter, the passage over the Strait is made in fine steamboats, owned and operated by the Dominion Government.

CONFORMATION AND AREA.

The Island is of an irregular crescent shape, some 140 miles long, and varies in width from four to thirty miles. Its area is about 2,184 square miles, containing about 1,397,991 acres of land, all of which, save a very small percentage, is splendidly adapted for agricultural purposes.

NAME.

The Island, up to the closing years of the Eighteenth Century, was called the Island of St. John, after St. John the Baptist, from the day in which it was said to have been discovered. There were a number of other St. John's in the Maritime parts of British North America, which caused confusion, and, to remedy this, the Legislature, in 1798, changed the name to Prince Edward Island, being so named after the Duke of Kent, (father of Her late Majesty, Queen Victoria), who had been for several years stationed at Halifax. Besides its official name, the Island has several other well-known familiar designations, all having reference to, and being derived from the noted fertility of its soil. The "Million Acre Farm" is a designation often applied to it. It is known far and near as THE Island. But, outside of its official name, the name by which Prince Edward Island is best known, and which was given it generations ago, is that of "The Garden of the Gulf."

HISTORY.

Prince Edward Island is the smallest of the Provinces which go to make up the Dominion of Canada. It is also by far the most densely peopled, having a population of nearly fifty to the square mile, or twice as many as any other Province.

It was not one of the original members of the Confederation, when the Dominion came into being in 1867, although the initial Conference on the question of Union of the British North American Provinces, was held in Charlottetown, in 1864. The Island did not come into the Confederation until 1873, or six years after the union of the four original Provinces, Ontario, Quebec, Nova Scotia and New Brunswick.

DISCOVERY.

Prince Edward Island was long believed to have been discovered in 1497, by John Cabot, who sailed on his first voyage of discovery, from Bristol, that year. The better opinion, based upon recent researches, now seems to be that it was discovered by Jacques Cartier, in 1534. He sighted the North Coast of the

4

COUNTRY SIDE, PRINCE EDWARD ISLAND.

HARVESTING.

western part of the Island, on 30th June of that year. 'Cartier, who was a most observant explorer, and who carefully examined and noted the appearance and peculiarities of every place visited by him, described the land as being "low and plain, and the fairest that may possibly be seen, full of goodly meadows and trees." He landed at a number of places along the coast, and made a close scrutiny of its features, and was most favorably impressed with its appearance. He described the trees as wonderfully beautiful, and the land as good and abounding in small wild fruits. The climate also was very pleasant. A like favorable impression was made on all subsequent explorers. Save that the forests have largely given place to cultivated farms, Cartier's description is as applicable to-day, as when it was made.

THE FRENCH OCCUPATION.

After the cession of Acadia to Great Britain, by the Treaty of Utrecht in 1713, the French, who retained this Island, recognized its importance, owing to the great fertility of its soil, as a base of supplies for their great fortress at Louisburg, and for their naval and military forces in their remaining North American possessions. Consequently, they settled the Island, which soon became their granary, and furnished the grain, meat and other provisions, required by their forces, They had a Governor here, and a couple of officers or commissaries, whose duty it was, to procure in the Island, the food supplies required by the forces.

CEDED TO GREAT BRITAIN.

Upon the fall of Louisburg in 1758, the Island was taken possession of by the British. By the Treaty of Paris in 1763, it was, with the remaining French possessions in North America, excepting a couple of islands, and some fishing privileges on the Newfoundland Coast, ceded to Great Britain, and annexed to the Province of Nova Scotia.

HIGHLY VALUED BY BRITISH GOVERNMENT.

Because of its fertility, the British Government placed a high value upon the Island. In consequence, it was proposed to levy a Quit Rent upon lands granted here, fifty per cent. higher than was imposed upon lands elsewhere in the North American Provinces.

Its fertility and great promise as an agricultural country, as well as its situation in the centre of the great salt-water fisheries of the Gulf of St. Lawrence, recommended the Island to the attention of many officers of the navy and army, as well as to such civilians as were then acquainted with this part of North

America, numbers of whom made applications to the British Government, for grants of estates in this Island. Of these, the Earl of Egmont, at that time first Lord of the Admiralty, applied for a grant of the whole Island, which he estimated at two million acres, with a view to establishing the Feudal System here. His scheme was a most elaborate one, and would have been up-to-date about the Twelfth Century. Fortunately for the Island, and possibly still more fortunately for himself, the Earl's application was refused.

DIVIDED INTO COUNTIES AND TOWNSHIPS

The British Government caused the Island to be surveyed and divided into three Counties, King's, Queen's and Prince, which division has been retained, and these Counties constitute the Electoral Districts or Ridings, for the election of Members to the House of Commons of Canada. Each County was also divided into parishes and townships. The township, usually called "Lot", is the unit generally considered, in describing the sub-divisions into which the three Counties are divided. Of these townships, there are sixty-seven in all, containing, approximately, nearly 20,000 acres each, though they vary somewhat in area.

TOWNSHIPS LOTTERIED.

The number of applicants for grants was so great, that there were not sufficient townships for all. Lots were drawn for grants, and the whole island, with the exception of a couple of townships, and some town and fishery reservations, was parcelled out in a lottery among the applicants.

LANDLORD SYSTEM ESTABLISHED.

In this way, the landlord system was established, and for a century, the farmers were mainly tenants, holding under leases, at a yearly rental. This system of land tenure, or landlordism, is not adapted to this country, where every man looks to own the freehold of his land, and it does not take firm root here. It was, however, the cause of a century of bitter struggle on the part of the people, before they threw it off.

LANDLORD SYSTEM ABOLISHED.

Finally, in 1875, the Legislature of the Island enacted a Compulsory Land Purchase Act, under which the system was abolished, and the Proprietors of the different estates compelled to sell to the Provincial Government, at a valuation fixed by a competent Commission appointed for that purpose. The Provincial Government, when the title to the land became vested in it, sold the different holdings to the tenants, on easy terms, at a low valuation, calculated upon the rentals. The vacant, unleased lands were sold to any BONA FIDE purchaser, who applied for them. Since that time, the lease-hold tenures have been converted into freehold, and now, with very few exceptions, the farms of Prince Edward Island are owned by their occupants in fee simple. Of the farms of the Island in 1901, the year of the last census, 97.23 per cent were so owned. Of the total area, 85.44 per cent, or 1,194,508 acres, are occupied, of which 726,285 acres are improved or under cultivaion.

SEPARATES FROM NOVA SCOTIA.

The union with Nova Scotia did not long continue. In 1769 the Island of St. John was erected into a separate Province, with a Government of its own. It was then almost entirely in a wilderness state, without roads or other public conveniences, and the population consisted, besides Indians, of 150 white families.

THE FIRST GOVERNOR.

Walter Patterson, Esq., was the first Governor appointed to rule over the new Province. He arrived in the autumn of 1770. Like all others, who have come to this Island and examined into its capabilities and resources, he was greatly impressed with its natural wealth, both of the soil and in the surrounding waters.

HIS REPORT ON SOIL.

Shortly after his arrival, he reported to Lord Hillsborough, one of the principal Secretaries of State, that "the soil appeared good and easily cultivated. It was free from rocks and stones." He never saw finer grass in his life, than grew every place where it was clear of woods. "It would," he said, "produce every kind of grain and vegetable common in England, with little or no trouble," and such as he had seen of the latter, "are much better of their kind than those at home, though raised in a very slovenly manner." A year later, in October, 1771, after having himself tested the capabilities of the soil, by raising various crops upon it the Governor again reported that his expectations had been fully realized, and in many cases, surpassed. He expressed the opinion that, with care, the Island would become the Garden of America.

The same good report of the wealth of the soil, has been repeated from the days of the old Governor, until now. From the time when the French, recognizing the great possibilities of its fertile soil, made this Island their granary and the provider of food supplies for their forces, until to-day, Prince Edward Island has sustained its splended reputation for the abounding productiveness of its soil. All visitors to its shores are struck by the fertility of its fields, the fresh green of its pastures and meadows, and the well-cultivated appearance of the farms.

GEOLOGICAL FORMATION.

The Geological formation is new red sandstone. With the exception of some few isolated granite boulders in the Western part of the Province, evidently brought there by glacial action in former ages, there is no rock or stone to interfere with agricultural operations. The soft red sandstone underlying the surface, is a handsome and valuable building material. It is easily quarried and dressed, and hardens after exposure to the air. It is now being used in the erection of public buildings, churches, etc.

Of coal and minerals, the Island cannot boast. That seams of the former underly some sections of the Province, seems beyond question, but no serious effort has been made to test them, as the nearness of the vast coal-producing areas of Nova Scotia and Cape Breton, assures the Island an unlimited and cheap supply of that article.

SCENERY.

The surface of the country is undulating, the highest land being about 300 feet in height, and, owing to the gentleness of the slopes, nowhere presents obstacles to cultivation.

While there is no grand scenery to be found in Prince Edward Island, there are innumerable vistas of surpassing beauty. The views over many a winding river or stream, about the numerous harbours and bays, with the snug farms, equipped with good, roomy, comfortable houses and buildings, with groves and scattered trees of spruce, birch, beech, maple and other woods, sloping down to the water's edge, give a park-like appearance to the face of the land, most fair to look upon, and give evidence of the natural wealth with which the soil is blessed.

NATURE OF SOIL—EXPERT OPINION.

The soil is a rich sandy loam, mixed with decayed vegetable matter, and is well adapted to the growth of the crops usually grown in Great Britain or in Canada. In this connection, Mr. J. P. Sheldon, Professor of Agriculture in the Wilts & Hants Agricultural College, who, with other delegates, visited Canada in 1880, for the purpose of inquiring into the Dominion's agricultural capabilities, and spent some time here, reported that:—

"In some respects, this is one of the most beautiful Provinces of the Dominion, and it has probably the largest proportion of cultivable land. The soil, generally, is a red sandy loam, of one character throughout, but differing in quality. On the whole, the grass land of the Island and the character of the sward, consisting, as it does, of indigenous clovers and a variety of the finer grasses, reminded me strongly of some portions of old England. Prince Edward Island is covered with a soil that is easy to cultivate, sound and healthy, capable of giving excellent crops of roots, grain and grass,—an honest soil, that will not fail to respond to the skill of the husbandman. The Island grows very good wheat, and probably better oats than most other parts of the Dominion. * * * * The Island is noted for its large crops of excellent potatoes, which not uncommonly foot up to 250 bushels an acre, of fine, handsome tubers. Swedes make a fine crop, not uncommonly reaching 750 bushels per acre, of sound and solid bulbs."

The late Sir William Dawson, F.R.S., K.C.M.G., Principal of McGill University, and some time President of the British Scientific Association, who had made a careful examination of the whole Island, on the same point says:—"The great wealth of Prince Edward Island consists in its fertile soil, and the preservation of this in a productive state, is an object of imperative importance. The ordinary soil of the Island is a bright, red loam, passing into stiff clay on the one hand, and sandy loam on the other. Naturally, it contains all the mineral requisites for cultivated crops, while its abounding in peroxide of iron, enables it rapidly to digest organic manures, and also to retain well their ammoniacal products."

Numerous brooks, rivers, and short arms of the sea, abounding in trout, smelt, bass, oysters, lobsters and other fish, intersect the country.

NATURAL MANURES.

Along many of the brooks and rivers are found large deposits of black mud, mainly consisting of decayed vegetable matter, and forming an excellent fertilizer in compost; while, in the rivers, bays and harbours, are found immense beds of mussel-mud, consisting of oyster and mussell shells, mixed with marine deposits of various kinds, forming a most valuable manure and soil stimulant, as it remains for years in the land, and supplies, not merely lime to the soil, but also the animal and vegetable matter, of which the mud is largely composed.

Large quantities of kelp and other sea-weeds are continually being cast up by the sea, and are much used in composts, forming a valuable fertilizer. In some sections, where lobster packing is carried on, an exceedingly rich manure is offered from the offal of

APPLE BLOSSOMS, PRINCE EDWARD ISLAND.

the lobster factories, but this is purely a local and very limited supply, depending entirely upon the prosecution of the lobster industry. To the farmer, the great deposits of black mud, mussel-mud, kelp and other sea-weeds, are invaluable.

Referring to the various manures, Sir William Dawson says: "The chief natural manures afforded by the Island, and which may be used in addition to the farm manures, to increase the fertility of the soil, or to restore it when exhausted, are (1) Mussel-Mud, or oyster-shell mud of the bays. Experience has proved this to be of the greatest value. (2) Peat and Marsh Mud and swamp soil (Black Mud). These afford organic matters to the worn-out soil, at a very cheap rate. (3) Sea-Weed, which can be obtained in large quantities on many parts of the shores, and is of great manurial value, whether fresh or composted. (4) Fish Offal. The heads and bones of cod are more especially of much practical importance."

Speaking of the Mussel-Mud, Professor Sheldon says:— "The Island possesses one advantage, which is unique and immensely valuable. I refer now to its thick beds of 'Mussel-mud,' or 'oyster-mud,' which are found in all the bays and river mouths. The deposit, which is commonly many feet thick, consists of the organic remains of countless generations of oysters, mussels, clams, and other bi-valves of the ocean, and of crustaceous animals generally. The shells are generally more or less intact, embedded in a dense deposit of mud-like stuff, which is found to be a fertilizer of singular value and potency. The supply of it is said to be inexhaustible, and it is indeed a mine of great wealth to the island. A good dressing of it restores fertility in a striking manner, to the poorest soils. Clover grows after it quite luxuriously, and, as it were, indigenously. * * * * It may be regarded as a manure of great value. * * * * Nor is it soon exhausted, for the shells in it decay, year by year, throwing off a film of fertilizing matter."

Owing to the quantity of lime in it, the mussel-mud is a powerful stimulant, and has a decided action on other food elements of the soil, rendering them available for plant food. Owing to its stimulating properties, it requires to be used with a certain amount of caution. The earlier users of Mussel-mud, before its properties were well understood, used it too freely, with the result that the land was over-stimulated, and its fertility temporarily impaired. Experience has taught farmers to guard against this result, and, with greater knowledge of its use and action, its value to farmers has gone on increasing.

CROPS COMMONLY GROWN.

The most common crops grown are wheat, oats, potatoes, turnips and hay, but very many others, such as mangles, beans, peas, vetches, flax, buckwheat, carrots, beets, sugar-beets, parsnips, corn (for fodder), and similar crops, give large yields, and the soil is well adapted for their growth, and many of them are being more extensively grown, year by year, with most satisfactory results.

Of these, wheat is grown only for home use, and not for export. The crop is a good one, but the hard wheat of Manitoba and Canada's other Western Provinces and territories, is better adapted for milling purposes and for export. Formerly, considerable quantities of wheat were exported, but it is doubtful if it ever was as profitable a crop for that purpose, as several others that are grown.

INTENSIVE FARMING.

Up to the last couple of decades, enormous quantities of the other crops, notably potatoes and oats, were shipped to Great Britain, the United States, the neighboring Provinces, and to Newfoundland. Immense quantities are still exported, particularly to the mining sections of the other Provinces, to Newfoundland, and to the United States; but, of recent years, the system of farming has changed very much. It has become much more intensive and more profitable. The farmers, instead of sending away their crops in the raw state; are, more and more, going in for converting their grains, hay, vegetables, etc., into the more finished products of beef, lambs, poultry, bacon, horses, eggs, butter, cheese, etc., etc., great quantities of which are yearly exported. The soil has benefitted by the changed system while the cost of disposing of the product, in its more concentrated form, has been reduced.

BUTTER AND CHEESE.

Twenty years ago, there were no cheese or butter factories in the Province. Now they are in operation from one end of the Island to the other. Twenty years ago, there was no export of cheese and very little of butter. In 1901, the last census year, the cheese output amounted to 4,457,519 lbs., worth $449,088.00, or over £90,000 Sterling. The total value of both the butter and cheese output, together, amounted to $566,824, or nearly £120,000 Sterling. There were 47 factories, of which 27 manufactured both cheese and butter, 15 made cheese only, and 5 made butter only. In 1891, only ten years previously, there were only four factories on the Island, and the total value of their output was only $8,448, or a trifle over £1,700 Sterling.

GRASS LANDS.

Prince Edward Island has splendid grass lands. Owing to its insular position, there is always some moisture by night, if not by day, to prevent the grass and hay fields from being parched up, as sometimes happens, during dry seasons, in more inland countries. In this respect, the appearance of the surface of the Island more closely resembles Ireland, than other places. It has the fresh green of the Emerald Isle, without the wet weather so prevalent there. Consequently, the Island is well adapted for the dairying industry, which should grow to large dimensions, and there is no reason whatever, why it should not rival Denmark or any other country, in the quality of its dairy products.

STOCK-RAISING.

Year by year, the farmers are devoting more care and attention to stock-raising, for which the Island is well adapted. The excellence of the pasture, the immense quantities of all sorts of roots, grasses and grains, required for feeding purposes, which, in a soil so fertile and easily tilled as the Prince Edward Island soil, are raised without difficulty, make stock-raising and fattening, profitable. This industry is also capable of enormous development, as there is always a demand for good live-stock and its products. For many years, good breeds have been introduced into the Province. The Provincial Government, both before and since Confederation, has devoted much care and attention to this subject. Private enterprise has immensely promoted the work, by importations from Great Britain and elsewhere, and the number of animals of the most approved breeds, are thus being increased, with a corresponding improvement in the quality of the stock.

The Island has long been famed for its horses. A large number are annually shipped away, to supply the needs of the lumber-woods, the mines, and the general traffic of the neighboring Provinces and of the New England States. This industry was founded here at an early period in the Island's settlement. Some of the early Lieutenant-Governors, then appointed from Great Britain, took a practical and active interest in this matter, and imported the best sires and dams that they could procure. In a country so favorable to animal life and development, the good results of these importations were soon seen, and the Island soon established a high reputation for its horses. Ever since, the raising of horses for sale has been a regular branch of the farmer's business, and it has been a successful one. The breeds are of all the usual ones deemed best suited to the Island, and embrace the Thoroughbred, the Clydesdale, the Shire, the Standard, and others. Frequent importations of sires keep the quality up, and infuse new blood into the business.

The same forethought and care, which, applied to horse-breeding, made the Island noted for its equine production, were equally instrumental in bringing good breeds of cattle, sheep and swine to the Province. Cattle-raising and fattening, forms a most important part of the work of the best farmers. The cattle, both beef and milch, are of excellent quality, and judicious importation of animals for breeding purposes, keeps up the high standard already established. The favorite breeds, for beef and milk, are the Shorthorn and Ayrshire. The other breeds, however, such as Polled Angus, Hereford, Holstein, Jersey and Guernsey, have their admirers, and considerable numbers of excellent animals of these breeds, are in the hands of farmers.

Besides supplying the local demand, live cattle and beef are largely exported, and find a ready and near market, in the neighboring Provinces and Newfoundland, as well as in St. Pierre and Miquelon, the headquarters of France's fishing industry in North American waters. The facilities for water carriage, give the Island a special advantage in supplying these important markets.

SHEEP.

Sheep are very successfully raised, almost every farmer having his flock. The Island mutton and lamb are of exceptionally fine quality, probably due to the superior pasture the fields afford. Large numbers of lambs are shipped alive, to supply the markets of the cities of the other Provinces, and to Boston and other cities of the New England States, where the demand is large. The dressed meat goes largely to the same markets as the beef. The wool also finds a ready sale. Shropshires and Leicesters, are the breeds most commonly found, though all the other standard breeds are well represented.

SWINE AND DRESSED MEATS.

Large and increasing numbers of hogs are raised and fat-tened. Every farmer raises some swine, and they are found to pay well, and always find a ready sale. In swine, as in other live-stock, Prince Edward Island is noted for its good breeds, which are always being improved by importations of new blood. Very many pure-bred animals are annually shipped from the Province, for breeding purposes. Thousands of the dressed

OAT FIELD, FARM HOUSE AND BARN.

TOURISTS AND END OF HAY-MAKING.

carcases are sent away, to supply the fresh pork markets of towns and cities outside the Island. Many more go to the pork-packing and bacon-curing establishments in the neighboring Provinces, while yearly, they are converted, in ever increasing numbers, into bacon, hams, and mess-pork, in the packing establishments here, whence the finished product is sent to the outside markets. A large quantity of bacon finds its way to Great Britain, where, owing to its excellent quality, Canadian bacon commands good prices, and, as it becomes better known, will command still better prices.

The pork and bacon industry is of great and growing importance. Like all the other industries of P. E. Island, it is capable of enormous development, possibly of greater development than any other. It is safe to say that it could be quintupled without difficulty. Moreover, there is always a large and constantly growing demand for hogs for manufacturing bacon and ham. The supply, not only in Prince Edward Island, but throughout Canada, is quite unequal to the demand. Packers are crying out for more hogs, and cannot get them. The Canadian pigs are so much better cared for, and fed on so superior a food, than those in the United States, that the Canadian bacon and ham are of a vastly better quality, than are produced in the Great Republic. In consequence, it commands much higher prices. In no part of Canada, are there better live swine, better cared for, or better fed, than in Prince Edward Island.

POULTRY.

Raising poultry for the home markets and for export, is a growing and profitable business. Until recently, it was of small dimensions, but, during the last few years, considerable attention has been devoted to it, and many persons are now interested in raising and fattening poultry, for local use, and for sale off the Island. The industry received a great impetus from the efforts of the Dominion Department of Agriculture, in giving practical instructions and illustrations, through poultry raising and fattening stations, in the rearing and preparation of the poultry for table use, as well as in methods of fitting them for export. Under the same direction, the best breeds of fowl are being adopted, which was not always the case.

EGGS.¹

The output is enormous. It brings in quite a large revenue to the farmers of the Island, or rather to their wives, eggs being generally recognized as a legitimate source of supply for pin money. They are shipped off the Island in great numbers. The eggs are either taken direct from the farms to the buyers, or are collected by the regular dealers, whose teams are sent over the routes as often as required, to bring in the eggs while perfectly fresh.   They are then carefully picked over, and packed in crates for shipping.   This industry has long been an important one, and, like its allied industry of poultry fattening, is capable of almost indefinite expansion.

HISTORY REPEATS ITSELF.

While the system of agriculture has much changed of recent years, and is still in a state of transition, the farmers adopting more intensive methods, feeding their raw material on their farms with great benefit to the soil, and with pecuniary advantage to themselves, enormous quantities of grain, hay and roots, as well as the finished products of the farms, are still sent away.   History is repeating itself, though with a difference.   During the French regime, a century and a half ago, the Island's fertile fields were called upon to furnish food for the naval and military forces of France, in Cape Breton and other parts of North America.   In the Twentieth Century, the Island is still called upon to supply the same regions, but now it is the industrial forces of Canada and Newfoundland, which draw upon its resources.   The vast development of the iron and steel industries, at Sydney, only a couple of hundred miles away, the immense coal mines of Cape Breton and Nova Scotia, employing their thousands of miners, with families dependent upon them, the hosts of lumbermen employed in the forests of the other Maritime Provinces, and of New England, the myriads of men engaged in or dependent upon the magnificent fisheries of the neighboring Provinces and of Newfoundland, and the families of these men, as well as those engaged in the fisheries prosecuted from France, with headquarters at St. Pierre and Miquelon, off the Newfoundland Coast, all look largely to Prince Edward Island for food.   The West Indies also are good customers, taking largely of the products of the Island's farms and fisheries.

AGRICULTURAL VALUES.

The total value of agricultural property in 1901, according to the census of that year, was $30,434,089.00, (about £6,500,000 Sterling), including land, buildings, stock, etc., etc., while the gross value of farm products amounted to an aggregate of $7,413,297.00.   Roughly speaking, two-thirds of this consists of crops, and the remaining third of animal products.

WATER-CARRIAGE.

An important advantage, in addition to its fertile soil, possessed by Prince Edward Island, is the fact that its export and import trade can be carried on by water. A look at the map will show that practically all parts of the Province have easy access to the sea. Steamship lines connect with the other Provinces and Newfoundland, while a fleet of other vessels ply between the Island and its markets. That this advantage is an important one, is evident, as water carriage is so much cheaper than that by land. Every part of Prince Edward Island being near a shipping port or harbour, the products of the farm can be shipped by steamships or schooners, direct to their destination. Owing to this, railway freights have been kept low, and the Prince Edward Island Railway, which ramifies over the Province, and is owned by the Dominion Government, affords excellent and cheap facilities for transporting freight over such short distances, as it may be more convenient to ship goods by land. It connects with the shipping at all the principal parts of the Island coast.

The water-carrying trade is a considerable one, affording employment to a large number of small vessels, in addition to the trade carried on by the steamship lines. The advantage derived by the Island, from its facilities for water-carriage to all its markets, cannot easily be over-estimated. Numbers of vessels, engaged in the produce trade, find their way, not only to the principal harbours, but up the many rivers which intersect the Island, and into the bays and inlets, collecting the crops of the surrounding country, or taking away the horses, live cattle, sheep, pigs and poultry, the dead meats, the vegetables and grains, from the farms, to feed the toilers of the sea, the workers in the mines, the factories and the woods, who look to Canada's Garden Province, for their food supplies.

In the shipping season, when the crops are being sent away, the sight of a lot of trim schooners, lying along the wharves, or alongside a bridge, at the upper reaches of some tidal stream, taking in farm produce as fast as it can be put on board, is an object lesson. It gives, in small space, ocular proof of what the productiveness of the surrounding country must be.

SHIPPING VILLAGES.

The villages at the shipping places do a thriving business. They not only export the products of the land, but they import all the goods required by the neighboring country. Excellent and up-to-date shops and general stores, are found in all these places, and the trade done is large. In these villages, there is a ready sale for all the farmer has to sell, and in them he can procure all the agricultural and other machinery he may require, at reasonable cost, as well as all other supplies which he may need, and his farm does not, in itself, produce.

FRUIT.

The idea long prevailed, that Prince Edward Island was not adapted for fruit growing. In what this absurd notion originated it is impossible to conjecture, yet it certainly was held, almost as a matter of faith, until a very recent period. In view of the excellent crops of fruit, of many kinds, now being produced, it is astounding that so erroneous an idea ever existed, or that it was not exploded a century ago. The splendid displays of home-grown fruit, shown at the Annual Agricultural Exhibition in Charlottetown, have proved veritable eye-openers, to those who doubted Prince Edward Island's fruit-raising capabilities.

APPLES.

The most valuable and marketable varieties of apples are produced. The Northern Spy, the Wealthy, the Fameuse, the Ben Davis, the Russetts, the Pippins, in a word, all the well-known and most prized apples for commercial purposes, are now grown. The famous Gravenstein, of delicious flavor, which does not ordinarily remain good beyond Christmas, here acquires a better keeping quality, and is in a good condition for a month or more, than when grown elsewhere.

PLUMS AND CHERRIES.

The Island excels in its plums, all varieties of which are raised in perfection. Fruit experts, brought to examine and judge the exhibits of fruit, have invariably extolled the superior excellence of this crop. Cherries also are largely grown, and of good quality.

THE SMALLER FRUITS.

The smaller fruits—strawberries, raspberries, gooseberries, currants, cranberries, etc., are grown in abundance for home use and for export. Owing to the season being somewhat later than in the United States and inland parts of Canada, some of these small fruits, particularly the strawberry, mature after the season is over there, and the Island berries come upon an empty market. Consequently, there is a ready demand for them in the cities of the New England States. Some farmers, who have made a specialty of strawberry culture, get handsome returns from catering to this market, and it is capable of very great expansion.

PASTURE.   CHURCH IN SPRUCE GROVE.

CONTENTMENT.   CHEWING THE CUD, IN PRINCE EDWARD ISLAND.

No factories for converting the abundance of fruit into jams, etc., have yet been established, and what is not utilized for home use, is sent away. In view, however, of the unlimited supplies of the very best raw material, that can be grown, with little difficulty and with small expense, it would seem but a matter of a short time, before a highly remunerative industry in this line must spring up.

THE FISHERIES.

While Prince Edward Island is wonderfully rich in its agricultural resources, and while the products of its soil have made the name of Canada's Garden Province a household word for abundance and excellence, yet its natural riches are by no means limited to the dry land. Situated in the Gulf of St. Lawrence, the Island is in the midst of the almost illimitable deep-sea fisheries of North America. The richness of her soil on shore, has to a large extent prevented the full development of her magnificent fishery resources. The people, as a rule, turn to agriculture, and look to the land rather than to the sea, for the rewards of their industry. More attention is given to fishing off the coast of Prince Edward Island by the fishermen from Nova Scotia, and from the Atlantic cities of the New England States, than by the Islanders themselves. That the sea about the Island could be made to rival the land in the wealth of its products, will be evident to those who look at the Island's position in the fish-teeming waters of this part of America.

Yet, although ranking in a very secondary position in comparison with agriculture, the fishing industries employ a large number of men, and bring a very considerable revenue into the Province. The varieties of commercial fish are numerous, and always command a ready market. Codfish, haddock and ling, are caught in large quantities, and, when dry-cured, form a valuable item in the Island's trade. They find their way into the United States, to Cuba, the West Indies and South America, as well as to the European markets. The fish are usually cured in the sun, but of late, the Department of Marine and Fisheries of Canada, have devoted much attention to the improved curing of fish, and are establishing artificial curing and drying establishments, which are proving most successful, the quality of the fish so cured being superior to the fish cured by the sun-drying process.

Formerly, people living along the sea shores sought to combine the two industries of farming and fishing, with very unsatisfactory results. It is found by experience, that men cannot successfully carry on both farming and fishing operations. Both are apt to require attention at the same time, and those who sought to carry them on together, usually failed in both. Now they are kept apart, and the fisherman is no longer a farmer. A farmer or his sons, living by the sea, may, and often does, go out for a day in his boat, to catch a supply of fish for his household, or to cure for winter use, but he no longer makes a business of doing so. The results from separating the two industries, are seen in the greater prosperity, which now attends the operations of each.

Mackerel abound off the Island shores, and are caught by seines and hand-lines. They are always sure of a ready sale, and the salted fish are a staple and valuable article of export. They command good prices, and are largely sold in the United States.

Herring, which frequent the harbours and coasts, are caught largely for local consumption, and for use as bait in the lobster and other fisheries. So far, they have not been utilized, as fully as might be, for commercial purposes. Recently, however, the Department of Marine and Fisheries have taken steps to instruct the fishermen in the best methods of curing and putting up herring for market. Experts have been brought to Canada, and practical instructions given the fishermen. The results from this policy, though only lately inaugurated, are already showing, in the demand for the fish, at excellent prices, which is springing up.

Of other valuable fish, halibut, bass, salmon (which are netted off the harbours, and are equal to the best Scotch salmon), gaspereaux or alewives (caught largely for bait), trout, smelts and other fish, are taken in the bays and rivers, and along the coasts of Prince Edward Island.

LOBSTER PACKING.

Lobster packing is prosecuted extensively all along the Island shores. The industry is a most valuable one, affording employment, during the season, to many people of both sexes. It is a profitable business, bringing a handsome revenue into the Island. It is carried on under stringent Government regulations as to seasons, cleanliness of packing, and the quality of the fish put up. So extensively and keenly was this business prosecuted, that it was feared the lobsters, which are caught by millions, would be exterminated. To remove any possibility of such a misfortune, the Fisheries Branch of the Marine and Fisheries Department has established lobster hatcheries, from which enormous quantities of the young lobsters are distributed in the different bays and fishing grounds. From one hatchery alone, at the mouth of Charlottetown harbour, over a hundred million young lobsters were hatched and distributed, to replenish the

23

fishing grounds, during the seasons of 1905 and 1906. Under careful management, this industry now promises to become greater and valuable as time goes on.

OYSTERS.

But the fish for which Prince Edward Island is best known, is the oyster. For the last half century, more than half of the annual catch of oysters in Canada, has been taken from the natural beds under the waters of Prince Edward Island. There are many qualities of oysters. They vary according to the locality in which they are caught. They are found in almost all the bays and tidal rivers of the Island. While some are of a coarse quality, the great bulk are of excellent flavor. The best Island oysters are the "Malpeques" and "Bedeques." The Malpeque oyster is known and famed as the oyster PAR EXCELL-ENCE throughout North America. The name "Malpeque" derived from the name of the place where the oysters are fished, is a synonym for the best oysters. To know that oysters are "Malpeques," is to know all that any North American lover of the oyster, wishes to learn about them. They are the best oysters in the world. They are taken from the natural beds in Richmond Bay, in the western part of the Island, off the shores of the Malpeque settlements. Up to the present, but little has been done in the way of cultivation. What little has been done, has demonstrated that the industry can be made most successful and profitable. There are many thousands of acres of land covered with salt water, in the harbours, mouths of rivers, bays and stretches of water along the coasts, on which there are no oysters at present, but where the conditions are eminently adapted for their successful propagation and culture. Interest is just beginning to be taken in this industry, as the natural beds are beginning to show the depleting effects of half a century of continuous fishing, without any steps having been taken to keep up their fertility. This business is capable of practically unlimited expansion. The available grounds are of enormous area. No other oyster can compare with the one grown in these waters. The demand all over Canada, and in the New England States, for the Malpeque oysters, seems insatiable. The demand is, and must continue to be, far greater than the supply. Prices are consequently good, and each year are getting better. With present day fast steamships and cold storage facilities, they can very easily be placed on the English market. There are fortunes to be made in this industry.

CLIMATE.

The climate of Prince Edward Island is good. The summer is particularly delightful. The temperature very seldom reaches 80 degrees F., while a night, not pleasantly cool, is a rarity. The extremes of heat and cold, found in other lands, are here moderated by the fresh breezes from the surrounding sea.

TOURIST TRAVEL.

The delightful, health-giving summer climate, makes Prince Edward Island a favorite resort for tourists from the towns and cities of Canada and the United States, who come for rest and recreation during the hot weather. The number of these visitors is yearly growing. Spare rooms in good farm houses, near the rivers and shores, are in demand. They are coming more and more into request, and bring many a dollar in profit to their owners. Numbers of people come, year in and year out, to spend a few weeks in this way. The change is pleasant and healthful. They enjoy the good, wholesome food the farm produces, or which they themselves may procure from the streams and neighboring waters. They are surprised at the keen appetites engendered of the ozone-laden air, from the Gulf of St. Lawrence, which give zest to life, almost compelling them to make away with the good things of the table, in quantities, and with an enjoyment they would think impossible at home. Then to the brain-wearied worker of the cities, the strong, fresh, bracing sea air brings new life, and, when the lights are put out, sleep comes without wooing. The tourists not only gain themselves, but they bring profit and variety of life to many a farm-house, as well as to the large summer resorts.

WINTER.

The winter is a bracing and pleasant time. It is the season when the farmer has most spare time on his hands, as he has little to do beyond attending to his stock, and generally keeping things in order. The temperature never goes to the depths it does away from the sea. The snow makes excellent roads, and sleighing is a most enjoyable way of getting over the neighboring country.

SPRING.

Spring is the most unpleasant season of the year, as it is long, and the snow is then melting, and the water is flowing off the fields. The skies are bright, but the air is raw from the melting snow. It is, however, the prelude to the glorious Prince Edward Island summer, a season unexcelled.

WATER AND FUEL.

Water is abundant and pure. Springs and brooks are found all over the country, while excellent water is procured at almost every farm-house, from wells sunk into the ground.

Throughout the greater part of the Province, wood, (beech, maple, birch, etc.), is the common fuel, but coal is coming more and more into ordinary use. The proximity of the splendid coalfields of Nova Scotia and Cape Breton, and the facility with which it can be brought by water to almost every section of the Island, . enables consumers to procure coal at a very cheap rate.

SYSTEM OF GOVERNMENT.

The Dominion of Canada is a Federal Union of the several Provinces and Territories, which make up its vast area of 3,745,-574 square miles. The seat of the Central Parliament, is at Ottawa, while each of the Provinces has its own local Government and Legislature, meeting once a year in the Capital of the Province, to attend to such matters as are assigned to it, under the provisions of the British North America Act, the Imperial Act which is the Charter of the Dominion.

The Central Government has to do with such matters as concern the Dominion as a whole. It has sole control of such subjects as the Customs, Excise, Postal Service, Militia, Trade and Commerce, Marine and Fisheries, Immigration, Patents, Quarantine, the Criminal Law. Speaking generally, it has to do with all subjects of Dominion importance, as distinguished from those of merely local or Provincial consequence.

The Provincial Governments and Legislatures have to do with subjects of local importance. These vary much in the several Provinces. In those having great mining, manufacturing and lumbering resources and industries, the matters managed by the Local Legislatures, are much more numerous and extensive, than they are in a purely agricultural community, such as Prince Edward Island. The Local authorities have charge of education, roads, bridges and ferries, devotes much attention to the conditions of agriculture within the Province, the Administration of Justice, (except the salaries of the Judges, which are paid by the Central Government), public works of only local importance, titles to land, etc. In a word, the Local Legislature is concerned with matters which are of importance to the Province alone, in contra-distinction to subjects which are wider in their scope, and affect the Dominion as a whole, or, at least, affect more than a single Province.

26

DUNK RIVER, PRINCE EDWARD ISLAND
FAMED FOR ITS TROUT.

AUTUMN.   SHIPPING FARM PRODUCTS.

EDUCATION.

Prince Edward Island has an excellent system of education. It is the result of continued efforts on the part of the people for generations. Beginning with the school of the pioneer days of the Eighteenth Century, when a teacher, paid his salary by the British Government, and having his residence in Charlottetown, ministered to the educational needs of the few people who composed the population of that time, it gradually grew with the growth of the community. During the first half of the Nineteenth Century, schools increased in number, and were spread over the Province, at considerable, though ever-lessening distances apart. They were mainly supported by voluntary subscriptions, and the teacher "boarded" from house to house, of the parents of his pupils. The Government gave what help it could, from the very meagre revenue of the struggling colony. The schools in the new settlements, were few and far between, yet the children, and even grown up men and women, trudged for miles over the bad roads, and through the woods and half-cleared fields, to attend, and gain such elementary education, as the little log school-houses, with their ill-paid teachers, offered. Many, whom the hard work of pioneer life on the farm prevented from attending in the day-time or in summer, would avail themselves of the "night school" of those years.

The results from these seemingly unpromising conditions, were often wonderful. Many a young man, who went out into the outside world, and prospered there, has cause to thank the humble school of his back-woods home, for the training which enabled him to succeed, and more than hold his own, in competition with those, whose early opportunities seemed so superior to his.

From the days of the earliest settlers, the people actively concerned themselves about education. It was the issue in many an Election contest. In 1852, a great advance was made. The colony was poor. Its resources were undeveloped. Its revenues were of the smallest. But the Island was possessed of men, who felt the need of education, some of them from personal experience, and they had the courage and foresight to grapple the question. In that year, a free school system was introduced, giving the benefit of an elementary education, to every child, whose parents or guardians wished it to attend the school. At once, a great improvement in the education of the youth of the Island began. From time to time, still further improvement was made, until, in 1877 was enacted the "Public Schools Act" of that year, which, with such amendments as experience has since suggested, is still in force.

Under that Act, every child is entitled to attend the common schools, where a good elementary education is afforded, free of charge. The Government pays the salaries of the teachers, applying a large proportion of the Provincial Revenue to that purpose. The country schools are at short distances apart, so that children can have no difficulty in attending them. The school districts, by voluntary assessment, provide their own school-houses, and usually provide a supplement to the teacher's salary, in addition to the Government grant, and provide for heating, etc. The system is undenominational, and works well. The schools, on the whole, are good.

PRINCE OF WALES COLLEGE.

In Charlottetown, at the apex of the system, is the Prince of Wales College, a very excellent institution, famed throughout Canada for the thoroughness of its work. It is supported by the Provincial Government. A very small annual fee of $5 (£1) for the child of a farmer or countryman, and $10 (£2) for a resident of Charlottetown, is the only charge made at the Prince of Wales College. Board for students in Charlottetown can be had at cheap rates, so that an industrious young man or woman, who passes from the common schools into this Institution, can obtain a really good education, at a very small cost. Agriculture is one of the subjects taken up in connection with the College, and is yearly becoming more of a feature of the work carried on. Much attention is given to the subject, and at the Local Government's farm, a short distance from Charlottetown, much experimental work is carried on. This work is to be extended. At the present writing, the Prince of Wales College, by the generosity of Sir William McDonald, himself a Prince Edward Islander by birth, is being nearly doubled in size, and in its accommodation for classes and the teaching staff, while the subjects of study are to be increased in number.

ST. DUNSTAN'S COLLEGE.

Situate a couple of miles out of Charlottetown, is St. Dunstan's College, a Roman Catholic Institution, which has been in prosperous existence for half a century. It is patronized, not only by the youth of Prince Edward Island, but draws numerous students from the other Provinces of Canada, as well as from the New England States. It does a good work, and sends forth its graduates well equipped for the battle of life.

CHURCHES.

The different Christian denominations are well represented in the Island. Of the total population of 103,259, in 1901, the Roman Catholics were the largest religious body, numbering 45,796. The Presbyterians came next, with 30,750. The Methodists numbered 13,402. The Anglicans were 5,976, and the Baptists 5,905. The several bodies of Christians are on the most cordially good terms, and are always ready to assist each other on their church building or other works, when occasion arises. Good churches have been provided by the respective denominations throughout the Province, and the people of Prince Edward Island, being a church-going and Sunday-observing community, make full use of them.

NATIONALITIES.

The population is almost entirely composed of persons of English, Irish, Scotch and French nationality, or rather descent, as the present inhabitants are nearly all born to the soil. Of these, the Scotch, or their descendants, are by far the most numerous, 41,753 claiming the "Land of the Heather" as their ancestral home. The English number 24,043, the Irish 21,992, while the French muster 13,866, these being almost entirely descendants of the French settlers, who remained on the Island after the cession of 1763.

TAXATION.

Outside the towns of Charlottetown and Summerside, which tax themselves for civic purposes, the taxation is so light as scarcely to deserve being called such. A land tax of one-fifth of one per cent (less than a halfpenny in the £1) is levied on real estate, the valuation being made by the owner, who makes a Statutory declaration of its value. Throughout the country, male persons, between the ages of twenty-one and sixty years, pay an annual poll tax of one dollar (4s. 2d.), and twenty-five cents for each horse, as a tax towards keeping up the public roads. There are some other small taxes levied, but these are the only ones that affect the farmer.

IMPROVED FARMS.

It will be evident that, in a Province so thickly settled, so conveniently situated, and already so well provided with the necessaries of agricultural life, there are no free grant lands to be had. The farms, with comparatively few exceptions, are improved, cultivated holdings, with dwellings and out-buildings. The farm-houses and out-buildings are almost invariably constructed of wood, that being the cheapest and most readily procurable material. It is also, on the whole, the most suitable for the country. They are usually roomy and comfortable. The size of the farms runs from the small one of fifty acres or less, up to those having an acreage of several hundreds. The averaged sized farm, in 1901, contained rather over ninety acres. The price of these farms varies according to location, quality of soil, and class of buildings. `From $15 to $50 (£3 to £10) an acre would be about the range of prices, although some farms in specially good locations, may command more. A good 100 acre farm, with buildings, can be had for from $25 to $30 (£5 to £6) an acre. This, it must be borne in mind, is for a title in fee-simple, free from encumbrances. And these are good farms, not worn out.

FARMS FOR SALE.

During the past few years, the "Lure of the West" has cast its spell over thousands of the young men of the older Provinces of Canada, and drawn them away from the comfortable old home-steads. The life in these old, long-settled Atlantic communities has grown too prosaic and matter-of-fact for them, and they have joined in the Westward march. There are great possibilities in the West. It is a new country with a magnificent future, and its attractions for young men frequently prove irresistible. And these young men from Canada's Eastern Provinces, make the best pioneers and settlers for a new country, the world can produce. Their early training on the home farms has well equipped them for the work of opening up and developing the vast, virgin plains of the Canadian West. They can turn their hands to anything. Hence, they readily adapt themselves to new conditions, and our great West draws them to the prairie lands. In consequence, numbers of farmers are moving towards the setting sun, and are offering their fine, well-equipped farms for sale. Because of this Westward migration, they can be purchased more cheaply now, than a few years ago, before the movement began, or than they can be purchased a few years hence, when it begins to slacken. These are the farms suited to the old-country farmer, with his knowledge of practical and scientific agriculture, and his experience and training under old and settled environments.

FOR FARMERS WITH A LITTLE CAPITAL.

It cannot be claimed that Prince Edward Island is a place for a farmer without any means, to seek a home. Farms are too valuable, and cost too much money, and such a man, while he would likely come out all right in the end, if he be industrious, and a good, practical agriculturist, would yet start with a burden of debt for the purchase and equipment of his farm, which would be a handicap for years. The Island is eminently adapted for a practical farmer with some capital, who can buy his farm, and start unburdened with debt. Such a man, with a few hundred pounds, can scarcely fail of success.

To such men, the Island offers a healthy and pleasant climate. It gives him an abundance of good water, and a very fertile soil. His taxes are so small as to be scarcely worth considering. He has sure markets, and has water-carriage to carry his produce cheaply to those markets. Accustomed to an old-country farm, which has been under cultivation for centuries, the change is to a comparatively new country, but it is not a violent change. He comes to an old, settled community, where his old-country experience stands him in good stead. The conditions he meets with are familiar to him, though in a modified and newer form. Compared with the land he has left, this Province is new. Here he will find neighbors on every side, within easy reach of his door. He finds a school for his children, in which a free elementary education is provided. Should he desire to give them a higher education, he can do so at small cost. Every settlement has its church. What the new comer requires on his farm, or in his house, he can readily procure. Compared with other new countries, the population is dense. Compared with Great Britain, or the countries of Continental Europe, it is sparse.

MARKET-HOUSE, POST OFFICE, PROVINCIAL LEGISLATIVE BUILDING
AND COURT HOUSE, CHARLOTTETOWN.

WESTERN CANADA

THE GRANARY OF THE
BRITISH EMPIRE

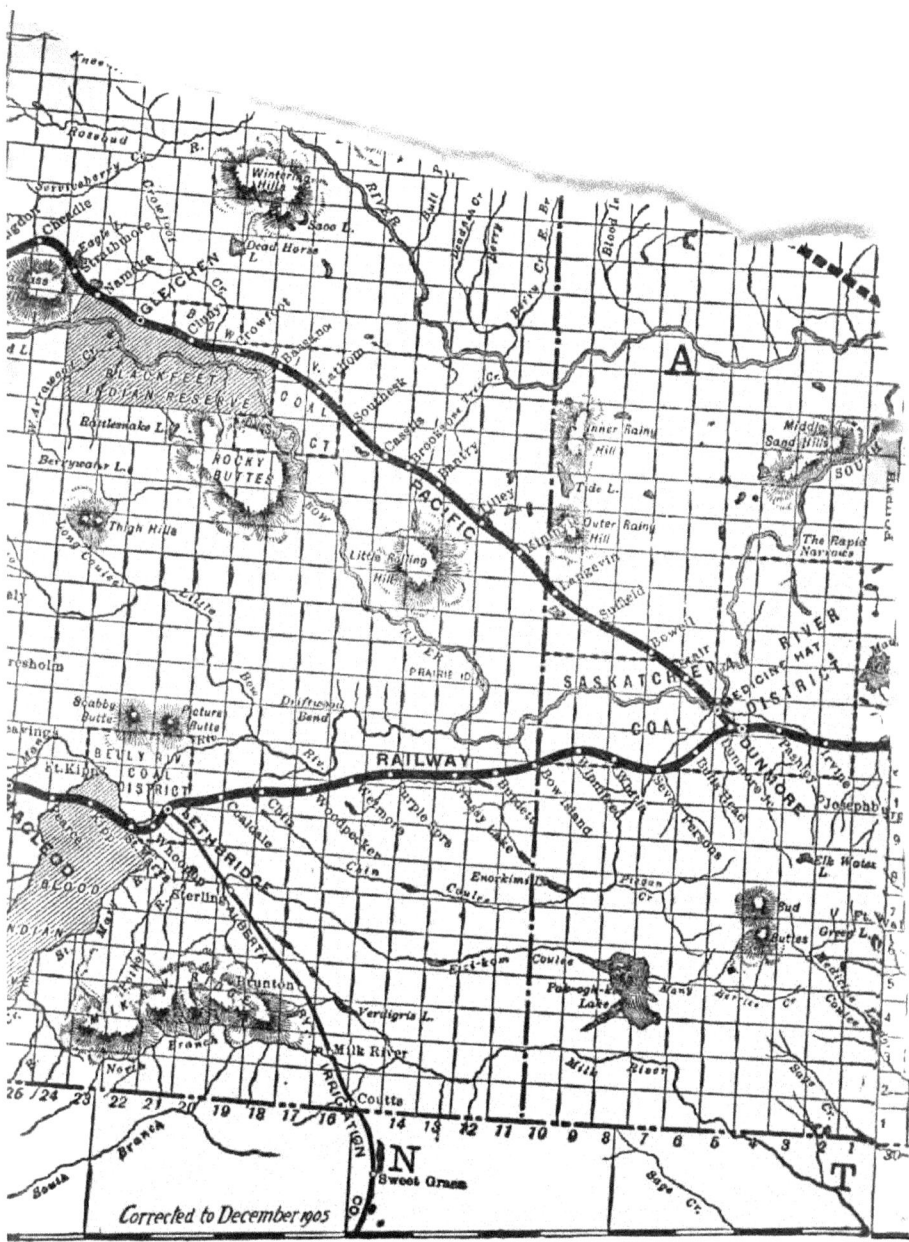

"Corrected to December 1905"

113°   112°   111°   - 11 0°

MAIN STREET, WINNIPEG, CAPITAL OF MANITOBA.

Population, 1874, about 2,000; 1907, 110,000. One of the most prosperous cities in the world.

WESTERN
CANADA

Manitoba
Alberta
Saskatchewan
and New Ontario

How to Reach It
How to Obtain Lands
How to Make a Home

A07

Western Canada

TABLE OF CONTENTS

WESTERN CANADA

EMBRACING

MANITOBA, SASKATCHEWAN, ALBERTA and NEW ONTARIO

THE INLAND EMPIRE

Speaking in Winnipeg in the year 1877, the late Lord Dufferin, then Governor-General of Canada, referred to Manitoba as the key-stone of that mighty arch of sister provinces which spans the continent from the Atlantic to the Pacific. It was bold, picturesque and eloquent language—it marked the thought of a great states-man, who, even at that early stage in the opening up of the West, could picture and predict other provinces yet to be created in that wonderful stretch of prairie land that lay for a thousand miles to the west of Manitoba, and that was some day to be the scene of ever-broadening harvests, multiplying towns and villages, and expanding pastures. Thirty years have now elapsed, and the pre-diction of the far-seeing statesman has been realized in that inland empire, which, with Manitoba, now embraces the new and vigorous provinces of Saskatchewan and Alberta, and which is the centre of an industrial, commercial and agricultural development that can scarcely be credited by those who have not been actual witnesses of its wonderful achievements.

In those thirty years it has been demonstrated beyond a doubt that Western Canada, with its illimitable dimensions, its wealth of resources, and in the strength of its material might, is destined to be the peer of any power on earth. A land of stupendous possibilities; a rich alluvial region, whose only limit seems to be an ever-receding horizon—truly has it been said that the develop-ment of this Kingdom of the North represents the genius of the twentieth century, for here a nation is to be builded that will play a most important part in solving more than one of the economical problems that now confront the statesmen and the philosophers in the older portions of the world.

Canada is a country of great distances. Extending from the Atlantic to the Pacific, it is more than equal in size to the United States, and, in fact, covers 3,614,000 square miles—one twelfth of the land surface of the earth. The Eastern Provinces of Canada are a land of woods and forests, of sea ports and harbors, lakes and valleys, corn lands and pastures, more extensive than half a dozen European kingdoms, practically all throbbing in some degree with the energy of strenuous commercial activity and rich in agriculture, timber and mineral resources.

The provinces which make up the agricultural region of Western Canada—Manitoba, Saskatchewan, Alberta, and incidentally may be included, New Ontario—comprise practically 600,000 square miles, and embrace within their combined limits the vast extent of the available agricultural region of the Great West.

It has been estimated that one-half of this area is well adapted to cultivation, and that nearly all the cultivable area will produce wheat. Explorations show that for a hundred miles or more at a time no bad soil is seen in any direction, and explorers confidently predict that wheat culture will yet extend to what even now is considered the remote north.

A very important consideration in this connection is undoubtedly the climate, and many wrong impressions regarding Western Canada frequently prevail on this point. It will interest and probably astonish many to be informed that Edmonton—a thousand miles northwest of Winnipeg—has as high an average annual temperature as St. Paul, in Minnesota, five hundred miles south of Winnipeg; but a glance at any map having climatic lines will show that this is true. Further, that Northern Michigan and Manitoba have similar temperatures, and that as we go north and west the influence of the winds from the Pacific have a marked effect in modifying the climate. The mean temperature for July in Winnipeg is 66, which is higher than in any part of England. The average diurnal range is also much greater than that in England, being from a maximum of 78 degrees to a minimum of 53 degrees. This high daily temperature during the growing months, with the long hours of sunshine, matures the crops quickly.

In Alberta remarkable characteristics of the climate are the light snow-fall and the warm chinook winds, and especially is this the case in the southern part of the province. Both cattle and horses can remain outside the entire winter, living and doing

exceedingly well on the sun-cured buffalo grass which covers the plain.

Three times the size of the German Empire, and five times larger than Great Britain and Ireland, Western Canada is a vast plain, watered and drained by three great river systems—the Red and the Assiniboine in Manitoba, the Saskatchewan in Southern Alberta and Saskatchewan, and the Peace and Arthabaska in Northern Alberta. With a gentle slope to the east and a slight tilt to the north, this plain stretches for fully a thousand miles from the Rocky Mountains on the west to the granite country of New Ontario on the east, and from the International Boundary on the south to a yet-to-be-determined point on the north, and the river systems make it one vast network of interesting valleys, the topographical features, as well as the climate, in a large measure accounting for the remarkable productivity of the soil.

The completion of the Canadian Pacific Railroad saw the beginning of the real development of modern Canada. It was then that Eastern Canada, emerging from her forests and fields, first gazed upon the rolling prairies and unexplored North-West, and learned of the vast and rich interior which had hitherto been scarcely considered as a factor in the economical problems of the Dominion. The announcement that in the Canadian North-West had been discovered more than two hundred million acres that would produce wheat and that this vast area could undoubtedly, if properly developed, cover many times over England's demand for wheat to meet her annual bread deficiency, naturally riveted the attention of the whole world, and a stream of settlement immediately began to flow into the country. It was not, however, until the dawn of the twentieth century that the real rush began, and since 1900, when the people of the republic to the south began to fully appreciate the agricultural possibilities of Western Canada, and to realize what a wonderful agricultural country lay immediately to the north of their own boundary, the influx has continued with ever-increasing volume.

Bringing with them capital, effects, and the necessary knowledge to develop the country, settlers are coming in thousands from the United States and from Great Britain, and every country of Europe, come other thousands to take advantage of the opportunities offered.

With regard to capital, the development of Western Canada has been unique, for with the twentieth century discovery of the West by those who come to make their homes in the country, there has been concurrently a corresponding awakening of the investing class and the capitalists to the opportunities offered them. The result has been a remarkable inflow of capital from England and European countries, and from the United States, and there is no lack of funds for either public or private improvements.

To sum up—what is the immediate outlook for Western Canada? It is a country that is now a long way beyond the experimental stage—a country that has been tried in the most exacting test to which a new country can be put, and has come triumphantly through the ordeal. There can be no doubt now that the settlers who are so rapidly peopling the Great West of Canada and making their homes here, are destined to be the wheat producers for the British Empire, and that they will also supply for all deficiencies that may arise in other countries. In this connection there seems to be no limit to the expectations that may reasonably be formed. For instance, what were once, in the imperfect knowledge of the country, supposed to be semi-arid districts are now, on thorough investigation, found to be capable of producing full crops, and providing opportunities for which no superior can be found elsewhere for cereal and garden roots, for dairying and for stock raising. In, other districts, which were once little thought of, winter wheat is revolutionizing the character of the whole territory, and elsewhere irrigation is proving an assured method of getting the best results.

Once it was thought that there were large areas of doubtful rainfall. The soil was known to be first class, but natural conditions were thought to be too uncertain to justify any attempt at settlement. Later explorations and practical experience has shown that conclusions in this regard have often been arrived at too hastily—and it now seems certain that there is very little really waste land in the whole vast territory.

For soil and climate Western Canada cannot be beaten—it is destined to be the bread basket of the world—the stockman's paradise, and the home of millions of happy, contented and pros-

CITY HALL AND UNION BANK BUILDING, WINNIPEG

A PROFITABLE CROP IN WESTERN CANADA.

perous people, living under the best form of government, and with natural advantages which are unsurpassed.

Growth of Population.

The recent census gives the population of Manitoba as 360,000, Saskatchewan 240,000 and Alberta 185,000. The total of these three provinces in 1901 was 419,512. The total in 1906 is 785,000, an increase in five years of 365,488.

All this has been accomplished without what one might call a boom. Business and production have kept pace with the advance of lands and other real estate. While there has no doubt been some speculation, the actual settler has established himself on the soil, and by actual work brought values up to and beyond the expectation of the most sanguine.

Soil and Climate of Western Canada.

Professor Thomas Shaw, now of the editorial staff of the Orange Judd Farmer, an eminent agriculturist, writer and lecturer, after a recent trip through Western Canada, spoke in the following terms on this subject:

"The contemplation of this great country is bewildering whether viewed from the standpoint of size or resources. In size it is an empire. Our party has been travelling over it as fast as the engine could carry us for the past sixteen days, and we have only seen a very limited portion of its entire area. Its resources are almost fabulous in the aggregate, whether viewed from the standpoint of minerals, timber or agricultural production. But beyond all question, the agriculture of this country will be its greatest industry through all the centuries.

"The first foot of soil in the three provinces of Manitoba, Saskatchewan and Alberta is its greatest natural heritage. It is worth more than all the mines in the mountains from Alaska to Mexico and more than all the forests from the United States boundary to the Arctic Sea, vast as these are. And next in value to this heritage is the three feet of soil which lies underneath the first. The subsoil is only secondary in value to the soil, for without a good subsoil the value of a good surface soil is neutralized in proportion as the subsoil is inferior. The worth of a soil and subsoil cannot be measured in acres. The measure of its value is the amount of nitrogen, phosphoric acid and potash which it contains, in other words, its producing power. Viewed from this standpoint, these lands are a heritage of untold value. One acre of average soil in the North-West is worth more than twenty acres of average soil along the Atlantic seaboard. The man who tills the former can grow twenty successive crops without much diminution in the yields, whereas the person who tills the latter must pay the vender of fertilizers half as much for materials to fertilize an acre as would buy the same in the Canadian North-West in order to grow a single remunerative crop.

"Next in value to the soil is the heritage of climate. No citizen of north-western Canada should be anxious to apologize for the climate of his country. Good as the soil is, it would never have brought supremacy in grain production in this country, had it not been for the climate. The blessing of the climate is threefold. It consists in the purity of the air, in the temperature of the same and in the happy equilibrium in the precipitation. Every one knows the value of the pure air in this country, viewed from the standpoint of health. But does everyone know as to the inestimable character of the blessing which pure air proves to the agriculture of the country? It prevents the rapid decay and transformation of the vegetable matter in the soil, and also the too rapid transformation of inert fertility, thus virtually preventing waste in the hand of nature. In this fact is found our explanation of the extraordinary fertility of the soil. The cool temperature of the summer nights is responsible for the large relative yields of the grain. Raise the temperature of the summer days and nights, and the yield of grain will be proportionately reduced. The relatively cool temperature is one of the agricultural glories of this land. The relatively light precipitation is also a great boon to the northwestern farmer. It grows his crops and does not destroy them when grown. Nearly every portion of these three provinces has a rainfall of 15 or 20 inches; enough to grow good crops of grain on farms that are properly tilled, and not enough to waste the fertility of the soil through cracking. In this, another reason is found for the wonderful producing power of these lands.

TURNING SEVEN FURROWS AT ONCE BY STEAM.

DELIVERING GRAIN FOR SHIPMENT.

Manitoba is the pioneer province of the West. As the Red River Settlement it stoutly maintained for years the possibilities of the fertile valleys of the Assiniboine and the Red. After the Dominion was formed it was taken in as a province.

It is the smallest of the western provinces, measuring but 65,000 square miles, yet it is as large as England, Scotland, and Ireland and has 27,000,000 acres of arable land, about one-sixth of which is under plow.

The Dominion census of 1906 gives a population of 360,000. Most of this population consists of emigrants from Great Britain, the United States and Eastern Canada. Manitoba being the most thickly settled portion of Western Canada conditions here may be regarded as a fair index of the whole.

In 1906 about 4,850,000 acres were under plow.

The average yield of wheat for 1904 and 1905 was about 20 bushels; the average price 85 cents and 65 cents respectively. The cost of seeding, harvesting and marketing being reckoned at $6.00 per acre, we have a balance ranging from $11.00 to $7.00 clear profit to the farmer. When it is remembered that land can be had for from $8 to $30 per acre, according to location and improvements, the financial end of Manitoba farming may be appreciated.

Manitoba is a thoroughly settled community, and in nearly every part the difficulties of the pioneer are a thing of the past. A glance at the map will show the excellence of the railway com-

munications. From Winnipeg the branches of the Canadian Pacific
Railway spread out like a fan. The main transcontinental line
passes through Winnipeg, and extensions are built as needed to
keep pace with, and sometimes even to anticipate the rapidly increasing population.

The southern half of the province is generally open prairie.
but the northern half is mostly covered with timber, consisting of
poplar, birch and spruce.

The climate of Manitoba is typical of the interior. The winters
are cold, but the air is dry and the days bright. Spring comes
early and suddenly. The summers are warm and the days long,
making the growing season equal to that of the states lying to
the south.

Telegraph lines connect every part of the Province with Winnipeg, and the telephone and electric light are found in all places
of importance. The postal service is thoroughly well organized
and reaches every part of the Province, while the public schools
are efficient and numerous. Statistics show that on January 1st,
1906, there were 1,360 organized school districts, 63,287 registered
pupils and 2,272 teachers. Of these schools 29 had been organized
within the year. 7 colleges (6 in Winnipeg and 1 in Brandon)
and a university in Winnipeg were maintained.

In Canada there is no established church, every religious body
being on an absolute equality in the eye of the law. Fraternal
orders and other benevolent associations are a prominent feature
of the social life of the country. Associations of importance from
a more strictly business point of view are the agricultural societies
with their annual fairs, and the farmers' institutes for the discussion
of practical questions.

Fruit and Ornamental Trees.

No better currants (black, red and white) can be grown than
in Manitoba. Gooseberries, raspberries and strawberries yield
regularly crops of the finest fruit and stand the climate well; crab
apples, too, are heavy bearers and when well sheltered hardy
varieties of the standard apples can be grown where the altitude
is not too high. Ornamental trees and shrubs do well and some
farmers now have their lawns very tastefully arranged with such
trees and flowering shrubs. The Dominion Government supply
from the Experimental Farms 1,500 trees to all applicants owning

APPROACHING THE FERRY.

farms in Western Canada. These are delivered in good condition at the nearest station free of cost. All that the farmer is expected to do is to take good care of the trees. Some of the early settlers now have groves of trees which will supply them with both shelter and wood for fuel for years to come. The seed of the box alder or soft maple can be gathered in the fall of the year in abundance. Trees of this variety are no more trouble to grow than a crop of turnips.

Mr. A. P. Stephenson, of the "Pine Grove" farm near Morden, Manitoba, has an orchard containing about five hundred apple trees, of which three hundred are bearing. Most of his trees are still very young, but the annual yield is considerable. Mr. Stephenson is quite optimistic as to the future of apple growing in Manitoba, and it is believed that in time the production in the province will fully supply the local demand.

At a convention of Manitoba market gardeners, held recently, Dr. Thompson, a successful fruit-grower, stated that he believed there was no country where small fruits could be grown with less trouble than in Manitoba. There were few insect pests or diseases to interfere with their growth. Dr. Thompson called the

APPLES ON A. P. STEPHENSON'S FARM, NEAR MORDEN, MANITOBA.

attention of the farmers to the fact that when more was grown than was wanted, they would find a very profitable local demand for it. He had, therefore, no hesitation in advising the farmers of Manitoba to grow small fruits.

Mixed and Dairy Farming.

For many years Manitoba was treated as almost exclusively a wheat-growing country, but this is changed now, and stock-raising and dairying are attracting much attention. On January 1st, 1905, the number of horses in the province was set down at 143,386; cattle, 306,943, of which 127,562 were milch cows; sheep, 18,228; pigs, 118,986. Cattle raising is especially profitable, as there is a splendid market close by. At least 80,000 cattle are required each year for home consumption, while the young cattle find a ready sale among the ranchers of the west.

Dairying is becoming a more important industry every year in Manitoba.

In 1896 the Provincial Government established a dairy school in Winnipeg, which has been a great success. It is fitted up in the most modern way, and has trained many of those now in charge of the creameries and factories throughout the Province. Any resident of Manitoba may attend without paying fees.

The dairy statistics for the Province of Manitoba for 1906 are:

	Pounds	Price per lb.	Value
Butter, dairy	4,698,882	18.3	$840,006.85
Butter, creamery.	1,552,812	22	342,495.48
Total.			$1,182,502.33

	Pounds	Price per lb.	Value
Cheese, factory.	1,501,729	13	$195,244.51
Total dairy products			$1,377,746.84

Dairy butter shows an increase in one year of about 39 per cent. and an increase of $1.33 per 100 pounds in the price.

Creamery butter shows a gain of 20 per cent., with an average price of 22c per pound.

The amount of cheese manufactured was 25 per cent. over 1905, and the advance in price was $3 per 100 pounds.

Manitoba has great advantages as a dairy country. The pasturage is very rich and nutritious, with an abundance of variously flavored grasses; the water supply is excellent, and ample both for watering the stock and for use in the dairies, streams of pure running water being often available.

The annual government report on the crops and agricultural conditions in the province for the year 1906 has been issued. The total wheat yield for the province was 61,250,413 bushels, being over 5,000,000 bushels increase over last year. Other crops show

A FARMER'S HOME, 3 MILES FROM WINNIPEG.

a proportionate increase. The whole report shows that the farming community of Manitoba is in a most flourishing condition.

The oat crop for the Province of Manitoba for the year 1906 was 50,692,977 bushels; barley, 17,532,553 bushels; flax, 274,,330 bushels; Rye, 100,000; Peas, 67,301 bushels; corn, 249,840 bushels; Potatoes, 4,702,595 bushels.

In 1905 the estimated amount expended in farm buildings in Manitoba was $3,944,101, while in the year 1906 it was $4,515,085.

Liberal Exemption Law.

Manitoba has a liberal exemption law; that is, the law protects from seizure for debt, where no mortgage exists, a certain number of horses, cattle, swine and poultry, some household effects and a year's provisions, so that if a settler who has not mortgaged his property is overtaken by misfortune, he cannot be turned out of his house and home.

Land for Immigrants.

The new comer has the choice of four ways of securing a farm; he may homestead, he may buy land from the Canadian Pacific Railway Company, or other holders, he may rent an already culti-vated farm, or he may buy an improved farm on the crop payment plan. The terms on which nearly all the farms are leased is the half-share plan. The owner of the farm provides the seed (and if he is wise, sees that it is clean and of the best quality); he also pays for one-half the threshing and half the twine, and in return gets one-half the crop put into the granary on the farm. The tenant does all the work and also the statute labor, which is generally five days for a half section, and he, too, gets one-half the crop. To buy a farm on the crop-payment plan the holder in most cases asks a cash payment of from $500 to $1,000. The purchaser delivers in the nearest elevator one-half the crop till the land is paid for. The price is agreed upon and six per cent. interest is charged on the unpaid principal. The purchaser, if the land is of good quality and near to market, runs no risk, as he always has a fair return for his labor and in a few years owns the farm.

Opportunities to rent farms in the older settled districts are not uncommon, and are often worth seizing. The farms are rented generally during the winter or early spring for a year or more, the rent depending largely upon the kind and value of the improvements.

Cheap Fuel.

Besides the large tracts of forests, both in and adjacent to Manitoba, there are vast coal areas contiguous to the province of such extent as to be practically inexhaustible. The Manitoba Legislature has effected an arrangement by which this coal is to be supplied at a rate not to exceed $2.50 to $5 per ton, according to locality.

At Banff, Alta., deposits of anthracite coal have been recently opened up by the Canadian Pacific Railway Company. The coal resembles that obtained from the famous Pennsylvania mines, and will be supplied to the whole of Manitoba and the North-West.

Cities and Towns in Manitoba.

Winnipeg, the capital of Manitoba, and the largest city in Canada west of Lake Superior, is about midway between the Atlantic and Pacific oceans. In 1870 its population was 215; in 1874 it was 1,869; in 1902, 48,411, and 1907, 110,000, and is steadily increasing. Winnipeg is naturally a centre for the wholesale and jobbing trade of the North-West and every branch of business is represented; all the principal chartered banks of Canada have branches here, and there are a large number of manufacturing

A MANITOBA PIONEER.

establishments. There are extensive stockyards, and immense abattoirs, arranged for slaughtering and chilling the meat for shipment to Europe and other markets. There is ample cold storage in the city for dairy produce, etc. It is an important railway centre, from which both the East and the West may be reached. Branch lines run to nearly every part of the Province and a branch of the Canadian Pacific connects with the Soo line at Emerson thus affording a direct and easy route to St. Paul, Minneapolis and Chicago.

The yards of the Canadian Pacific Railway at Winnipeg are the largest in the world operated by one company, and contain 120 miles of track.

Winnipeg is the political as well as the commercial centre of Western Canada. The Legislative and the Departmental buildings of the Manitoba Government and the chief immigration, lands and timber offices of the Dominion Government for the west are located here. The Canadian Pacific Railway Company has its chief offices in the west in Winnipeg, and also the head offices of its land department, where full information regarding the company's land can be obtained.

The largest towns in the province outside of Winnipeg are on the main line of the Canadian Pacific Railway:—Portage la Prairie, 56 miles west (population about 5,500), and Brandon (population 11,000), 133 miles west are important railroad junction points and centres for a considerable area of grand farming country.

There are many other important towns, with populations ranging from 3,000 to 5,000. Grain elevators have been erected at nearly every railway station. Stores will be found in every town facilitating the business of the neighboring settlements.

ON LAKE WINNIPEG.

SASKATCHEWAN

The Province of Saskatchewan contains 229,229 square miles. It is 700 miles from south to north and a little more than 400 from east to west. It, therefore, of necessity has a variety of climates. The northern half is largely unexplored. The southern half may be divided into the agricultural and grazing sections.

The eastern portion, for a distance of some 120 miles west from its eastern boundary, is practically a continuation to the westward of the grain-growing areas of Manitoba. The soil is a friable loam, easily worked, and producing excellent crops of wheat, coarse grains, and vegetables. The winter climate answers all requirements, both as to degree of cold and as to sufficiency of snowfall, for the production of the No. 1 hard wheat for which Western Canada is now noted. This district will one day be one of the greatest wheat-producing sections of the American continent, and for the following reasons: 1st—It has a soil particularly rich in the food of the wheat plant. 2nd—It has a climate that brings the plant to maturity with great rapidity. 3rd—On account of its northern latitude it receives more sunshine during the period of growth than the country to the south. 4th—Absence of rust due to dryness of climate. 5th—Absence of insect foes.

These conditions are especially favorable to the growth of the hard, flinty wheat so greatly prized by millers all the world over, and commanding a higher price than the softer varieties grown elsewhere.

The summers leave little to be desired in an agricultural country, cyclones or violent storms being thus far unknown. In most parts good water can be obtained at a reasonable depth.

Coal in abundance is found in the south, in the district drained by the Souris River. Sufficient wood for all purposes for many years to come is to be found along the rivers and in the Moose Mountains.

The possibilities of Southern Saskatchewan are shown by the averages of tests made at the experimental farm at Indian Head, where eleven varieties, of the most suitable wheat, sown on April the 15th, were cut in 130 days, and yielded 4,314 pounds of straw and 43 bushels and 2 pounds of grain per acre.

This area embraces the Regina and Moose Jaw plains, nearly every acre of which is first class wheat land, the celebrated Indian Head district and the favored Moose Mountain settlement. Only a few years ago hundreds of thousands of acres of these lands could be bought from the Railway Company at $3.00 per acre, and a very large proportion of the free grant lands were vacant. Now the free homesteads are exhausted and lands sell freely at from $8.00 to $20.00 per acre. This part of the province is well served by the main line and branches of the Canadian Pacific Railway.

North of the Qu'Appelle River, along the Pheasant Hills and the Manitoba and North-Western branches of the Canadian Pacific Railway are excellent mixed farming tracts.

Southwestern Saskatchewan, embracing that section of country lying between the South Saskatchewan River and the International boundary, and west of the Moose Jaw district, has hitherto been regarded as semi-arid, with here and there localities with sufficient rainfall to insure average crops four years out of five. The rainfall at Swift Current, however, for the past ten years compares favorably with that of the best wheat growing districts in Canada.

From Swift Current to the Alberta boundary is mainly a ranching country. Two features have peculiarly fitted it for cattle and sheep. The first is the "buffalo grass." The plains are covered with a short, crisp herbage, which, though it turns brown at mid-summer, remains green and growing at the roots. On this cattle and sheep thrive the whole year round, and there is little need to provide other fodder at any season of the year. The other natural advantage is the "chinook" wind. This blows from the mountains and licks up the snowfall in winter with wonderful rapidity. The severity of the climate is greatly mitigated thereby, and cattle and sheep face the winter with little or no artificial protection. The Cypress Hills are especially adapted for stock-raising.

Further north, between the south and north branches of the Saskatchewan River, are immense stretches of open prairie land

CATTLE IN SASKATCHEWAN.

particularly suited to wheat growing on an extensive scale, and it is probable that these plains will within a few years compare favorably as a grain-growing section with the southeasterly portion of the province. The Moose Jaw Branch of the Canadian Pacific Railway, and also the Pheasant Hills Branch from Saskatoon to Wetaskiwin, will assist in developing this territory.

North and east of the arable part of Saskatchewan stretch extensive tracts of the finest spruce timber. From this the settler now gets, and will get for all time, cheap building material. It will also be very valuable when the opportunities it gives for pulp and paper, industries have been recognized.

The series of lakes north of the Saskatchewan River are well stocked with fish. Lake trout, pike, pickerel, sturgeon and white fish abound, and are available for export as well as local consumption.

Dairying in Saskatchewan

On the open prairie portions of the province farmers devote their energies largely to exclusive wheat growing, but in the eastern and northeastern districts conditions for mixed farming and dairying are eminently suitable. The natural growth of grass

is abundant and affords ample pasture of excellent quality for stock. In addition to this the country is well watered. Where streams and small lakes do not exist a sufficient supply of good well water is easily obtainable within from fifteen to thirty feet of the surface. Furthermore, the settlements suitable for mixed farming are comparatively well wooded with bluffs of trees which serve as shelter for stock during the warm summer months and also from the cool autumn winds, as well as affording relief from flies. The trees in most districts are of such a size that they can, in many cases, be used in providing buildings for comfortably housing stock during the winter months, thus permitting the settlers to engage in mixed farming without any great outlay of money when they first arrive in the country.

With the exception of one creamery, which commenced operations in 1906, all the creameries in the province are co-operative institutions and are under the direct control of the Department of Agriculture, Regina. The local matters requiring attention are supervised by a Board of Directors appointed each year for that purpose, and the Department assumes the work of finding a market for all produce beside keeping the books for all creamery companies. Detailed certified reports are sent weekly by creamery managers and from these patrons and companies receive a monthly advance on their butter and at the close of the season, when all butter has been marketed, the balance remaining, after deducting

THRESHING IN SASKATCHEWAN.

manufacturing expenses, is distributed by the Department. This system tends to establish confidence among the dairy farmers as the financing is altogether under Government control, and monthly payments are promptly made regardless of any sales of butter. While the creamery work is but in its infancy, it is estimated that upwards of 300,000 pounds of butter will be manufactured during this year. In addition to this, the make of dairy butter is estimated at over 700,000 pounds.

Official Report for 1906

Figures have been compiled indicating the agricultural progress of the Canadian West during the past year, showing that the grain crop of the three provinces totalled 201,020,148 bushels. There are 1,200 interior elevators, and yet these are insufficient to handle the grain crop.

The cattle industry also is a very important factor in the country's wealth, $4,029,639 net having been paid to ranchers alone last year, the prices ruling fairly high, an average of over $47 per head for export steers being paid. Some 130,000 head of cattle were received at the Winnipeg stock yards, and nearly 86,000 were carried to the seaboard over the Canadian Pacific Railway, an increase of 27,000 compared with the previous year.

The supply of hogs was altogether inadequate to meet the demands of the market, and an average price of $7.11 per hundred-weight prevailed.

The supply of sheep is much below the requirements of the market.

Towns in Saskatchewan

Regina, formerly the territorial capital and now the capital of the province, has a population of about 7,000, is on the main line of the Canadian Pacific Railroad, and is the terminus of the Arcola branch from the southeast.

Prince Albert, the oldest town of size in the province, with a population of 4,000, is located on the Saskatchewan River, near the centre of the province.

Moose Jaw, population 7,000, is a divisional point on the main line of the Canadian Pacific Railway; it is an important business centre and is situated in one of the best wheat sections of the province. It is the point where the Soo line, running to St. Paul,

Minneapolis and Sault Ste. Marie, connects with the main line of the Canadian Pacific Railway. A branch line is being built north-westerly and will open up immense tracts of finest wheat lands.

Saskatoon, with a population of about 2,500, is a thriving town on the Pheasant Hills branch, and is the junction point of that line with the line running from Regina to Prince Albert.

Weyburn, on the Soo line, is becoming a very important business centre.

There are many other important towns, and at nearly every station are elevators, stores, and all the business facilities which the settlers require.

HIS FIRST HARVEST.

Alberta

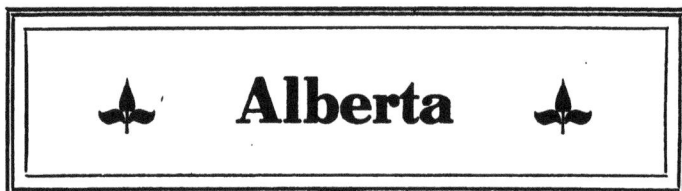

North of the International Boundary line and immediately east of the Rocky Mountains lies the Province of Alberta—a land blessed with all that is necessary to happiness and prosperity.

Between the 49th and 60th parallel of latitude between the 110th and 120th meridians lie 281,000 square miles of possibilities. No other political division of the great virgin empire of the North-West contains more that is necessary to modern civilization and less that is useless. Here are mountains and plains, foothills and valleys, rolling prairies with wooded stretches between, dense forests and grassy meadows, clean-shored, timber-girded lakes and winding brooks, cold mountain streams and navigable rivers, and a soil rich in the alluvial and vegetable accumulations of centuries. And as if not content with these outward signs of her favor, Nature hid beneath the surface vast deposits of coal and other minerals; she filled the subterranean reservoirs with gas and oil, and sprinkled the sands of the mountain streams with gold. That no living thing should go athirst she gathered together the waters of the mountains and brought them to the plains to be directed by the ingenuity of man to the use of the grazing herds and the planted fields. Then, to crown her effort and leave nothing incomplete, she brought the Chinook wind, warm with the breath of May, to temper the north wind.

One of the first things that strikes the casual observer in looking over the map of the Canadian North-West is the apparent remoteness of Alberta. Leaning against the eastern shoulder of the Rocky Mountains it seems a long way from the great-centres of civilization. It must be remembered, however, that to be remote from one place is to be near another and that isolation itself has advantages.

On reaching Alberta you feel at once the sundering of eastern ties—even the middle west is out of your mental range—you are in another world, the world of the Pacific. You are east of the mountains, yes, but you are within the sphere of the commercial west. The markets of the agricultural products of Alberta will ultimately be to the west—to the Orient, as it is now to the mining districts of British Columbia. Her coal will go down grade to the east—to the plains of Saskatchewan and the prairies of Manitoba, but her agricultural products will seek a nearer market. It is over two thousands miles to Montreal. and but six hundred to Victoria.

Alberta stands in the very gateway of railroad development. Through its territory lead the great river valleys, and beyond are the mountain passes. From Calgary on the main line of the Canadian Pacific a branch extends north to Edmonton, and another

DAYSLAND—A ONE-YEAR-OLD TOWN OF ALBERTA.

south to Macleod, where it connects with the Crowsnest branch running into the Kootenay mining country.

Central Alberta is an immense area of the most fertile land, well timbered and well watered. The soil consists of from one to three feet of black vegetable mould, with little or no mixture of sand or gravel. In ordinary good seasons a yield of oats of 100 weighed bushels to the acre has not been uncommon; less than 60 bushels is considered below the average, 70 to 85 bushels averaging 50 pounds to the bushel, being an ordinary yield; the barley will yield 60 bushels and wheat over 40, and potatoes of from two to three pounds weight are not a rarity.

Live stock of all kinds is raised extensively. including horses of all grades, from heavy draught to Indian ponies, horned cattle, sheep, pigs and poultry. Native horses do well without stabling all the year round, but good stock of whatever kind requires good treatment to bring it to its best, when it is most profitable.

A HOME IN WESTERN CANADA.

There are few summer or winter storms. As a consequence, a fine class of cattle can be raised very cheaply and with less danger of loss in this district than in some other parts. The advantages which tell so heavily in favor of the district for cattle raising, tell as heavily in favor of dairying.

Native fruits—wild strawberries, raspberries, gooseberries, saskatoon and cranberries, cherries and black currants—grow in profusion almost everywhere, and tobacco is successfully cultivated. Splendid vegetables are raised, and Wetaskiwin is noted for its turnips. All through the country small game, principally mallard and teal, prairie chicken and partridge, is very plentiful, and deer may not infrequently be found.

To accommodate the great increase of settlement north of Calgary the Canadian Pacific Railway Company are constructing two branch lines in an easterly direction from the Edmonton Branch. They start from Wetaskiwin and Lacombe, running through very fertile districts, and will connect with the Pheasant Hills branch.

Southern Alberta, between Macleod and Calgary, has attracted many settlers lately. Four years ago the sixty miles from Macleod to High River were given up to ranching, and there was hardly a house to be seen; now there are thriving towns eight to twelve

miles apart. The soil is very fertile, water may be easily obtained; unlimited supplies of timber await the lumberman in the mountains, and grey sandstone for building purposes is plentiful. Fall wheat is grown successfully. Southern Alberta is level, open prairie in the eastern portion, but is much broken along the western side by the foothills of the Rockies. The live stock industry is still the chief one, although the conditions are fast changing the large herds to smaller ones, which can be more easily handled and cared for. Large numbers of young beef cattle are usually imported from the east to be fattened on the Southern Alberta ranges, and are again profitably shipped as matured beef to European and eastern markets and to British Columbia and the Yukon. Mixed farming is now extensively carried on in Southern Alberta, and is very profitable. With a rapidly extending system of irrigation, this and other farming operations will develop very quickly.

Chief Towns.

Calgary is a busy city of 15,000 population, which is rapidly increasing. It is situated at the confluence of the Bow and Elbow Rivers, about 70 miles east of the Rocky Mountains. It is the centre of the northern ranching districts of Southern Alberta, and supplies many of the smaller mining towns in the west. It is built principally of grey sandstone. It is the junction of the Calgary & Edmonton branch with the main line of the Canadian Pacific Railway, being a divisional point, with machine shops, etc. It is also the headquarters of the British Columbia Land and the Irrigation Departments of the Canadian Pacific Railway Company.

Edmonton, on the north bank of the Saskatchewan, is the provincial capital and the market town for the farmers, traders, miners, etc., on the north side of the Saskatchewan, and for the trade of the great Mackenzie Basin. It is a well built and prosperous town with a population of about 12,000.

Strathcona, on the south bank of the Saskatchewan (population 3,500), and the present northern terminus of the Edmonton branch of the Canadian Pacific Railway, is another rising centre.

Medicine Hat has a population of 3,500 and is situated on the banks of the South Saskatchewan River. Natural gas wells supply the town with the material for heating and lighting.

There are a large number of other important towns and villages throughout Alberta, which are growing both in wealth and population.

Horse Raising.

The Alberta horse has already become noted for endurance, lung power and perfect freedom from hereditary and other diseases. Thoroughbreds from Great Britain and Kentucky, Clydesdales from Scotland, Percherons from France, and trotting stock from the United States have been imported at great expense, with the result that the young horse of Alberta will compare with any in Canada, and finds a ready market. Good three-quarter bred Clydes and Shires which at maturity will weigh 1,400 to 1,600 lbs., have been selling at three years old readily from $100 to $125. Good quality of other classes bring from $60 to $100.

ALBERTA HORSES.

Sheep, Hogs and Poultry Raising.

For sheep there are thousands of acres of rich, well watered grass lands, adapted in every way to produce first-class mutton and fine fleeces. Sheep mature early owing to the excellent quality of the grass. The popular breeds are Shropshires and Downs. in some cases crossed with Merinos.

The favorite breeds of hogs are Berkshires, small Yorkshire Whites and Tamworths. Hog raising may be increased indefinitely, as the demand exceeds greatly the supply. At present Manitoba, Saskatchewan and Alberta do not produce 50 per cent. nor British Columbia 25 per cent. of the ham and bacon they consume.

ALBERTA SHEEP.

The Orient will also take a large quantity. As things are the Eastern Provinces and the United States meet the demand, but there is no reason why the West should not raise its own hogs.

One of the most profitable branches of farming in the Canadian West is the production of eggs. During the winter months prices range from 30c. to 60c. a dozen. There is a ready demand for fowls for home consumption, the supply not nearly equalling the demand. This climate cannot be surpassed for the rearing of turkeys, the dryness and altitude being especially favorable for this profitable bird. Geese grow to a large size on the rich pasture without very much care or extra feeding.

Cattle Raising.

There are countless herds of fat cattle on the ranges of Southern Alberta, which at any season are neither fed nor sheltered. Shorthorns, Herefords and Polled Angus (black and red of the latter), are the chief breeds. There are some Holsteins and Ayrshires, but they are not generally used, except where dairying is the desideratum. For the small stock breeds, where dairying and beef producing must go hand in hand, probably a good milking strain of Shorthorns will be the most profitable.

The ranching industry in Southern Alberta seems, however, to be undergoing a radical alteration. The rancher is giving way to the mixed farmer. Some of the larger men are realizing on their property, and are being replaced by farmers, who have some of their land under crop, but keep a herd of cattle as well.

Crops of 1906

The crop areas and yields of 1906 in Manitoba, Saskatchewan and Alberta, according to the Provincial official returns, were as follows:

	Bushels.	Acres.	Average.
Wheat. . .	94,119,626	4,614,827	20.39
Oats.	86,216,627	2,024,127	42.59
Barley.	20,779,734	591,393	35.13
Flax.	695,180	49,372	14.08

Game in Alberta.

Alberta is an attractive country for the sportsman. Wild duck of all varieties, geese, prairie chicken, blue grouse, snipe, partridge and all other game are usually plentiful, while in the north and the mountain regions of the south, deer, moose, and other large game are by no means uncommon. Bands of antelope are also often seen on the plains in the south. Trout of several species abound in most of the streams and lakes of Southern Alberta.

Alberta Winter Wheat.

The development of winter wheat cultivation in the Province of Alberta has been so rapid and successful, that those who have not been favored with an opportunity of seeing the growing crops, the reaping, threshing and marketing of same, are apt to discredit the published reports.

The area sown in 1902 was only 3.444 acres and this yielded in

1903, 82,420 bushels, an average of 23.86 bushels per acre. In 1903 some 8,300 acres were sown, yielding in 1904, 152,125 bushels. The crop of 1905 was very large. In 1906 the spring conditions were not as favorable as in the previous year; the acreage under crop was 43,660 and the estimated yield 907,421 bushels.

In Alberta are millions of acres of virgin prairie land as good as that which has already been cultivated and it is only natural to predict a tremendous influx of farmers to take advantage of the conditions which prevail.

Irrigation and Irrigation Development.

The Canadian Pacific Railway Company has now under construction one of the largest irrigation schemes on the American Continent. It embraces an area of some 3,000,000 acres lying east

of Calgary between the Bow River and the Red Deer River. Of this area the Company expects to be able to supply water to irrigate about 1,500,000 acres. Canals have already been completed which are capable of furnishing water for irrigating 110,000 acres.

The district comprised within the Company's irrigation block is at the present time the largest unoccupied block of good land in the West, and with the introduction of the irrigation system, will afford a first-class opportunity for ranchmen who desire to obtain ranges for grazing purposes, to which are attached lands on which fodder may be raised every year by irrigation. It will also attract the immigrant who desires to obtain a small holding where he can combine ranching on a small scale with dairy or mixed farming.

Full information, maps and pamphlets descriptive of this scheme may be obtained on application to J. S. Dennis, Superintendent of Irrigation, Calgary, Alberta.

Many individual owners or ranch companies have undertaken irrigation of their own land, and there are some large corporations carrying on the work on a large scale. The Calgary Irrigation Company has 35 miles of main ditch to the west of Calgary, and the Alberta Railway and Irrigation Company, with headquarters at Lethbridge, have constructed 130 miles of ditch. As a result a prosperous beet sugar factory has been established at Raymond, and three good settlements have sprung up.

Beet Sugar Industry in Alberta.

The character of the soil and climate of Southern Alberta has for many years indicated the suitability of the district for the growth of sugar beets. It was not, however, until actual experiments in connection with their growth were made in the Raymond District, south of Lethbridge, that it became evident that both the soil and the climate were specially adapted to the growth of these roots. Following that experiment a beet sugar factory was erected at Raymond and has been in operation for four years. The beets raised in that district are of very exceptional quality both as to purity and percentage of saccharine matter, and the factory is turning out a large quantity of first-class sugar. Following these, experiments have been made at certain points adjacent to the main line of the Canadian Pacific Railway Company between Calgary and Medicine Hat in the growth of sugar beets and the results

obtained indicate that that district will produce beets of first-class quality. The sugar beet, while necessitating considerable labor in its cultivation, gives a first-class return to the farmers and it now seems certain that within a short time large areas in Southern Alberta will be devoted to the production of sugar beets and in the near future several large factories will be in operation producing sugar on an extensive scale.

Dairying.

There are now nineteen Government creameries in operation, of which number probably eight will be running all winter. Besides these, there are private creameries located at Didsbury, Knee Hill, Berrydale, Bowden, Red Deer, Pine Lake, Mayton, Neapolis, Content, Ponoka, Valley City, Highland Park, Lamerton, Leduc and Cardston. A number of others are under project.

The main creameries established on the co-operative principle and operated by the Provincial Department of Agriculture are situated at Calgary, Olds, Markerville, Red Deer, Evarts, Blackfalds, Lacombe, Earlville, Wetaskiwin, Clover Bar, Innisfail, Beaver Hills, Stony Plains, Martins, Ferry Bank, Rosenroll, Pine Lake, Stettler and Crossfields.

The main creameries are equipped with first-class cold storage rooms and other modern improvements. A regular semi-weekly refrigerator service is furnished by the Canadian Pacific Railway, which makes it practicable to ship perishable food products to the markets in the pink of condition.

Good prices are obtained for the output of butter, which finds a ready market principally in British Columbia, and for shipment to the Yukon territory, and the markets for creamery butter in China and Japan are, though limited, increasing satisfactorily, and shipments are going forward by nearly every steamer leaving Vancouver for the Orient.

Poultry Raising

Not enough attention is given to the business of raising fowls. Intent upon larger things, the Alberta farmer neglects the incidentals. The climate is wholly favorable and the business needs only close attention and ordinary judgment. There is a large field in Southern Alberta for the industrious poultry raiser. A few acres and a few hundred chickens will yield a good income.

With eggs at 25 to 50 cents per dozen and dressed poultry 15-22 cents per pound on the Calgary market, little need be said about this valuable side issue of the Southern Alberta farm. An enormous market exists in the Province of British Columbia for poultry products and this market is increasing every year. An egg gathering station is maintained at Calgary by the government, where the highest market price is paid for eggs, and from which periodical shipments are made to western points. No less than $367,950 worth of poultry and eggs were imported to Calgary by jobbers alone during 1905 for distribution in Alberta and British Columbia points. It only remains for the farmers of this district to go into the poultry business on a larger scale, in order to have this money circulated in Alberta. The climate is ideal for poultry raising and the market the best in Canada. Many a Southern Alberta farmer keeps his grocery bill square with the products from the poultry yard; and yet little attention is given to improving his breeds, or to housing or feeding their domestic feathered friends. A fine opening awaits those who will undertake poultry raising on a scientific basis, and put really fine fowls upon the market. Alberta needs enthusiastic poulterers as much as breeders of pure-bred horses, cattle and swine. One thing the poultry raiser there has, is a good market all the year round. There is money in the business when rightly managed, and comfort, cleanliness, a little grain and a good alfalfa field to range over, about comprise the requisites.

There is always a demand for chickens and eggs and many have found the business profitable, but whether it be on a large scale or not, the farmer's wife and daughters can always make good pin money with poultry.

Turkey raising has come to be an industry of importance. In parts of Southern Alberta, where range is good, thousands of these birds grow and fatten for market in the coast cities and thousands of dollars are brought into the country every year through this business alone. Where large areas of wheat stubble may be utilized for forage ground, the expense of putting turkeys upon the market is reduced to a minimum.

Minerals.

For years past gold in paying quantities has been found on the banks and bars of the North and South Saskatchewan and in the Pembina, Smoky, Macleod and Athabasca rivers. Veins of galena have been located, which are pronounced by experts to contain a large percentage of silver.

Vast areas are underlaid with rich deposits of anthracite, bituminous, semi-bituminous coal and lignite. The coal mines already discovered are of sufficient extent to supply Canada with fuel for centuries. Lignites are now mined at Medicine Hat, Cypress Hill, Red Deer, Otoskwan, Edmonton, Sturgeon River and Victoria, and are obtained at the pit's mouth at from 65c. to $2.50 per ton. The semi-bituminous is mined at Lethbridge (where $1,500,000 have been invested), Taber, Pot Hole, Milk River Ridge, Woodpecker, Crowfoot and Knee Hill Creek, and is obtained at from $1.50 to $3.00 per ton. The true bituminous is mined at Waterton River, Pincher Creek, on each of the South, Middle and North Branches of the Old Man River, on High River, Sheep Creek, Fish Creek, Bow River and Canmore, and fetches similar prices to the semi-bituminous. The most important anthracite deposit is near Banff, where at Bankhead the Canadian Pacific Railway Company is developing a mine of splendid quality. It is the only one now operating in Canada, and will supply the country from Winnipeg to Vancouver with a hard coal equal to that shipped from Pennsylvania.

System of Land Survey.

The land is divided into "townships" six miles square. Each township contains thirty-six "sections" of 640 acres, or one square mile each section, and these are again sub-divided into quarter-sections of 160 acres. A road allowance, one chain wide, is provided for between each section running north and south, and between every alternative section east and west.

The following is a plan of a township:

Township Diagram.
N.
SIX MILES SQUARE.

31 C.P.R.	32 Gov.	33 C.N.W. or C.P.R.	34 Gov.	35 C.P.R.	36 Gov.
30 Gov.	29 Schools	28 Gov.	27 C.P.R.	26 H.B.	25 C.N.W. or C.P.R.
19 C.P.R.	20 Gov.	21 C.N.W. or C.P.R.	22 Gov.	23 C.P.R.	24 Gov.
18 Gov.	17 C.P.R.	16 Gov.	15 C.P.R.	14 Gov.	13 C.N.W. or C.P.R.
7 C.P.R.	8 H.B.	9 C.N.W. or C.P.R.	10 Gov.	11 Schools	12 Gov.
6 Gov.	5 C.P.R.	4 Gov.	3 C.P.R.	2 Gov.	1 C.N.W. or C.P.R.

W ... E

Each Square is 640 acres, and a quarter section 160 acres.

A Section contains 640 acres, and forms one mile square.

S.

Government lands open for homestead (that is for free settlement).—Sections Nos. 2, 4, 6, 10. 12, 14, 16, 18, 20, 22, 24, 28, 30, 32, 34, 36.

Canadian Pacific Railway Lands for Sale.—Sections Nos. 1, 3, 5, 7, 9, 13, 15, 17, 19, 21, 23, 25, 27, 31, 33, 35.

School sections.—Sections Nos. 11 and 29 are reserved by Government for school purposes.

Hudson's Bay Company's Land for Sale.—Sections Nos. 8 and 26.

Free Homestead Regulations.

Any even-numbered section of Dominion Lands in Manitoba or the North-West Territories excepting 8 and 26, which has not been homesteaded, reserved to provide wood lots for settlers, or for other purposes, may be homesteaded upon by any person who is the sole head of a family. or any male over 18 years of age, to the extent of one-quarter section of 160 acres, more or less.

Entry must be made personally at the local land office for the District in which the land to be taken is situate, or with any Sub-Agent for the District. A fee of $10 is charged for homestead entry.

A settler who has been granted an entry for a homestead is required by the provisions of the Dominion Lands Act and the amendments thereto to perform the conditions connected therewith, under one of the following plans:—

(1) At least six months' residence upon and cultivation of the land in each year during the term of three years.

It is the practice of the department to require a settler to bring 15 acres under cultivation, but if he prefers he may substitute stock; and 20 head of cattle, to be actually his own property, with buildings for their accommodation, will be accepted instead of the cultivation.

·(2) If the father (or mother, if the father is deceased) of any person who is eligible to make a homestead entry under the provisions of the Act, resides upon a farm in the vicinity of the land entered for by such person as a homestead. the requirements of the Act as to residence prior to obtaining patent may be satisfied by such person residing with the father or mother.

(3) If a settler was entitled to and has obtained entry for a second homestead, the requirements of the Act as to residence prior to obtaining patent may be satisfied by residence upon the first homestead, if the second homestead is in the vicinity of the first homestead.

(4) If the settler has his permanent residence upon farming land owned by him in the vicinity of his homestead, the requirements of the Act as to residence may be satisfied by residence upon the said land.

NOTE.—The term "vicinity" used above is meant to indicate the same township or an adjoining or cornering township.

A settler who avails himself of the provisions of Clauses (2), (3) or (4) must cultivate 30 acres of his homestead, or substitute 20 head of stock, with buildings for their accommodation, and have besides 80 acres substantially fenced.

The privilege of a second entry is restricted by law to those settlers only who completed the duties upon their first homesteads to entitle them to patent on or before the 2nd June, 1889.

Application for patent should be at the end of the three years, before the Local Agent, Sub-Agent or the Homestead Inspector. Before making application for patent the settler must give six months' notice in writing to the Commissioner of Dominion Lands at Ottawa of his intention to do so.

Synopsis of Regulations

For Disposal of Minerals on Dominion Lands in Manitoba, Saskatchewan and Alberta.

Coal.

Coal lands may be purchased at $10.00 per acre for soft coal, and $20.00 for anthracite. Not more than 320 acres can be acquired by one individual or company.

Permits to mine coal for domestic purposes may be issued on application to the Agent of Dominion Lands for the district in which the lands are situated for an area not exceeding three acres, which area must previously have been staked out by planting a post at each corner. The frontage must not exceed three chains or the length ten chains. Rental $5.00 an acre per annum, and royalty 20 cents per ton for anthracite coal, 15 cents per ton for bituminous coal and 10 cents for lignite coal. Sworn returns of the quantity mined under a permit to be made monthly. No rental to be charged if the permittee is the owner of the surface.

Placer Mining and Dredging in the Rivers.

Placer mining claims generally are 100 feet square; entry fee $5.00, renewable yearly. On the North Saskatchewan River claims are either bar or bench, the former being 100 feet long and extending between high and low water mark. The latter include bar diggings, but extend back to the base of the hill or bank, but not exceeding 1,000 feet. Where steam power is used, claims 200 feet wide may be obtained.

A Free Miner may obtain only two leases of five miles each for a term of twenty years, renewable at the discretion of the Minister of the Interior.

The lessee's right is confined to the submerged bed or bars of the river below low water mark, and subject to the rights of all persons who have, or who may receive entries for bar diggings or bench claims, except on the Saskatchewan River, where the lessee may **dredge** to high water mark on each alternate leasehold.

The lessee shall have a dredge in operation within one season from the date of the lease for each five miles, but where a person or company has obtained more than one lease one dredge for each fifteen miles or fraction is sufficient. Rental $10.00 per annum for each mile of river leased. Royalty at the rate of two and a half per cent., collected on the output after it exceeds $10,000.00.

Timber on Dominion Lands

In Manitoba, the North-West Territories, and within the Railway Belt in the Province of British Columbia.

A license to cut timber can be acquired only at public competition. A rental of $5.00 per square mile is charged for all timber berths excepting those situated west of Yale in the Province of British Columbia, for which the rental is at the rate of 5 cents per acre per annum. In addition to the rental, dues at the following rates are charged:—Sawn lumber, 50 cents per thousand feet B.M.; railway ties, eight and nine feet long, 1½ to 1¾ cents each; shingle bolts, 25 cents a cord; all other products, 5 per cent. on the sales.

A license is issued as soon as a berth is granted, but in unsurveyed territory, no timber can be cut on the berth until the licensee has made a survey thereof.

Homesteaders having no timber of their own are entitled to a permit free of dues to cut the following quantities:—3,000 feet of building logs, not to exceed 12 inches at butt end, or 9.250 feet board measure. If the timber is cut from dry trees, 3,000 lineal feet of any diameter may be taken; 400 roof poles; 500 fence posts; 2,000 fence rails.

Homesteaders and all bona fide settlers whose farms may not have thereon a supply of timber, or who are not in possession of wood lots or other timbered lands, will be granted a free permit to take and cut dry timber for their own use on their farms, for fuel and fencing. A permit fee of 25 cents in each case is charged.

As the regulations governing the cutting of timber on Dominion lands are undergoing important changes, it would be advisable for settlers interested to communicate with the Crown timber agents in regard to dues, etc.

Grazing

In Manitoba, Saskatchewan and Alberta.

A settler in the vicinity of unoccupied hay lands may obtain a lease for an area thereof not exceeding forty acres. The term of the lease is five years and the rental twenty-five cents an acre per annum payable in advance. Leases for hay purposes of not more than 640 acres and not less than 160 acres of school lands may be issued upon payment in advance of the rental at the rate of twenty-five cents an acre per annum. Applications for permits to cut hay are made after the first day of January in each year to the agent of Dominion Lands in whose agency the land containing the hay is situated, and permits are issued on and after the first day of April following, upon payment of a fee of fifty cents and the dues hereinafter prescribed. If before the 1st of April, more

AN ALBERTA RANCHER'S HOME.

than one application is received for a permit covering the same tract of land, the agent, if he cannot arrange a division of the land to suit the applicants, may post a notice in his office calling for tenders for the purchase of the hay, and the permit is awarded to the person offering the highest cash bonus. No hay shall be cut prior to a date to be fixed each year by the Minister of the Interior. The dues chargeable for permits to actual settlers who require the hay for their own use are ten cents an acre or ten cents per ton, and 50 cents office fee, and to all other persons the rates are fifty cents an acre or fifty cents per ton, payable in advance.

Information for Settlers.

Newly arrived immigrants will receive at any Dominion Lands office in Manitoba, Saskatchewan or Alberta information as to the lands that are open for entry in that district, and from the officers in charge, free of expense, advice and assistance in securing lands to suit them. Full information respecting the land, timber, coal and mineral laws, as well as respecting Dominion Lands in the Railway belt in British Columbia, may be obtained on application to the Superintendent of Immigration, Department of the Interior, Ottawa; the Commissioner of Immigration, Winnipeg, Manitoba; the Deputy Commissioner of Agriculture, Regina, Sask. The Dominion Lands Agents in Manitoba, Saskatchewan or Alberta can furnish information only regarding land in their respective districts.

For disposal of the public lands by free-grant the Dominion has established the following agencies, at which all the business in relation to lands within the district of each must be transacted:

Government Land Offices.

(Figures are inclusive.)

Winnipeg District—Includes all surveyed townships, Nos. 1 to 25 north; ranges—all east of 1st meridian, and ranges 1 to 8 west; also townships 1 to 4, ranges 9 to 14, and townships 5 to 7, ranges 9 to 12 west. Agent, Winnipeg.

Yorkton District—Townships 15 to 20 inclusive, in range 23; townships 15 to 21 inclusive, ranges 24 and 25; townships 15 to 22 inclusive, range 26; townships 15 to 24 inclusive, range 27;

townships 15 to 26 inclusive, range 28; townships 17 to 26 inclusive, range 29; townships 17 to 38 inclusive, ranges 30 to 34, all west of 1st meridian; townships 19 to 38 inclusive, ranges 1 to 6; townships 22 to 38 inclusive, ranges 7 to 9, west of 2nd meridian; townships 24 to 38 inclusive, ranges 10 to 20, all west of 2nd meridian.

Brandon District—Townships 8 to 12 inclusive, ranges 9 to 12; townships 5 to 12 inclusive, ranges 13 to 14; townships 1 to 12 inclusive, ranges 15 and 16; townships 1 to 18 inclusive, ranges 17 to 22; townships 1 to 14 inclusive, ranges 23 to 28; townships 1 to 16 inclusive, ranges 29 to 34, all west 1st meridian.

Dauphin District—All townships lying to the north of the Brandon district and north of that part of the Yorkton district lying east of range 13, west of 2nd meridian and west of the Winnipeg district.

Alameda District—Townships 1 to 9, ranges 1 to 30, west 2nd meridian. Agent, Alameda.

Regina District—Townships 10 to 18, ranges 1, west of 2nd to 30 west of 3rd; townships 19 to 21, ranges 7 west of 2nd to 29 west of 3rd; townships 22 and 23, ranges 10 west of 2nd to 29 west of 3rd; townships 24 to 30, ranges 2 west of 2nd to 29 west of 3rd; townships 31 to 38, ranges 2 west of 2nd, to 10 west of 3rd. Agent, Regina.

Lethbridge District—Townships 1 to 18, ranges 1 to 24 west of the 4th meridian; townships 1 to 12, range 25 west of the 4th meridian to B. C. Agent, Lethbridge.

Calgary District—Townships 19 to 30, ranges 1 to 7, west 4th meridian; townships 19 to 34, ranges 8 to 24, west 4th meridian; townships 13 to 34, range 25, west of 4th meridian to B. C. Agent, Calgary.

Red Deer Sub-District—Townships 35 to 42, range 8, west 4th meridian to B. C. Agent, Red Deer.

Edmonton District—Townships north of and including township 43 from range 8, west of 4th meridian to British Columbia. Agent, Edmonton.

Battleford District—Townships north of and including township 31, range 11, west of 3rd meridian to 7 west of 4th meridian. Agent, Battleford.

Prince Albert District—Townships north of and including township 39, range 13, west of 2nd meridian to 10 west of 3rd meridian. Agent, Prince Albert.

At the offices in the districts, detailed maps will be found showing the exact homestead lands vacant.

Labor registers are kept at the Government Land and Immigration offices and may be made use of free of charge, by persons seeking employment as well as by farmers and others seeking help of any kind.

Railway Land Regulations.

The Canadian Pacific Railway Lands consist of odd-numbered sections along the Main Line and Branches, in the Lake Dauphin District in Manitoba and in Central Alberta and Saskatchewan. These are for sale at the various agencies of the Company in Manitoba, Saskatchewan and Alberta, at prices ranging from $8.00 to $25.00 per acre.

Maps showing the lands in detail have been prepared and will be sent free to applicants.

Terms of Payment.

If land (not exceeding 640 acres) is bought for actual personal settlement within one year, the aggregate amount of principal and interest is divided into ten instalments; the first to be paid at the time of purchase; one year's interest to be paid at the end of the first year; and the remainder of the instalments annually thereafter.

The following table shows the amount of the annual instalments on a quarter section of 160 acres at different prices:—

160 ACRES	CASH PAY'T	1ST YR'S INT.		
At $ 8.00 per acre..	$191.70..	$65.28	and nine instalments of	$160.00
9.00 "	.. 215.70..	73.46	" "	180.00
10.00 "	.. 239.70..	81.62	" "	200.00
11.00 "	.. 263.60..	89.78	"	220.00
12.00 "	.. 287.60..	97.96	"	240.00
13.00 "	.. 311.55..	106.10	"	260.00
14.00 "	.. 335.60..	114.32	"	280.00
15.00 "	.. 359.50..	122.44	"	300.00

Purchasers, who do not undertake the settlement conditions, are required to pay one-sixth of the purchase money down and the balance in five equal annual instalments with interest at six per cent.

Interest at six per cent. will be charged on overdue instalments.

General Conditions.

All sales are subject to the following general conditions:—

1. All improvements placed upon land purchased to be maintained thereon until final payment has been made.

2. All taxes and assessments lawfully imposed upon the land or improvements to be paid by the purchaser.

3. The Company reserves from sale, under these regulations, all mineral and coal lands, and lands containing timber in quantities, stone, slate and marble quarries, lands with water power thereon, and tracts for town sites and' railway purposes.

4. Mineral, coal and timber lands and quarries will be disposed of on moderate terms to persons giving satisfactory evidence of their intention and ability to utilize the same.

Liberal rates for settlers and their effects are granted by the Company over their railway.

Towns.

The Company offers for sale at its Land Office in Winnipeg lots in the various towns and villages along the Main Line and Branches.

The terms of payment for these lots are:—One-third cash, balance in six and twelve months, with interest at eight per cent.

For full information apply to

F. T. GRIFFIN,
Land Commissioner of C.P.R. Co., Winnipeg.

Information as to prices and terms of purchase of railway lands, may be obtained from all station agents along the Company's main line and branches. In no case, however, is a railway agent authorized to receive money in payment for lands. All payments must be remitted direct to the Land Commissioner at Winnipeg.

The Canada North-West Land Company.

This Company owns 550,000 acres of selected lands in Manitoba and Saskatchewan. These lands are on sale at the various land agencies of the Canadian Pacific Railway Co. For maps and further information application should be made to the office of the Land Company at Winnipeg.

British Columbia.

For descriptive pamphlet of British Columbia and particulars of lands, town lots, and timber areas for sale or lease by the Railway Company, in that province, write to J. S. Dennis, B.C. Land Commissioner, Calgary, Alta.

Stop-Over Privileges.

Intending settlers are given the privilege of stopping over at stations where they wish to inspect land. If stop-over is desired, application should be made to the Immigration Office of the Company at Winnipeg, in case the settler's ticket does not specifically provide for stop-over privileges.

Settlers' Effects.

Freight Regulations for their Carriage on the C. P. R.

1. These rates are subject to the general notices and conditions of carriage printed in the Company's form of Shipping Receipt, and will apply only on shipments consigned to actual settlers, and are entirely exclusive of cartage at stations where this service is performed by the Railway Company's Cartage Agents.

2. *Carloads* of Settlers' Effects, within the meaning of this tariff, may be made up of the following described property for the benefit of actual settlers, viz.: Live Stock, any number up to but not exceeding ten (10) head, all told, viz.: Cattle, calves, sheep, hogs, mules or horses; Household Goods and personal property (*second-hand*); Waggons, or other vehicles for personal use (*second-hand*); Farm Machinery, Implements and Tools (*all second-hand*); Softwood Lumber (Pine, Hemlock or Spruce—only), and Shingles, which must not exceed 2,000 feet in all, or the equivalent thereof; or in lieu of, not in addition to the lumber and shingles, a Portable House may be shipped; Seed Grain; small quantity of Trees or Shrubbery; small lot Live Poultry or pet animals; and sufficient feed for the live stock while on the journey. Settlers' Effects rates, however, will not apply on shipments of second-hand

Waggons, Buggies, Farm Machinery, Implements or Tools, unless accompanied by Household Goods; and will not apply on Automobiles, Hearses, Omnibuses, or similar articles, under any circumstances.

3 *Merchandise*, such as groceries, provisions, hardware, etc., also implements, machinery, vehicles, etc., if new, will not be regarded as Settlers' Effects, and, if shipped, must be charged the regular classified tariff rates. While the Canadian Pacific Railway is desirous of continuing to give liberal encouragement to settlers, both as to the variety of the effects which may be loaded in cars, and the low rates thereon, it is also the duty of the Company to protect the merchants of the North-West by preventing as far as possible, the loading of merchandise of a general character in cars with personal effects. *Agents, both at loading and delivering stations must personally satisfy themselves that contraband articles are not loaded, and see that actual weight is charged for when carloads exceed 24,000 lbs.*

4. *Top Loads will not be permitted.*—Agents must see that nothing is loaded on top of box or stock cars. This manner of loading is dangerous and is absolutely forbidden.

5. *Passes.*—One man will be passed free in charge of full carloads of settlers' effects when containing live stock, to feed, water and care for them in transit. Agents must fill out the usual live stock form of contract.

6. *Settlers' Effects*, to be entitled to carload rates, must consist of a carload from one point of shipment to one point of destination. Carload shipments will not be stopped in transit for completion or partial unloading.

7. *The minimum carload weight* of 24,000 lbs. is applicable only to cars not exceeding 36 feet in length; larger cars must not be used for this business. If the actual weight of the carload exceeds 24,000 lbs. the additional weight will be charged for at the carload rate.

8. *The minimum charge* for less than carload shipments will be 100 lbs. at regular first-class rates.

9. *Should a settler wish to ship more than ten head* of live stock, as per clause 2, agent will apply to his General Freight Agent for rate.

10. *Less than carload shipments* will be understood to mean only Household Goods (*second-hand*), Waggons, or other vehicles for

EXPERIMENTAL FARM, BRANDON, MAN.

personal use (*second-hand*), and second-hand Farm Machinery, Implements and Tools. Settlers' Effects rates, however, will not apply on shipments of second-hand Waggons, Buggies, Farm Machinery, Implements or Tools, unless accompanied by Household Goods; and *will not apply* on automobiles, hearses, omnibuses, or similar articles under any circumstances. *Less than carload lots must be plainly addressed.*

11. *Shipment of settlers' effects from connecting lines* will be charged from the Canadian Pacific junction point the settlers' effects rates from that point.

12. *Car Rental and Storage of Freight in Cars.*—Under this tariff, when freight is to be loaded by consignor, or unloaded by consignee, one dollar ($1.00) per car per day or fraction thereof, for delay beyond 48 hours in loading or unloading, will be added to the rates named herein, and constitute a part of the total charges to be collected by the carriers on the property.

General Information.

The question, " How much money is necessary? " is a difficult one to answer. It depends upon circumstances. Very many men have gone into Western Canada without any capital and have prospered.

Generally, it may be said that a settler commencing on a half section will need four good horses, which will cost from $600.00 to $700.00; harness, $65; one breaking plow, or a combination plow, $27.00; one set of harrows, $25.00; one waggon, $75.00 to $80.00, if new, and if second-hand, $45.00; one seeder, $85.00; one mower and rake, $95.00; two cows, $80.00; provisions for himself and family, about $200.00. A habitable house, 18 by 20, one-story high can be built for $200.00. It will, of course, have to be added to for the winter. He should also have one brood sow, $15.00; forty or fifty hens, $15.00. With this outfit he will be in a position to commence comfortably, and will be much better off than most of the early settlers were twenty years ago. Some of those who had scarcely any capital are now in independent circumstances. The outfit mentioned will cost about $1,500. When the first crop is ready for harvest a binder will be required, but it can be paid for out of the proceeds of the crop.

A young man entering for his homestead, say, in May or June, for which he pays the Government agent $10.00, can with prac-

tically no capital, start for himself. If he is willing to work and understands horses and general farming he can earn from $160.00 to $180.00 for the summer season. He can employ a neighbor to break ten acres on his land, and in November can put up a cheap house at, say, from $40.00 to $50.00 and live on his land during the winter months, when the wages are not as high as in the summer season, thus complying with his settlement duties. He can do this for three years, and at the end of that time will be entitled to a recommendation for his patent. He will then be in a position to borrow sufficient capital on the security of his homestead to purchase the outfit necessary to enable him to devote his whole time to the cultivation and improvement of his farm. A settler with a family old enough to work can follow the same course. To enable a settler with a young family to start comfortably on a quarter sec· tion of free grant land, he should have at least $500.00 to $1,000.00 capital.

Customs Regulations.

Settlers' Effects.

The following is an extract from the Customs Tariff of Canada, specifying the articles that may be admitted free as Settlers' Effects:

"455. Settlers' Effects, viz: Wearing apparel, books, usual and reasonable household furniture and other household effects; instruments and tools of trade, occupation or employment, guns, musical instruments, domestic sewing machines, typewriters, bicycles, carts, wagons and other highway vehicles, agricultural implements and live stock for the farm, not to include live stock or articles for sale, or for use as a contractor's outfit, nor vehicles nor implements moved by mechanical power, nor machinery for use in any manufacturing establishment; all the foregoing if actually owned abroad by the settler for at least six months before his removal to Canada, and subject to regulations by the Minister of Customs: Provided, that any dutiable articles entered as settlers' effects may not be so entered unless brought by the settler on his first arrival, and shall not be sold or otherwise disposed of without payment of duty until after twelve months actual use in Canada."

"A settler may bring into Canada free of duty live stock *for the farm,* on the following basis if he has actually owned such live stock abroad for at least six months before his removal to Canada, and has brought them into Canada within one year after his first arrival, viz.:

If horses only are brought in, ... 16 allowed.
" cattle   "   "   16   "
" sheep   "   "   160   "
" swine   "   "   160

If horses, cattle, sheep and swine are brought in together or)art of each, the same proportions as above are to be observed.

Duty is to be paid on the live stock *in excess* of the number above provided for.

For custom's entry purposes a mare with colt under six months old is to be reckoned as one animal; a cow with a calf under six months old is also to be reckoned as one animal.

Cattle and other live stock imported into Canada are subject to Quarantine Regulations."

The settler will be required to fill up a form (which will be supplied him by the customs officer on application) describing the kind and value, etc., of the goods and articles he wishes to be allowed to bring in free of duty. He will also be required to take the following oath:

I, ..., do hereby solemnly make oath and say that all the Goods and Articles hereinbefore mentioned are, to the best of my knowledge and belief, entitled to Free Entry as Settlers' Effects, under the tariff of duties of Customs now in force, and that all of them have been owned and in actual use by myself for at least six months before removal to Canada; and that none of the goods or articles shown in this entry have been imported as merchandise or for use in any manufacturing establishment, or as a contractor's outfit, or for sale, and that I intend becoming a permanent settler within the Dominion of Canada, and that the "Live Stock" enumerated and described in the entry hereunto attached is intended for my own use on the farm which I am about to occupy (or cultivate), and not for sale or speculative purposes, nor for the use of any other person or persons."

Cattle Quarantine.

Settlers' cattle should when possible be accompanied by a certificate signed by a veterinarian of the United States Bureau of Animal Industry or a State Veterinarian stating that no contagious disease of cattle (excepting Tuberculosis and Actinomycosis) has existed in the district whence they have come, during the period of six months immediately preceding the date of their removal therefrom.

Cattle for breeding purposes, or milk production, six months old or over, must be accompanied by a satisfactory Tuberculin chart signed by a veterinarian of the United States Bureau of Animal Industry, or will be quarantined until tested.

Cattle re-acting to the test will be returned to the United States or slaughtered without compensation.

Swine.

All swine are quarantined for thirty days and will not be admitted unless accompanied by a certificate signed by a veterinarian of the United States Bureau of Animal Industry that neither Swine Plague nor Hog Cholera has existed within a radius of five miles of the premises in which they have been kept for a period of six months immediately preceding that date of shipment.

Swine found to be suffering from contagious diseases will be subject to slaughter without compensation.

Sheep.

Sheep are admitted subject to inspection at port of entry and should be accompanied by a certificate signed by a veterinarian of the United States Bureau of Animal Industry or by a State Veterinarian, stating that no contagious disease of sheep has existed in the district whence they have come, during the period of six months immediately preceding the date of their removal therefrom.

These regulations also apply to goats.

Immigration Statistics.

The number of immigrants into Manitoba, Saskatchewan and Alberta has been increasing steadily for the last few years, a marked feature being the number of settlers from Great Britain, Ireland and the United States. The official figures for the years, July 1, 1900, to June 30, 1905, are:

	British.	American.	Continental.	Total.
1900-01	11,810	17,987	19,352	49,149
1901-02	17,259	26,388	23,732	67,379
1902-03	41,792	49,473	37,099	128,364
1903-04	50,374	45,171	34,785	130,330
1904-05	65,359	43,652	37,255	146,266
1905-06	86,796	57,796	44,472	189,064

Of the 86,796 British immigrants of 1905-06, 65,932 were from England and Wales, 15,846 from Scotland, and 5,018 from Ireland.

Educational Facilities.

A school district in Saskatchewan and Alberta comprises an area of not more than twenty-five square miles, and must contain not less than four resident ratepayers, and twelve children between the ages of five and sixteen inclusive. Any three qualified rate-payers may petition for the formation of a school district, and upon its proclamation the ratepayers therein may establish a school and elect trustees to manage it. These trustees have power to erect and equip buildings, engage certificated teachers, levy taxes and perform such other acts as may be necessary for the proper conduct of a school.

The classes of schools established are denominated Public and Separate. The minority of the ratepayers in any organized public district, whether Protestant or Roman Catholic, may establish a separate school therein, and in such case the ratepayers establishing such Protestant or Roman Catholic separate school shall be liable only to assessment of such rates as they impose upon themselves in respect thereof. Schools are maintained by Legislative grants and by local taxation.

The school year for which grants may be paid does not exceed 210 teaching days. The Legislative grant is paid as follows:— To rural districts: for each day a school (with an average attendance of at least six pupils) is open, $1.20, and for each additional day over 160 days, 40 cents per day, provided that these additional days shall not exceed 50 in number. For a teacher holding a first-class certificate, 10 cents per day for each day such teacher is actually employed in the school; to each school, according to its percentage of attendance, a sum not exceeding 25 cents per day. The grants to village and town schools are similar to the above, except that the grant made for each day a school is open is 90 cents. High Schools receive a special grant of $75.00 per term for each department. Salaries average about $50.00 per month. In the programme of studies provision is made for teaching the elementary subjects and such additional subjects as are required for teachers' examinations and university matriculation. The last half-hour of school may be devoted to such religious instruction as the trustees may determine.

In the Province of Saskatchewan for the year ended December 31st, 1905, there were 726 schools in operation, with 821 teachers engaged. The estimated expenditure for fourteen months ending the 28th of February, 1907, is $250,000.

In Alberta the number of schools in existence January 1, 1905, was 484; January 1, 1906, 602; the pupils enrolled January 1, 1906, were 24,254; the teachers employed January 1, 1906, were 628; the total grants paid the schools in 1905 were $120,723.61, while the estimate for 1906 provided $165,000 for the same purpose.

Harvest Hands.

So bountiful are the harvests that it is now necessary to bring in from Eastern Canada and elsewhere, from 20,000 to 25,000 farm laborers to work in the wheat fields. These earn good wages and many remain and become actual settlers themselves. Cheap rates are offered to points in Manitoba, Saskatchewan and Alberta, and special trains run for their accommodation. Those who go are given certificates and when they have them properly filled out and signed by the employer to the effect that the holder has done one month's farm work he is returned to his home at a low fare. Agents meet each train en route with maps of the province.

Milling in Western Canada.

Wheat-flour milling is the most important manufacturing interest in Western Canada, and the product not only finds a ready market throughout the whole Dominion, but is exported to Great Britain, Newfoundland, South Africa, China, Japan and Australia. Mills are located at different points throughout the country, one at Keewatin, having a daily capacity of 4,000 barrels, and another at Winnipeg of 3.800 barrels; another mill has recently been completed at St. Boniface with a capacity of 4,000 barrels. Other mills are in course of erection. There are also oatmeal mills in operation at Winnipeg, Portage la Prairie, Brandon, Pilot Mound and Strathcona.

Mills and Elevators.

The grain elevator system throughout Western Canada is perfect, the facilities now existing being sufficient to handle, if necessary, 125,000,000 bushels of grain in less than six months' time. The rapid increase in the storage capacity is one of the best indications of the continuous development of the country's agricultural re-

sources. In 1891 the total storage capacity was 7,628,000 bushels; in 1901, 18,879,352; in 1902, 23,099,000; in 1903, 30,356,400. For the year ending June 30, 1904, the total storage capacity was 41,186,000.

The Canadian Pacific Railway terminal elevators at Fort William have a capacity of 8,493,400 bushels; "D" containing 3,000,000. The seven terminal elevators have a capacity of 15,000,000 bushels; the 912 public country elevators and 64 warehouses, 27,214,000 bushels, the average capacity of these being about 28,000 bushels.

The following is a summary:—

Canadian Pacific Railway:	Bushels.
Ontario	12,217,000
Manitoba	14,078,500
Saskatchewan and Alberta	8,614,000
	34,909,500
Canadian Northern Railway:	
Ontario	6,467,000
Manitoba	4,780,400
	11,247,400
Grand total	46,156,900

CANADIAN PACIFIC GRAIN ELEVATORS, FORT WILLIAM.

How to Reach the Canadian West.

Colonists having arrived in Canada at Quebec or Montreal in summer, or Halifax or St. John, N.B., in winter, travel to new homes in Ontario, Manitoba, Saskatchewan and Alberta or British Columbia by the Canadian Pacific Railway direct. Settlers from the Eastern States travel via Montreal, Prescott or Brockville, and thence by the Canadian Pacific; but if from Southern and Western New York and Pennsylvania via Niagara Falls, Hamilton, Toronto and North Bay, thence Canadian Pacific Railway; those from the Middle States either by Toronto, or St. Paul and Emerson, Man., or Minneapolis and Portal via St. Paul; from the Middle Western States by Portal (or, if for Manitoba, by Emerson, Man.); from the Pacific Coast States by Vancouver or Sumas, or through the West Kootenay mining regions and Canadian Pacific from Rossland and Nelson.

On the same fast transcontinental trains with the first-class cars are colonist cars, which are convertible into sleeping cars at night, having upper and lower berths constructed on the same principles as those of first-class sleeping cars, and equally comfortable as to ventilation, etc. No extra charge is made for this sleeping accommodation. Second-class passengers, however, must provide their own bedding. If they do not bring it with them, a complete outfit of mattress, pillow, blanket and curtains will be supplied by the agent of the company at the point of starting at a cost of $2.50—ten shillings.

The trains stop at stations, where meals are served in refreshment rooms, and where hot coffee and tea and well-cooked food may be bought at reasonable prices.

All trains are met upon arrival at Winnipeg or before reaching that city, by the agents of the Government and Canadian Pacific Railway Company, who give colonists all the information and advice they require in regard to their new home.

Special round-trip explorers' tickets can be obtained by newly arrived settlers at the Company's Land Office at Winnipeg, the full price of which will be applied on account of purchase money if the holder buys from the Company 160 acres or more.

⚜ New Ontario ⚜

The Rainy River District.

Before reaching Manitoba, the traveller on the Canadian Pacific Railway passes a fertile belt estimated to contain about 600,000 acres of good agricultural land, which lies in the valley of the Rainy River. Fort Frances, the principal town on Rainy River, has a saw mill and several flourishing stores and industries; its population is about 1,400. The region is reached during the season ot navigation by steamer from Kenora, on the main line of the Canadian Pacific Railway. All the cereal and grass crops common to Ontario grow there, and garden crops flourish exceedingly. The country is well wooded with pine, oak, elm, ash, basswood, soft maple, poplar, birch, balsam, spruce, cedar and tamarac. Lumbering operations are extensively carried on and there are well-equipped sawmills on Rainy River, Rainy Lake and at Kenora. There are gold mines now being worked on the Lake of the Woods, Rainy Lake and Seine River, and elsewhere mining operations are being carried on. The mining and lumbering industries combined afford the settler the best markets for his produce at prices considerably higher than can be secured in Eastern Ontario. The lands are owned and administered by the Government of Ontario (Department of Crown Lands, Toronto), and are open for settlement in 160 acre lots free, with conditions of residence, cultivation of ten acres for every 100 located and erection of buildings.

Any person may explore Crown Lands for minerals and mining lands may be purchased outright or leased at rates fixed by the Mines Act. The minimum area of a location is forty acres. Prices range from $2 to $3.50 per acre, the higher prices for lands in surveyed territory and within six miles of a railway. The rental charge is at the rate of $1 per acre for the first year and from 30 cents to 15 cents per acre for subsequent years, according to distance from a line of railway and whether the land is situated in surveyed or unsurveyed territory; but the leasehold may be converted into freehold at the option of the tenant, at any time during the term of lease, in which case the first year's rent is allowed on the purchase money. At the expiration of ten years, if all conditions have been complied with, the lessee is entitled to a patent without further cost and free from all working conditions. A royalty of not more than three per cent. is reserved, based on the value of the ore, less cost of mining and subsequent treatment for the market, but not to be imposed until seven years after the date of the patent or lease.

The Wabigoon Country, Rainy River District.

North of the country, bordering on the Rainy River, described above, and directly on the line of the Canadian Pacific Railway, is a section to which the Wabigoon River gives its name. The land was thrown open for settlement in the spring of 1896, and has been rapidly taken up. The settlers consist almost entirely of a good class of Ontario farmers, and the development of the country is being pushed forward with energy. The little towns of Dryden, on the Canadian Pacific Railway, and Wabigoon are the business centres of the district and have steamboat communication via Lake Wabigoon with the mines and fishing and shooting regions in the vicinity.

The land is not free grant, but is sold to actual settlers only, at fifty cents per acre (subject to settlement regulations), one-fourth down and the balance in annual instalments. How much agricultural land there may be available at this point has not as yet been definitely ascertained, but it is estimated at two million acres. The land, although not a prairie, is easily cleared. Some stretches are entirely destitute of timber, having been swept by forest fires, and require only a little under brushing before the plough starts to work. Elsewhere the growth is light, and may be cleared with much less labor than is required in heavily timbered countries. At the same time, sufficient large timber for building purposes is to be found here and there, so that it will be seen, the advantages of a prairie and of a timbered country are here combined to a large extent. The country is well watered, and possesses a good soil and a good climate. It is adapted to mixed farming, but particularly to dairying and stock-raising.

Thunder Bay District.

In the vicinity of Port Arthur and Fort William, two important points on Thunder Bay, Lake Superior, there are a number of townships of good agricultural land similar to that of the Rainy River Valley, besides a country rich in gold, silver and iron. The land here is given as free grants, subject to settlement duties, and is attracting a good many settlers from the United States. The principal movement of settlers to this district is occurring in the Slate River Valley, the White Fish Valley, south and south-west of the two towns, and the township of Dorion, east of Port Arthur, on the main line of the Canadian Pacific Railway.

The Dominion Government maintains a Settlers' Home at Port Arthur, and an agent of the Department of the Interior, Mr. R. A. Burriss, is located at this point.

Algoma and Nipissing.

At Sault Ste. Marie, at the junction of Lakes Superior and Huron, another stretch of country adapted for settlement is reached. The country to the north of Lake Huron is known as the Algoma District, and includes St. Joseph and Great Manitoulin Islands. It contains a large proportion of fertile land, but sparsely settled, yet considerable development has already taken place. A fine stretch of agricultural land containing at least 200,000 acres has recently been discovered north and west of Chapleau Station on the main line of the Canadian Pacific Railway. It surrounds Trout Lake and is due east of the Michipicoten iron district. Already there are thriving settlements not only on the large islands of St. Joseph and Manitoulin, but here and there along the north shore also, from Goulais Bay, about twenty or twenty-five miles northeast of Sault Ste. Marie, to the valley of the French River, some two hundred miles to the eastward, and elsewhere.

Sault Ste. Marie is the central point of the Algoma District The town is easily reached either from older Ontario or the United

States. It is situated on the "Soo line," a branch of the Canadian
Pacific, connecting with St. Paul and Minneapolis in the west, and
Boston in the east. In addition, several steamship lines call there.
Large pulp and paper mills, iron smelters and other industries are
making Sault Ste. Marie an important industrial centre. The Algoma Central Railway, now running from Sault Ste. Marie north-
wards, will aid materially in the development of the country.

The land, while good, is not in an unbroken continuous
stretch, as is the case of the southern portion of Ontario. It may
be described as an undulating plateau some 600 or 1,000 feet above
the sea level, covered for the most part with a vigorous growth of
forest. Between the ridges and protected by them, stretches of
arable land, often unbroken for thousands of acres, wind in and
out. As a dairy, stock and sheep-raising country, it has all the advantages of cheap land, good transportation facilities, rich soil,
good water and cheap building material, while its climate is unexcelled for the production of vigorous stock and vigorous men.

The Algoma and Nipissing districts are known to be rich in a
variety of minerals. Gold, silver, copper and iron have been discovered to the north of Lake Huron, and this region contains the
most extensive nickel deposits in the world, which are now being
worked in the vicinity of Sudbury. It also contains the richest
silver and cobalt mines in the world. New discoveries of mineral
are being made daily. The farmer has the best of markets at his
door for all produce. Cleared farms can be bought at very reasonable rates, with buildings ready for occupation.

The Timiskaming Country.

Another agricultural section in the northern part of the province is the Timiskaming country, which borders on Lake Timiskaming, a broadening of the Ottawa River. It is in the Nipissing
District, and about two hundred and fifty miles west of Toronto
in a direct line. It is reached from Mattawa on the Canadian
Pacific Railway, partly by railway and afterwards by steamboat
on Lake Timiskaming.

The whole country is overlaid by a rich, alluvial soil, level in
character, and equal in fertility to any in the province. The land
is thickly timbered with a somewhat small growth, but for the
most part may be cleared without excessive labor; 600,000 acres

nave been placed on the market at fifty cents per acre. The country is attracting quite a number of settlers from the older parts of
Ontario, and is well worthy of attention. The region of the Upper
Ottawa is to-day one of the most important lumbering districts in
Canada, and affords the settler an excellent market for the products of the farm, while the market for pulp wood, with which the
country is covered, furnishes the new settler a source of income.

A pamphlet giving full particulars regarding New Ontario may
be obtained on application to the Department of Crown Lands,
Toronto, Ontario.

A HOME IN WESTERN CANADA.

SETTLERS REPORTS,

HOW MEN AND WOMEN FROM VARIOUS COUNTRIES HAVE SUCCEEDED IN WESTERN CANADA.

A Glasgow Man's Success.

Moose Jaw, Nov. 20th, 1906.

"I came from Glasgow, Scotland, in 1887. The cash and effects at that time of myself and brother, who accompanied me, were less than $1,000. I homesteaded a quarter section and pre-empted another quarter. My brother Thomas homesteaded a quarter section. We worked together. Since that time we have bought three quarter sections and now have a section and a half. We have a crop of 300 acres each year.

"We have 24 horses, 80 cattle, a number of hogs and poultry. We have always made considerable butter each year. Our profits have been put into improvements from year to year. We have 200 acres fenced, have a new barn, which cost $3,000 and a house which cost $2,700. with other out-buildings. We would not sell our land at $30 per acre. The total value of land, stock and implements is easily $40,000

"I was a bank clerk in Glasgow, and my brother Thomas was a lawyer. We knew nothing about farming—never handled horses before coming here. Many dollars were lost the first years of our farming in experimenting, until we learned how to farm and how to care for our live stock."

Wm. Bennie.

How $200 Became $20,000.

Jas. McMillan, located on Sec. 30, Tp. 15, R. 25, W. of 2nd meridian, about eight miles south-east of Moose Jaw, says:

"I came here from London, Ont., in 1883. My cash on hand and effects did not exceed $200. I homesteaded a quarter section, and have since bought another quarter section.

"For the last five years I have cropped from 150 to 200 acres. I keep stock as well as raise grain. I have, at present, 20 horses, 60 cattle, a number of hogs and poultry. I do considerable dairying, and always find a good cash market for everything I have to sell at Moose Jaw.

"I have a fairly good house, good stables and granary. If $20,000 were offered me, cash, to-day for farm, stock and implements, I would not take it."

ON THE BANKS OF THE SASKATCHEWAN RIVER, NEAR EDMONTON.

Well Satisfied with Saskatchewan

NEUDORF, P.O., Sask., Sept. 29th, 1906.

I arrived here in 1894 and selected my homestead, being at that time 22 miles north of Grenfell. I had $1,100.00 in cash, and was, therefore, in a position to purchase the necessary implements and cattle. Having a large family—eight children—and the first few years not being very good ones, I made only slow progress. I was, however, able to purchase an additional 160 acres in 1899 and another 160 last year, at $10.00 per acre. On these lands I have cultivated 220˙acres and will thresh this fall about 1,800 bushels of oats and 2,600 bushels of wheat. I have 11 horses, 28 cattle, all implements, pigs and domestic fowl. My implements, horses and cattle are all paid for. I have married four of my children, ·
giving each of them the necessary cattle, wagons and plows, and have assisted them to establish themselves on their lands. I am very well satisfied with my new home, and only wish that there was more land in my vicinity to buy for my younger children.

JOHANN HUBENIG,
From Czernowitz, Bukowina, Austria.

A Worker in this Land can Reach Independence.

STOCKHOLM, Oct. 2nd, 1906.

Through reports from successful countrymen living in the Esterhazy Colony, I emigrated with my family from Hungary in 1901. As I could not find any homestead in the Esterhazy Colony, I went eight miles further west, and located on my present homestead, half a mile from Stockholm Station, which was placed here at the building of the Pheasant Hills Branch, two years ago.

My two sons have homesteaded still further west in the neighborhood of Lipton. The capital I brought with me was $200.00. I have cultivated 70 acres of land, and expect to thresh 1,200 bushels of wheat. I have two horses, two teams of oxen and twenty-two head of cattle. I have all my machinery paid for except my binder, and have good buildings on my farm, which I would not sell for less than $30.00 per acre.

I am very well satisfied with my success, which testifies to the fact that any one who is willing and able to work on land in this country, can reach an independent future.

JOHN SIVAK,
From Szabolco Komitat, Hungary.

Glad He Came.

STOCKHOLM, Sask., Oct. 1st, 1906.

Looking back over the 17 years I have spent in Canada, I cannot but feel really satisfied with my present position. On my arrival here I had very little money, only sufficient for my homestead entry fee and one cow. By exchanging work with my neighbors, I acquired an ox-team, and gradually the most necessary implements. As I am a good carpenter I was employed in building houses within our colony during the first three years, and could thereby care for my large family and gradually put my land under cultivation. The long journeys to our nearest town, Whitewood, made the first years rather difficult, but by sticking to it we were well rewarded by being prepared to take the greatest advantage of the new railway, which, after a long wait, was constructed six miles north of my farm.

I and my son own now 480 acres of land with a good living house, necessary stables and barns, 26 horses, 80 head of cattle, necessary implements, in short, all that is necessary on a well conducted farm. This year we have threshed 3,500 bushels of wheat and 1,000 bushels of oats and have a good yield of potatoes. On an average we have threshed 3,000 bushels of wheat the last five years. On account of advanced years I am leaving the care of the farm to my oldest son. In conclusion, I may say that I am very well satisfied with the result of my labors here.

A. P. SJOSTROM,
From Dorothea, Sweden.

A Report of Progress.

EDMONTON, Alberta, 29th Oct., 1906.

I and my whole family are well satisfied with the results from our labors since we came here in the spring of 1899—now seven and a half years ago. We have been carrying on a mixed farming business. We came here with about $6,000 all told. We bought 500 acres of land at a little less than $6 per acre. We have now 15 horses, 40 head of cattle, 145 hogs, and plenty of implements and machinery to run the farm, a good log house with seven rooms, a log barn, 28x56 feet, two small granaries, and barley and oat bins in the barn, a small horse stable for colts and breeding mares, 14 x 24 feet, wood house, 8x6 feet, smoke house, 10x10 feet, and a hen house, 14x18 feet; besides a considerable amount of cheap stabling and sheds for stock cattle and for hogs. Our horses and cattle are good stock; the cattle are about one-third pure-bred Shorthorn; horses are Morgan and Cleveland bay, of good type and quality, and the hogs are Poland China and Berkshire, and three pure-bred Berkshire sows. Have had plenty to eat and· wear, good health, and good schools for our children to attend, our oldest son attending college in Edmonton. Have had good crops. Our wheat was frosted one year, but made 60c. per bushel by feeding to hogs; it was also damaged by hail once, but we again fed that to advantage to hogs and cattle. We have 164½ acres broken, 37 of which is in Timothy, the rest being in pasture and timber. The land is all fenced, and field fenced from pasture land. We live three miles from Edmonton.

D. W. WARNER.

The Bryce Farm.

One of Western Canada's farmer magnates, W. H. Bryce, of Arcola, Sask., has on hand an enterprise for the advancement of his large stock-raising interests. Mr. Bryce, who is a cousin of Rev. Dr. Bryce, the historian of Manitoba, was one of the pioneer farmers in the Arcola district, where he began with small capital in 1882, before the days of railway communication in that part of the country. He has now a magnificent farm of 3,000 acres, and an elegant home eight miles from Arcola. His family are enjoying at "Doune Lodge" refinements which popular fallacy is accustomed to associate only with the city life, and which many sons and daughters of the soil foolishly leave home to seek, but never find.

Mr. Bryce has unbounded faith in Western Canada and in farming, using this word to include both stock raising and grain growing.

He is noted among Canadian horsemen for his importation of Clydesdales. Last year his importations cost some $20,000, and his success in the horse raising industry was fully attested at the Winnipeg exhibition by the number of prizes won by his exhibits. His first prize horse, "Perpetual Motion," was a magnificent animal. Mr. Bryce believes that in the next ten years there will be a considerable demand for first class stock, and he is taking timely steps to prepare to aid in supplying this demand. He considers the climate and soil here well adapted to horse raising and especially favorable for developing good feet and bone. He also gives attention to cattle raising and has forty head, twenty head of grade stock and twenty head of registered. Shorthorns are his favorites in cattle, and each year bring him prizes and medals at agricultural and stock shows.

His is one of the finest farm ranches in Canada, the barn alone is worth $10,000. Mr. Bryce's original capital of $500 has worked marvels, for, to-day, the original "homestead" has grown to 3,000 acres, with a cattle run in Moose Mountain of 640 acres more. The foundation of Mr. Bryce's fortune was laid when he bought 1,000 acres of Canadian Pacific Railway lands at $2.50 an acre— land which to-day is rated at $30 an acre. In 1906 10,000 bushels of wheat and 5,000 of oats were raised. He is fond of experimenting, and this year has seeded 160 acres with an improved variety of English clover, and is testing the capabilities of an automobile thresher.

GROWN ON THE BRYCE FARM.

Made the Hit of His Life.

KILLARNEY, Manitoba, August 18th, 1906.

So many people coming to this country, wonder how the old-timers got on. I think they should be told that it was by hard work and sticking to the job. My own experience proves this.

I arrived here in 1882 with $75 cash. I came from Iowa, where ague got into my bones, and, hearing about Western Canada I struck out for the open plains. The beginning was uphill work, but the ague left me, and I felt well and was able to work, and work it was, for I had to haul my grain to Brandon, 60 miles away, and was glad to get 45 cents a bushel for it. Since then I have sold wheat as high as $1.30, and the biggest yield I ever got was 50 bushels to the acre, but the average yield, year in year out, gives 25 bushels to the acre.

My $75 proved a good investment in Western Canada, for to-day, standing on my 200 acres (under cultivation) with 320 acres rented ground for pasturage and hay lands, I am an independent man; horses, cattle, swine, poultry, etc., implements, and here on "Sunny Heights" I live with my family in comfort and peace.

I consider I made the hit of my life the day I started out for Western Canada with my little capital of $75. I came against the advice of my own family, but to-day they admit they wish they had struck out for the land where a poor man can succeed on less than one hundred dollars.

FRED. FINKBEINER,
Killarney, Manitoba, Box 505.

SHIPPING SHEEP AT MAPLE CREEK, SASKATCHEWAN.

Percy's Place Not for Sale.

ARCOLA, Saskatchewan.

Perhaps a man who has seen considerable of farm life both in Canada and the United States may be allowed to give an opinion on Western Canada as a productive country. I came here with a single shilling in my pocket, and claim to have "made good" on that amount. True, I had always worked on a farm, and I had the asset of experience, but against that I had again the liability of a new country, far away market and no railway. Since then

I have seen the Canadian Pacific Railway spread out. like the fingers on a hand all around me; markets spring up at the door; a rush of settlers following after, and to-day you can't find an unbroken acre betwixt my place and Arcola town on the C. P. R.

Five years ago I took a trip to Kansas, saw farming operations carried on there; spent a winter and—well, here I am back in Canada West, and mighty glad to call it home!

I consider there is no country under the shining sun to equal Western Canada. Good prices for everything that grows; everything will grow that you can stick in the ground, and as for health, my boy, who walks two-and-a-half miles to school, winter and. summer, hasn't missed a day since he began. That, I think, speaks for the health of the place as well as the weather. Winter is a moderate season, my horses running. out all winter in the hills. They tell me my place would bring in fifty dollars an acre to-day if I sold, but "Percy's place" isn't for sale!

THOS. LEES PERCY.

Sept. 19, 1906.   (Percy Dist.)

A Story of Success

MUNICIPALITY OF DALY,
BRANDON, Manitoba, October 2, 1906.

It may interest you to know how a Huron County man, who came to the Canadian West on borrowed capital, got on; 1881 found me a squatter on the present place, some nine miles from the city of Brandon, and it may please you to know that the land I purchased from you at $4 an acre to-day you couldn't get back at $40, supposing you were aching to buy. "Sunnyside," my place, has at the time of writing (a busy time it is) 24 stacks of grain representing the summer's yield in grain. I have 1,160 acres of land; 600 under cultivation; 100 head of stock, 17 horses, 35 swine, and poultry that refuse to be counted. My milk cows are thorough-bred, and each spring I market, say, $1,000 worth of beef on foot.

This year I will market 8,000 bushels of wheat, 5,000 bushels of barley, and almost double that amount in oats, while my wild hay lands (160 acres) will give in returns $7 per ton. I put $2,000 this year in the ground, counting the expenses of plowing, seeding, hired help to garner my crop, and the last penny paid to the threshers, and will reap $6,000 clear cash from the venture. This gives me $4,000 profit, and if my bones ache a bit from the effort to gain this, why that's all right, for without boasting, I think I may say "Sunnyside" can hold its own against any farm anywhere!

As I said, I began life on this place in 1881 with a single pair of oxen, and I owed the money that went to buy them; but I've squared that debt many a day ago, and I guess Tom Clark blesses the day he hit Manitoba.

In the 25 years I've farmed here I never had a failure. I've suffered from want of farm help—the great drawback of the West ,—and although I'd pay $300 a y ar round, and give board to a good man (harvest times paying $2 a day), I can't get one. Every man that comes to the country "moves on after a season and buys an acre for himself; and I don't blame him, for a man should be his own master in the great west .

I am secretary-treasurer of the Hunter School District, and while I am telling you of my own success I might mention the names of a dozen neighbors who've done just as well. From where I sit I can look out on the homes of Venning, Gray and Hunter, all of them wealthy men, and they, like myself, made it on the farm.

THOMAS CLARK.

Where a Young Man Gets a Chance.

PORTAGE PLAINS, MANITOBA,
Sept. 1st, 1906.

As a Scottish immigrant, who has lived in the country since 1873, I may, after twenty-eight years, be able to say whether I believe Western Canada makes good its promises to the settler.

Western Canada, in my opinion, is the best colony in the Empire for Scottish people to come to. The conditions of the climate, soil, customs of the people—the latter congenial to a high degree —suits our countrymen and women. Scotch people need feel no hesitation in coming out to Canada West; every kindness and every friendly help is given; and my own experience may illustrate this.

When I went home to Scotland, a delegate farmer, speaking at a public meeting in Cheshire, one of my audience rose up and asked, "What is the price of a homestead in Western Canada?" I told him it would cost him an entry fee of two pounds ($10). "But," he asked again, "what if I found myself in Canada without the two pounds, what then?"

I answered his practical question in this way: "Well, sir, you might do as I did under these very circumstances; when I landed in Western Canada I wanted a homestead, and I hadn't two pounds to pay the entry fee, so I went to work and dug a well for which I was paid two pounds. I put the money into that homestead, and I have since that time refused $8,000 for it!"

I have been asked regarding the climate: On our place we have never had to quit work on account of the mercury dropping. All winter we haul our grain, wood, rails, etc., and the sunshine is beyond compare. It is the healthiest country I know of; a country where a young man gets a chance—where industry finds reward— and if a man who is sober, industrious, and capable comes here, even though his hands and pockets be empty, it is all the capital he requires. I have proved this myself and so has my brother.

WILLIAM FULTON.

An Orkney Island Man Speaks.

"HARRAY ACRES," YORKTON, Sask.
October, 1906.

I am not an old-timer of the plains, but it may interest Orkney friends to know how a man can get along on a Western Canadian prairie farm, who comes to this new world without money. I am a practical farmer from Pomona, Orkney Islands, and after trying fortune in New Zealand concluded to test some of the statements made about your wonderful West. To say every fact I ever heard stated has been abundantly verified is little, for my own experience has proved that too much cannot be said for this country.

I arrived in Winnipeg in 1898, and sitting in one of the hotels met a Yorkton man, Francis Bull, whose pioneer experiences, told in a little circle of chums, interested me greatly. I decided to ask Bull what chances for a fellow lay in Yorkton. His answer gave me some misgivings, but I decided to go. "You can do well enough," said Bull, "if you are willing to work and wait." I had worked in New Zealand and "waited"; and having but five dollars in my pocket knew I had to get to work pretty soon. I went to Yorkton, taking a job with a farmer, and worked for wages one year, saving up $150. This was all right, I thought, so I hired for two more years and got onto the swing of Canadian farming. I then bought 160 acres, paying a deposit of five dollars—two dollars an acre the price—improved it some—sold within three years for $1,000, taking the crop of seventy acres myself. This crop I took off paid for every dollar's worth of improvements done. I then purchased a half section at $3.50 per acre, broke 100 acres, fenced, sold it two years later at $9 an acre, and had one crop off it for sale. I then bought this farm I live on at $15 an acre, it is 480 acres in size, and I only wish it were twice the size.

Off 275 acres in 1906 I took 12,000 bushels of grain, 4,500 being best grade wheat. In a granary, 16x24, to-day I have stored 4,500 bushels of wheat, 190 bushels being of Red Fyfe, taken from a patch of five bushels sown as a test. This reaped 51 bushels to the acre.

Am I satisfied? I think my statement answers that. Being a new-comer perhaps my experiences may be of use to another man who fears to come without capital. With capital I cannot begin to say how a man can make money on a Canadian farm. It simply overwhelms one, the simple facts of this wonderful country.

I have nothing to say against New Zealand. It's a fine climate —but you can't eat climate, and a fellow has to live. In Canada you get climate all right, and you get what's more, cash returns for your labor! JOHN HOWRIE.

ROYAL MAIL S.S. EMPRESS OF BRITAIN

One of the new palatial steamships for the Canadian Pacific Railway Company's Atlantic service on the Montreal, Quebec and Liverpool route. Length 570 ft., breadth 65 ft., displacement 20,000 tons, 18,000 horse power, and makes the passage between Liverpool and Quebec in less than a week. Accommodation for 300 1st cabin, 350 2nd cabin, 1,000 3rd class passengers.

THE FINEST AND FASTEST STEAMSHIPS IN THE CANADIAN TRADE.

TO THE

PACIFIC COA

THE ORIENT, THE TROPICS, AND
THE ANTIPODES

THE BEST, CHEAPEST AND QUICKEST

——————— TO ———————

NEW ONTARIO	MANITOBA	SASKAT
ALBERTA	BRITISH COLUMBIA	PUGE
THE YUKON	ALASKA	CALIFORN
JAPAN	CHINA	PHILIPPINES	NEW Z
	HAWAII	AUSTRALIA

——————— OR ———————

AROUND THE WORLD

IS BY THE

CANADIAN PACIFIC RAIL

(Read before the Royal Horticultural Society, London, England,
January 31st, 1899.)

Revised 1907 by C. H. Hooper and Professor F. C. Sears.

A Year Among the Orchards

OF

NOVA SCOTIA.

ILLUSTRATED.

By Cecil H. Hooper (Member Royal Agricultural College,
Fellow Surveyors' Institution), of the South Eastern
Agricultural College, Wye, England.

Issued by the Nova Scotia Government Office,
57a, Pall Mall, London, S.W.

An Apple Orchard in Nova Scotia.

Gathering, Sorting and Packing Early Apples.

A YEAR AMONG
THE ORCHARDS OF NOVA SCOTIA.

The Peninsular of Nova Scotia is situated on the east side of the Dominion of Canada, and south of the entrance of the Gulf of St. Lawrence. The climate here is very pleasant, in spite of a long and rather severe winter. The country is remarkably well supplied with water by its countless little springs and its numerous lakes, as well as by the heavy falls of snow in winter and frequent and heavy showers during the summer months, the latter generally falling at night, leaving the days bright, warm and cloudless. The scenery is beautiful, the abundance of native trees rendering it particularly attractive; the most common of these are spruce, fir, pine, larch, birch, maple, ash, alder and oak. The acacia tree is often seen, and also in some parts the French willow and English elm have been introduced, and thrive well. Nova Scotia is said to have the largest variety of flowers, mosses and ferns of any country; wild eatable berries are also very plentiful; they include strawberries, raspberries, blueberries, huckleberries, blackberries and cranberries.

THE CORNWALLIS AND ANNAPOLIS VALLEY.

The Cornwallis and Annapolis Valley is the principal fruit-growing district in Nova Scotia.* It is one continuous valley of about 100 miles in length, and varying in width from 6 to 11 miles, situated between two nearly parallel ranges of hills about 600 feet in height. The North Mountain shelters the valley on the northwest, and from the strong winds off the Bay of Fundy; the South Mountain, which is a little higher, bounds it on the eastern and southern side, and runs N.E. to S.W. In the middle of the valley there is a watershed, the Annapolis River running S.W., the rivers of the Cornwallis District running N.E.

---

*The fruit belt is however being extended, as owing to the establishment of experimental orchards by the Nova Scotia Government, it has been found that apples can be successfully cultivated in practically any part of the Province.

These rivers are small, but owing to the great rise and fall of the tide (60 feet) the salt water runs up far inland, carrying with it enormous deposits of alluvial mud or silt, and allowing ships to go several miles inland.

Near the mouths of the rivers there are salt marshes which are over-flown by the tide and grow salt hay, which is eaten by the cattle during winter. Higher up are the dyked marsh lands reclaimed from the sea, forming very rich meadow land. Grand Pré, the district rendered famous by the expulsion of the Acadians, lies in the eastern part of the valley, on the shores of the Basin of Minas, across which Cape Blomidon, the termination of the North Mountain, is clearly seen. Owing to the beauty of the country, its historic interest, and the cooler temperature, it attracts many visitors from the United States during the summer months. Apples and plums are grown throughout the valley, and in the centre, near the towns of Middleton, Aylesford, and Berwick, cranberries, raspberries blackberries, and strawberries are grown, also some peaches and a few grapes ; one farm I visited had six acres of strawberries. Most of the soft fruits are sent to Halifax and Boston. In the centre of the valley there is a large area of bog land, which, it has been found, is well adapted to cranberry growing. This industry is rapidly increasing.

The greater part of the valley was originally covered by forest, which has been cleared, save at the foot and sides of the mountains.

THE SOIL.

The soil of the valley is partly formed from the disintegration of the Trap Rock of the North Mountain, partly from the syenitic granite of the South Mountain, together with the red loam and coarse-grained sand of the new red sandstone in the valley, which abounds in oxide of iron, lime and gypsum, forming a fertile soil admirably adapted to the cultivation of apples, plums and various fruits, as well as of potatoes, swede turnips, oats, maize, pumpkins, beans, etc. Wheat growing and beef production have lately decreased, owing to the competition of the western provinces, but more attention is now being paid to the rearing of stock by the farmers. The dairying industries are also increasing.

THE FARMS.

The farms are, almost without exception, occupied by their owners, most of them small, compared with the average size of English farms, and still smaller, of course, compared with many farms in the west of Canada

and the United States.   The labour is largely performed by the farmer and
his sons, with but little hired help.   The farmhouses and buildings for the
most part are neat, comfortable, and give the impression of prosperity ; they
are almost all constructed of wood, painted white ; they are generally
situated near the high road, and, as the farms are frequently long and narrow
extending often back into the woods and down through the marsh land to
the river, the farmhouses are frequently within a quarter of a mile of one
another, which enables life there to be of a social nature if desired.   Prohibition of intoxicating drinks is rigidly enforced throughout Nova Scotia,
with the exception of a few towns ; there is consequently very little drunken-
ness.   Roughly speaking, the area of these farms varies from 20 to 120
acres, about equal parts of grass and arable land, the latter including one to
five acres of apple orchard.   There are a few farms with as many as 60 or
more acres of orchard ; but a large proportion of this has been planted within
the last ten years, and is not yet in full bearing.   Many orchards are 50
years old, and a few apple trees remain which were planted by the French
more than 150 years ago.   The apple tree certainly thrives here, and the
orchards are generally neatly laid out and well cared for ; the growth of the
trees is more rapid, and they attain a larger size than is common in England.
The fruit is usually large, well coloured and abundant, and of pleasant flavour,
particularly the Gravenstein.   Owing, I suppose, to quicker growth and shorter
season, I do not think the flavour generally is quite as nice as that of
our best English apples.   Although the shape of the trees, the cultivation
and the fruit in the best English orchards equal, I think, anything I
saw in this valley, the average of the two countries is much in favour of
Nova Scotia.

Throughout the valley there is a telephone system which connects railway
stations, shops, doctors' dwellings, and many of the farmers' houses.   The
charge is five pounds for installation, £2 10s. od. yearly.

On one occasion I sent a cable to England from the sitting room of the
farmer's house in which I was staying, and received one back.

Co-operative cheese and butter factories stud the valley every few miles,
These encourage the keeping of dairy cattle, which industry profitably accom-
panies fruit growing.   The local agricultural societies own pedigree cattle
for the improvement of native stock.   At Canning, in the Cornwallis Valley,
there is a vegetable evaporating factory (Kerr's), which was busy drying
vegetables for the soup of the miners at Klondyke.   It has in former years
fulfilled contracts to the satisfaction of the English Government for naval and
military supplies.

PRUNING AND TRAINING.

The trees are, as a rule, well shaped, as the farmers begin early in the life of the tree to shape them. They like their trees to have a central leader with the main branches distributed evenly about it. The height of the branches from the ground is regulated so as to allow horse cultivation under them.

It is found by experience that it is best to saw off the branches as close against the trunk as possible. If it is necessary to remove a large limb, they commence by sawing in a short distance from below, upwards, in order to avoid splitting the wood and tearing the bark. Large wounds grow over best when the edge is smoothed off with a knife and then covered with some substance to exclude moisture, and thereby prevent decay. Gum shellac dissolved in wood-alcohol is found to be the best substance for this purpose, though white lead paint or grafting wax are both good.

Generally speaking, summer pruning, of which a good deal is done, promotes fruitfulness; winter pruning tends more to wood growth. Pruning in Nova Scotia is chiefly done at the end of winter, whilst snow is still on the ground. When the trees are bursting into bloom is found to be a good time, though the opinion is that pruning may be done any time during the winter without disadvantage to the trees, the discomfort being that of the man who prunes.

In case of a tree being split at the forking of the branches, a hole is bored with an augur right through the tree, at right angles to the split, and the parts are drawn together by an iron screw bolt and nut; which damages a tree less than binding together with a hoop of iron.

Again. on Mr. Ralph Eaton's farm (Kentville), in order to train the young trees to grow upright, in case of the trunk bending, a screw-hook is screwed into the tree, and by means of a wire attaching the hook to a peg in the ground. the tree is drawn into the desired position. These hook-eyes and wires are also used to train the branches into correct position where necessary.

FERTILIZING OF ORCHARDS.

Rotation in the fertilizers applied to the orchard is recommended as advantageous. For example: stable manure one year, chemical fertilizer another. Farm-yard manure greatly benefits old, neglected orchards requiring nitrogen; but its use should be discontinued where trees run too much to wood and leaf without fruit, and some fertilizer containing potash and phosphate would probably prove more beneficial.

Green manuring or cover cropping is much employed in Nova Scotia to supply vegetable matter.

In Canada, wood ashes are the best possible manure. They are applied at the rate of twenty to forty bushels per acre; those from hard-wood being better than those from fir trees. The ashes contain about five to seven per cent. potash, two per cent. phosphoric acid.

As the available supply of farm-yard manure and wood ashes is very limited, commercial fertilizers are largely used; the two in most common use are finely ground bone meal at the rate of five to eight cwts. per acre to supply phosphate and nitrogen, and muriate of potash at one to three cwts. per acre to supply potash.

In the adjoining valley of the Gaspereau there is a bone mill, to which farmers take bones to be ground.

Nitrate of soda is not, from what I noticed, much used in the Nova Scotian orchards, save sometimes to give young or old trees increased vigour. Nitrate of soda at 95 to 96 per cent. purity equals 15.6 to 15.8 per cent. of nitrogen.

In a paper on fertilizers for orchards in Nova Scotia, published in " The Farmers' Advocate," the following ingredients were recommended:

For small fruits (strawberries, raspberries), per acre: 150 lbs. nitrate of soda; 250 lbs. muriate of potash; 800 lbs. bone meal.

For apple orchards: 100 lbs. nitrate of soda; 200 lbs. muriate of potash; 550 lbs. bone meal.

For orchards with clover: 200 lbs. muriate of potash; 250 lbs. Thomas phosphate. (Basic slag.)

E. B. Voorhes, of New Jersey Experimental Station, said:

" To provide vegetable matter and to improve the physical quality of poor " soils, apply yard manure once in four years, in fall or winter, at the rate of " from five to ten tons per acre. To aid in the decomposition of vegetable " matter, and to insure a sufficiency of lime as plant food, apply lime at the rate

" of twenty-five bushels per acre once in five years. To provide, in addition, an
" abundance of all forms of available plant food at the times needed for the
" development of the tree and fruit, apply annually chemical fertilizers in the
" following proportions :

Nitrate of soda	100 lb.
South Carolina rock superphosphate	100 lb.
Ground bone	200 lb.
Muriate of potash	200 lb.

" The amounts to be applied depend upon the character of the soils, the
" kind of fruit and the age and vigour of the tree ; these given perhaps mark the
" minimum. In a number of best orchards the quantities applied are very much
" larger than is here indicated, and the larger application is believed by the
" growers to be proportionately profitable."

Frank T. Shutt, Chief Chemist of the Dominion Experimental Farms,
wrote :

"Assuming the leaves of a full-grown apple tree to weigh 50 lb., and
" reckoning 40 trees per acre, the manurial value contained in the 2,000 lb. of
" leaves is equal to :

Nitrogen	17.74 lb.
Phosphoric acid	3.88 lb.
Potash	7.84 lb.

The leaves are returned to the soil, but the fruit is exported. This, in
the case of an orchard 25 years old, producing 160 barrels per acre—
equal to, say, 20,800 lbs. of apples—there is a loss to the soil of
approximately,

Nitrogen	8.9 lb.
Phosphoric acid	5.3 lb.
Potash	32.8 lb.

The following is given as another useful formula for manuring
orchards :—

Good rotten barn-yard manure, 10-15 tons per acre.

Kainit (13 per cent. potash)	300-700 lb.
Or muriate of potash (50 per cent. potash) ...	100-200 lb.
Bone meal (fine ground)	100-200 lb.
(2 to 3 per cent. nitrogen, 22 per cent. phosphoric acid.)	
Or superphosphate	125-250 lb.
(16 per cent. phosphoric acid.)	

A.
Apples from adjoining Rows of Apple Trees B.

A. Sprayed with copper sulphate and lime.
B. Not sprayed, and worthless through apple scab fungus.

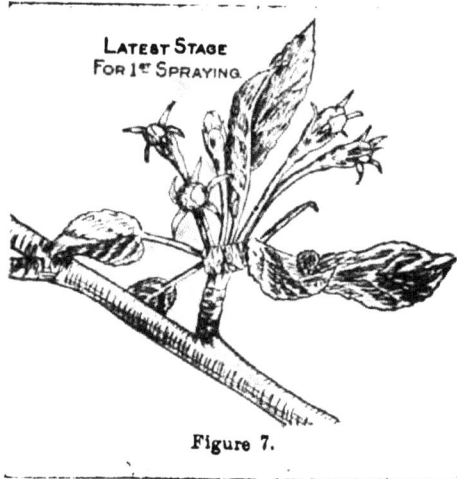

Figure 7.

1st Spraying before| blossoms open, to destroy spores of apple scab and other fungi and caterpillars of winter and other moths.

2nd Spraying, petals fallen, but calyx still open; the object being to kill the young caterpillars hatching out from eggs laid in the open flowers by the Codlin Moth.

ORCHARD TILLAGE.

The apple trees are planted 33 to 40 feet apart ; in a few instances with plum trees between, in one direction of the lines.

For the first few years the ground is ploughed deeply (8 inches), in order to break up the soil and encourage the roots to grow down to a sufficient depth to escape injury in the case of drought and to be below the reach of the plough. The whole surface of the orchard is tilled from the beginning. In ploughing the plough is turned partly out when within a few feet of the trees and runs shallower (4 inches deep), as the roots near the butt are closer to the surface ; immediately round young trees the surface is generally carefully forked over. Between young trees potatoes are frequently grown, using bone meal and muriate of potash as fertilizer. The deep ploughing needs only to be kept up for a few years in order to establish root growth.

The kind of plough in general use has no wheels, but has a sharp curved mould board, which, although it increases the draft, the difference is more than compensated for by the better condition the soil is left in. The ploughing is done either in the fall or early spring ; in Canada fall ploughing is not recommended for clay land as it tends to puddle it and make it become hard and stiff, the frost consequently enters to a greater depth, and root injury may result. In ploughing one aim is to obtain a level surface, thus one year the soil is ploughed from the trees, the next towards them, one year east and west, the following north and south. Most of the farmers whose land run down to the river bank, dig and haul the salt marsh mud on sleighs during winter and spread it on the orchard land ; this is disintegrated by frost and more carefully spread in spring. The mud has a fertilizing value, and also the salt in it probably aids in keeping the land moist. Early tillage saves the moisture accumulated during winter and early spring, and puts the soil into fine condition to warm up and get the trees quickly to work. As thorough cultivation renders plant food available, and is the best conservator of moisture, tillage is begun early by ploughing as soon as the snow has thawed, and the land is sufficiently dry to be worked. Harrowing follows, which stirs the ground thoroughly to the depth of about three inches. This is performed about every two weeks until late in the summer ; the drier the soil the oftener it should be done. The varieties of harrow used include the spring tooth, the spike tooth, the disc and acme. If the wood growth of trees is too luxuriant, it may be checked by lessening the tillage and by withholding nitrogenous manure. As the orchard trees stop growing about midsummer vigorous tillage then ceases, so that the new growth may ripen sufficiently to stand the cold of winter, and as the trees can now spare considerable moisture, catch crops are with advantage sown, such as ares and buckwheat.

THE CROPPING OF ORCHARD LAND.

Young orchards, say for the first 12 years, generally have some crop grown in them such as early potatoes, maize, beans, and other hoed crops. These crops need cultivation during the early part of the season and are removed about the middle of July or first week in August. Buckwheat and oats are also grown but are not as satisfactory, as the land cannot be thoroughly worked. Some space is allowed around the trees so as not to grow crops directly over the roots of the trees. For these crops bone meal and muriate of potash are usually sown in the drills.

In the older orchards the land is generally uncropped, but frequently harrowed. I have seen orchards almost weedless, due to this frequent cultivation. In some cases after the trees are about 12 years old clover is sown and left down for three or four years and mown for hay. This is however, exhaustive unless some fertilizer is used.

ORCHARD COVER CROPS OR GREEN MANURING.

The object of cover crops or green manuring is to gather and return vegetable matter to the soil, and to protect the ground from the effects of severe frost, especially when unprotected by snow.

For this purpose the orchards are very thoroughly cultivated during the earlier part of the season, and after the close of active growth, about July or August, the surface is sown with some crop which will grow quickly and be large enough to protect the soil during winter.

Crimson Clover (Trifolium incarnatum) in parts of Canada and the United States, where it stands the winter, is found admirably adapted to supply nitrogenous vegetable matter to orchards at a little cost; it is, however, rather a risky crop in Nova Scotia.

Mammoth clover is found to be the next best, sowing 14 lbs. per acre ; tares, lucerne, common red clover, peas, buckwheat, rye and oats are also employed for this purpose. A fair growth will be obtained the same season which is ploughed in early the following spring, in order not to retard the spring growth of the trees. These crops help to keep down weeds, and where successfully grown enrich the soil at less cost than with farmyard manure. Where clover and other plants of the leguminous family are grown, nitrogenous manures may be omitted, as these plants have special power to take up nitrogen.

Summer vetch is also excellent and when nitrogen is not important, and weeds (especially couch grass) are to be smothered out, rape and buck-wheat are excellent.

SPRAYING FOR FUNGI AND INSECTS.

In Canada both insects and fungi, where they do exist, appeared to me to be more plentiful and more destructive than at home. Canker in apple trees is, however, very rare. Woolly aphis is not common, and it is said to have come from Europe, and that Europe did not derive it from America.

For horticultural purposes insects may be divided into two classes: (1) Those that chew their food, such as caterpillars, and (2) those that feed by sucking the juices, such as scale insects and aphides.

The chewing insects may be destroyed by distributing poison over those parts of the plant upon which they feed.

In Canada spraying is almost universally practised; useful pumps are manufactured for the purpose, provided with a paddle to agitate the liquid. The pump is usually fitted into a paraffin barrel, the hose pipe is 10 to 15 feet long, often lengthened by a light bamboo tub rod, 6 to 10 feet long, in order to reach high up into the trees. The spraying nozzles are mostly of the Vermorel pattern. In the past year or two a new style of nozzle is being used, in appearance much like a large Vermorel. It has the advantage of being lighter to handle and less likely to catch on the branches of trees as it is being moved about through the top in spraying the interior of the tree. It is also less likely to clog. Several manufacturers have nozzles of this style on the market, notably the Goulds Company of Seneca Falls, New York; the Friend Company of Gosport, New York; and the Deeming Company of Salem, Ohio. The barrel is mounted on a cart or low waggon ("sloven") and drawn by a horse through the orchard, taking two or three persons to drive, pump and direct the spray. In some cases orchards are sprayed quite early in spring to clean the bark of the trees from scale and moss, using about 6 lbs. caustic rock potash to 50 gallons of water, or this may be applied to the trunks with a vegetable fibre (not hair) limewash brush, either at this time or in June. For spraying with stronger potash solutions, men sometimes wear old mackintoshes, rubber gloves and strap a mackintosh over the horse. This potash spraying is strongly to be recommended.

Powdered caustic potash is sold at most grocery shops in Canada, as it is used for soap making, a 1 lb. tin costing about 5d. These are very convenient for spraying purposes.

Spraying for fungi, Black Spot or Scab on apples (Fusicladium dendricum) may be done before the blossom or foliage opens, in order to kill the spores; for this, copper sulphate alone (2 to 4 lbs. to 50 gallons of water) can be used, without lime, as there is no foliage to burn. The first spraying jointly for insects and fungi is done when the fruit buds begin to unfold, but before the flowers expand, and should cease before the blossoms open. This is to destoy the spores of apple scab fungus and to kill the larvae of winter and bud moth and apple weevil. The second spraying is given immediately after the petals have fallen; this is specially to destroy codlin moth caterpillar, and if caterpillars are numerous or black spot shows on young fruit or leaf, the spraying should be repeated say once every two or three weeks.

The mixture in common use is the Bordeaux mixture, as the fungicide, Paris Green as the insecticide.

To save time, for the former, stock solutions of sulphate of copper and lime are made separately, the Paris Green is added direct.

The formula commonly adopted is :

	For apple.	For plum and peach.
Copper sulphate	4 lbs.	3 lbs.
Quicklime	4 lbs.	3 lbs.
Water	40 gallons	40 gallons.
Paris Green	4 ozs.	3 ozs.

Copper sulphate is readily soluble in cold water, and still more so in hot. The solution is made by banging the crystals, contained in a sack or basket, in a barrel of water, near the surface, so that it is just covered by the water. Vessels of wood or earthenware should be used for dissolving the sulphate. Dissolve 1 lb. of copper sulphate per 1 gallon of water for a stock solution : thus, take 40 lbs. for a 45 gallon barrel.

Take one barrel of lime, which should be freshly burnt, place it in a barrel, and pour, say, four gallons of water on it to slake it, afterwards add enough to make it into a creamery mixture like putty, then pour on a little more water to exclude the air and prevent change in character, then cover the mouth of the barrel to prevent evaporation.

For making up a 40 gallon barrelfull of the spraying mixture fill the barrel half full with water, then take four gallons of the copper sulphate solution (4 lbs.), dilute with four or five times the quantity of water. Next take some of the lime putty, mix it with water, and add it to the solution, straining it through a funnel-shaped box with a copper wire mesh strainer at bottom. In order not to add an unnecessary amount of lime, thereby risking the clogging of the machinery, it is advisable to test, so as to ascertain whether the sulphuric acid has been neutralised by the lime. For this purpose a solution of ferrocyanide of potassium (half ounce in half-pint of water) is used. Take a spoonful of the mixture from the barrel, in a white saucer or a glass, and add a few drops of the ferrocyanide solution. If a reddish brown colour appears the mixture needs more lime ; if there is sufficient lime no discolouration takes place.

Next weigh or measure out about ¼ or ½ lb. of Paris Green, put it in a cup, and make into a paste with water; add this to the mixture in the barrel.

In Nova Scotia, Paris Green is sold at most of the hardware stores in cardboard boxes, containing 1 lb., costing about 9d. It may be tested for its purity by ammonia, which should dissolve it completely, producing a deep blue liquid.

When Paris Green is used alone without the Bordeaux mixture it is always advisable to add an equal or double quantity of lime, for the purpose of taking up the soluble arsenic which may exist and might injure the foliage.

To spray twenty-year-old trees, planted forty to the acre, costs for materials 2s. to 4s per acre per application, and takes about 1½ to 3 gallons per tree to spray thoroughly on both sides. It takes double the quantity to spray when in full leaf t hat it does before the blossoms open. About four to six applications are generally needed.

Professor L: H. Bailey tells us that with a 300-gallon tank drawn by two horses, with three men, one drawing and pumping, the other two standing on the rear platform two or three feet above the tank directing the spray, each with a hose pipe, the pump having an automatic stirrer ; with this rig five acres of full-grown apple trees can be thoroughly sprayed in a day.

For currant and gooseberry caterpillar, white hellebore is used either as powder or if used as a liquid 1 oz. to 3 gallons of water is recommended.

The trunks of old trees are sometimes scraped with a short-handled triangular hoe or box-scraper, to clean off loose bark and moss, the dwelling places of the bark lice, and the winter quarters of the Codlin moth; this is usually done about April.

Insects, the food of which consists only of the sap or juice of the plant, and which thrust their beaks through the epidermis of the plant before they begin to suck in their food, are unharmed by any poision on the outside of the plant. This class of insect, to which scale and aphis belong, can only be destroyed by some substance which is applied to the insect itself, which enters the pores of the body and so kills it.

For apple bark scale (Mitylaspis pomorum) the following mixture is used, either in winter or about the middle of June, when the young lice are hatching out :

Paraffin	...	...	...	...	...	...	2 gallons.
Rain Water	...	...	...	...	...	...	1 ,,
Hard Soap	...	...	....	...	...	...	½ lb., or
Soft Soap	...	...	...	...	...	...	1 quart.

The soap and water are boiled together, then paraffin is added. The mixture is well stirred. For use, one part of the mixture is added to nine of water and applied to the bark, either as a spray before the leaf opens or with a brush in summer.

Dr. Fletcher recommends a solution of washing soda, so strong that no more will dissolve in the water, then dilute the soap to the proper consistency.

To counteract mildew on the leaves of gooseberry bushes Potassium Sulphide, 8 ozs. to 25 gallons of water, is employed.

FRUIT TREE BANDING.

The placing of bands of sticky material to prevent the ascent of the female winter moth, in America called the Canker Moth, is practiced to some extent in Nova Scotia, though it is generally considered that if spraying is thoroughly done at the right periods, grease banding is not necessary.

The substances chiefly used are bands of tarred roofing paper painted with printer's ink, or castor oil and resin applied direct to the tree. Professor Craig recommended, for winter use, 2 lbs. castor oil to 3 lbs. of resin, warmed together but not boiled, applied warm ; and for spring use, 2 lbs. castor oil to 4 lbs. resin. These mixtures are applied after scraping off loose bark, either direct or on the surface of paper.   The band is placed about two feet from the ground and about six inches wide, and is put on at the end of October or early in November.

Grease banding seems specially useful in the case of large trees, which it is difficult to spray thoroughly.

APPLE PICKING AND PACKING.

The kind of ladder commonly used for apple picking is one fairly broad at the base, but the sides of which at the upper end terminate in a point. They are liked, as they can be conveniently placed among the branches. The baskets used are of a rounded shape, with swing handle, holding a little more than a peck. These are convenient for emptying the fruit gently into the barrels, in which they are removed from the orchard without sorting. Fallen apples are picked up and sent to Halifax.

The barrels of apples are placed either in a special apple packing house, or more generally in the cellar under the barn or house, ready for sorting and repacking. The apples are sent over to England, chiefly between the end of September and the end of March. For sorting, the apples are poured out of the barrels on to a table, usually about five feet long and three feet wide, with a ledge all round four to five inches high, covered with carpet, felt or sacking. The sorting divides the apples into :

1st, of good size and quality.
2nd, smaller, but of good quality, both shipped to England.
3rd, scrubs, which are sound, but scabbed, ill-formed or otherwise defective, sent to local market.
4th, rotten, for pigs.

Sometimes the extra good apples are picked out as specially " Selected."

The Fruit Marks' Act which has come in force since I was in Canada, provides for four grades:—Fancy, No. 1, No. 2, and No. 3 and defines the size and quality of the first three.

The barrel most commonly used in Nova Scotia is made of fir staves with birch hoops, holding about 120 lbs. fruit and costing 10d.   Barrels with staves of maple or elm with elm hoops are used to a lesser extent, but are commonly used in Ontario and the United States.   These hold about 150 lbs. of apples and cost 1s.

In packing, the bottom and bilge hoops are first nailed, then a thin layer of wood wool called "Excelsior" is placed at the bottom, next a sheet of white paper the same size as the end of the barrel. A layer of "headers" is then laid; these are apples of average size, those best coloured being chosen. These are placed stem downwards. The barrel is then filled by carefully emptying in the fruit, using hinge-handled baskets. The barrel is gently shaken each time fruit is added, so as to pack the fruit closely. It is filled one to two inches above the rim. In order to get the apples tightly packed so that they do not move after packing and become "slack" in travelling, a round board lined with sacking or saddler's felt is placed, padded side downwards, on the top of the apples ; the barrel is rocked on the floor if of cement, or if not, on a heavy plank. The apples thus shaken and pressed sink to about the level of the rim ; any spaces are filled with small apples, then another piece of white paper is placed on the apples, the chine hoops are knocked up to loosen them, to allow the lid to enter the mouth of the barrel, then the lid is laid on, and the screw or lever press is applied to the barrel to press the lid into position ; followed by the tightening of the chine hoops and nailing of the head and hoops.

The name of the apple, together with the owner's name and address are stencilled on the top of the barrel ; the name of the salesman on the bottom.

The cost of sending over to England—London or Liverpool—from the Annapolis Valley, via Halifax, is about:

Rail	...	...	(60 miles)	...	...	1s. 3d.
Steamer	...	...	(2,450 miles)	...	...	2s. 6d.

3s. 9d per barrel.

The Salesman's commission in London is usually five per cent.

CRANBERRIES.

Around Berwick, Waterville, Auburn, Aylesford and Cambridge, in the middle of the valley, cranberry growing is fast increasing ; here the soil is moist but not stagnant; it consists generally of one or more feet of peaty soil over sand. In 1896 there were about two hundred acres of cranberry bog, and in 1897 about 2,500 barrels were raised. They keep well in barrels for fully nine months ; when required for shipment they are sorted and cleaned by hand and by machinery and sell in London at 25s. to 35s. per barrel. The cranberry beds take four years to come into bearing and are said to last about forty years. The crop is said to average about 40 barrels per acre. The rows are planted 15 to 24 inches apart, plants four to five inches apart.

Spraying tall Apple Trees with extension rod. (Photograph
lent by Prof. Allwood of Virginia.)

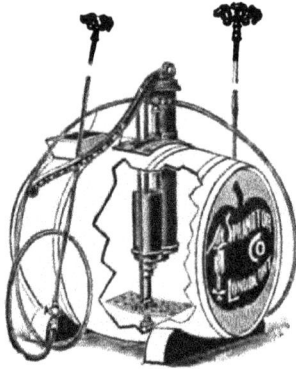

Pump with Agitator, 2 hose pipes, extension rods and nozzles
for Orchard Spraying.

Codlin Moth Caterpillar in an apple;

to prevent damage by this insect the trees are sprayed with a poisonous
insecticide directly the petals have fallen to destroy the young
caterpillars.

Once in three years the bogs are sanded about ½ inch deep, taking 50 two-horse loads per acre. During the winter the plantations are in many cases flooded, in order to destroy insects. One company owning a large area of cranberry bog which had been badly infested with "fire worm," at the recommendation of the Canadian Department of Agriculture, sprayed the plantation with arsenate of lead (½ oz. arsenate of soda in 1 qt. water, ¾ oz. acetate of lead in 1 qt., pouring the two together and adding 5 gallons of water). This insecticide has been found very effective and may take the place of Paris Green in orchards, as in a trial in the Central Experimental Farm, Ottawa, in 1895, the percentage of wormy fruit (Codlin Moth) was even less than where Paris Green was used.

For useful information as to cranberry culture I would recommend purchasing the Report of the Nova Scotia Fruit Growers' Association for 1897. I believe cranberries would be a remunerative crop on moor and heather land in England and Scotland.

Mr. S. C. Parker has written to me saying that upon the recommendation of Messrs. Nothard & Lowe (Tooley Street, London), cranberries have been generally shipped this season in boxes holding 10 lbs. This package gave good results, netting about 2s. 6d. per box.

NOTES OF THE YIELDS OF FRUIT.

At Cornwallis Rectory there is an old French Nonpareil apple tree probably 150 years old, measuring 10 feet girth one foot from the ground. The Rector, the Rev. F. J. H. Axford, told me that during the 20 years he has been there, the amount of fruit from it has varied from two barrels up to 16, the average being nine barrels.

At Wolfville, in 1896, from Mr. Elliott Smith's orchard, 20 barrels were gathered from three Gravenstein trees 25 years old. In the same year from the farms on Canard Street, 50,000 barrels of apples were gathered within a distance of 3½ miles along the road.

In the Gaspereaux Valley 1,700 barrels were gathered from 13 acres of orchard belonging to Mr. Gurtridge.

Strawberries, 100 bushels or 3,200 quarts per acre, I was informed, was an average yield, 5,000 being exceptionally good. Price usually 5d. to 7½d. per quart, but sometimes as low as 2½d. Strawberries are grown on the "matted row" system.

Raspberries, 2,000 quarts average, 5,000 very good ; price 4d. to 6d. per quart.

Blackberries 2,000 average, about the same price as raspberries.

Cranberries, 50 to 100 barrels of 150 lbs.   Price 25s. to 30s.

Price paid for picking all berries, ½d. per quart.

Potatoes are extensively grown and exported to Cuba.   Burbanks, early and late Rose and Chilis are among the varieties most grown.   The average yield is about 200 bushels per acre, price varying from 1s. 3d. to 3s., according to season.

VARIETIES OF FRUIT CHIEFLY GROWN IN NOVA SCOTIA.

APPLES.   Gravenstein—and Banks' Red Gravenstein are fit for shipping about the middle of September.

Baldwin, the most widely grown of any variety.

King of Tompkins County, very high quality, but rather shy bearer.

Nonpareil—commands a very high price, which is in its prime in the following May or June.   It is mostly grown in Annapolis County.

Ribstone Pippin—the best apple of English origin.

Golden Russet – commands the highest price, but shy, needs high feeding.

Ben Davis—still sells fairly well, but is poorer than the Western Ben Davis.

Northern Spy.

The Bishop Pippin or Belle Fleur, introduced by Bishop Inglis. the first Bishop of Nova Scotia, is much grown for home use, but does not stand shipping.

Blenheim Orange—one of the most popular.   Both leaves and fruit very free from " scab."

Fallawater.

Rhode Island Greening.

PEARS.   Bartlett or Williams' Bon Chretien, Clapp's Favorite, Flemish Beauty—as standards. (In Ontario Duchesse d'Angouleme is much grown as a dwarf.)

PEACHES.   Early Crosby and Early Alexander.

PLUMS.   Moer's Arctic, Lombard, Greengage, Magnum Bonum, and Bradshaw.   The wood of the European and American varieties are subject to a very destructive fungoid disease, " Black Knott"

(Plowrightia morbosa), which I trust we may never get in England, but from this disease the Japanese varieties are almost entirely free. Burbank, Wickson, and Abundance are the most popular Japanese sorts.

GOOSEBERRIES AND CURRANTS are but little grown. The American varieties of gooseberry are small, the best being the Downing. English varieties of gooseberry are so susceptible to the American gooseberry mildew that they are very little grown. Red currants fruit well, but black currants do not thrive.

RASPBERRIES. Cuthbert, Marlborough, both of American origin.

BLACKBERRIES. Snyder, of American origin.

STRAWBERRIES. Wilson's and Crescent Seedlings are mostly grown in alternate rows; Parker Earle, a new variety, is highly recommended. European varieties do not thrive on the American Continent.

CHERRIES are largely grown around Bear River and Digby; it is a district in which wild cherry and oaks grow naturally in the woods. Black and White Hearts and French are the varieties mostly grown.

CRANBERRIES are obtained wild from the North Mountain and near the lakes.

ORCHARD IMPLEMENTS. Some of the best orchard implements I saw in use in Nova Scotia were:

Pomona Pump (Gould's Manufacturing Co., Seneca Falls, New York).

Eclipse Pump (Morrill & Morley, Benton Harbour, Michigan). Both of these pumps have hard brass working parts complete, fitted on a 40-gallon barrel, with hose, bamboo rod, and double Vermorel nozzle, £3 15s. od. ; carriage to London, England, singly, about 24s.

The Spray Motor Pump, manufactured by the Spray Motor Co., of London, Ontario, is probably the most largely used of any pump now on the market in the Province. Several power sprayers made by this Co., and run by a gasoline engine, are also in use.

Orchard Spring Tooth Harrow (Syracuse Chilled Co., Syracuse, N.Y.), £3 15s. od.

As an example to us in co-operation amongst farmers and Government aid to agriculture, I will give a short account of the work of the Nova Scotia Fruit Growers' Association, its School of Horticulture, the Nova Scotia Office of Agriculture, and the Canadian Government Department of Agriculture.

THE NOVA SCOTIA FRUIT GROWERS' ASSOCIATION

was formed in 1863, when the acreage under fruit was probably about 2,500 acres. In 1893 it was estimated that there were 12,000 acres of bearing orchard, with 8,000 more of young trees. In 1871 apples were first sent to England. In 1896, a good year, it is estimated that 750,000 barrels were produced, of which 450,000 were exported to England, and 500,000 barrels in 1904. It is estimated that there are two million apple trees in Nova Scotia, and that not 10% of the land suitable for apple culture has yet been planted. The Association has for its objects :

The increase of cultivation of the various kinds of fruit.

The spreading of information as to the best methods of cultivating, packing, and shipping fruit.

The discussion of subjects of mutual interest, as freight rates by rail and steamer, ventilation on steamer, handling of barrels on embarking and disembarking, to prevent damage, condition of fruit on arrival in markets, possible new markets.

Once a year, in February, it has a three days' session at Wolfville, at which papers are read and discussed. I attended it in 1898, and thought it must be very valuable to the farmers, who attend in large numbers. Each member receives the printed report.

The yearly subscription is a dollar (4s. 2d.).

The Secretary is S. C. Parker, Berwick, Nova Scotia.

At the Halifax Industrial Exhibition in October there was a fine exhibition of apples, plums, cherries, peaches and grapes from Nova Scotia and New Brunswick under the management of the Association.

THE NOVA SCOTIA SCHOOL OF HORTICULTURE

was established at Wolfville in 1894 by the Nova Scotia Fruit Growers' Association, assisted by a Government grant. Horticulture, Botany and Microscopic Botany are taught by an able professor, F. C. Sears. The School consists of a class-room with a good collection of English, Canadian, and American books on Horticulture, Horticultural journals, about twelve good microscopes, and a collection of pressed wild plants. Beneath the class-room is a potting-shed, or workshop, and adjoining a glass house with economic and ornamental plants and flowers, in which grafting, budding and propagation can be taught during winter. There is also a root-cellar, in which apple stocks for root-grafting during winter are kept. Surrounding the school are ornamental grounds, with a nursery of young fruit and other trees and plants close by.

The horticultural course is at present confined mostly to the propagation of plants and to fruit-growing, dealing with wind-breaks, protection from frost, setting out and planting, tillage, manuring, cover crops, renovation of old orchards, grafting, budding, the life histories of fungi and insects, spraying, harvesting and packing of the fruit, cold storage, etc. The School of Horticulture is attached to the Wolfville University.

The classes are held during the winter months from November to the end of April. Having myself attended the course, I cannot speak too highly of it.

The course is free, and farmers are invited to come and look round at any time, attend any lecture, and bring any questions. Although there are such exceptional advantages, there are but few who attend regularly.

The following were the Text Books used at the School of Horticulture in winter, 1897-8 (they are excellent books).

"The Principles of Fruit Growing," by Prof. L. H. Bailey, of Cornell University.

"The Spraying of Plants," by E. G. Lodeman.

"The Nursery Book," by Prof. Bailey.

"The Pruning Book," by Prof. Bailey.

All published by the Macmillan Co., New York and London. Price about 5s. each.

Since I was in Canada the School has been removed from Wolfville and now forms a department of the Government Agriculture College at Truro.

THE NOVA SCOTIA OFFICE OF AGRICULTURE

in 1897 issued free to all farmers and others interested in the subjects a useful pamphlet, " Practical Hints to Fruit Growers," upon insects, fungi, spray machinery, insecticides and fungicides.

The Provincial Government has an Agricultural and Dairy School with Model Farm at Truro, which deals also with the growth and care of fruit. Lectures on agriculture, dairying and horticulture are given throughout the province, with field demonstrations, which appear to be well attended, judging from the one on spraying which I attended on Mr. Ralph Eaton's farm, near Kentville.

THE CANADIAN DEPARTMENT OF AGRICULTURE

issues at frequent intervals valuable bulletins on agricultural and horti-cultural subjects, sent free to farmers. It has an experimental station in each province for trial of different varieties of corn, vegetables and fruits. That of Nova Scotia is at Nappan ; the Central Experimental Station is near Ottawa, where soils and manures are analysed free to farmers ; insects are identified and advice given on matters connected with Agriculture and Horticulture.

In 1897 it sent over an expert, Mr. J. E. Starr (himself a large apple-grower) to report on the condition of the fruit arriving in England and Scotland from various parts of Canada, to see whether any improvements in packing, cold storage (for grapes, plums, peaches, etc.), steamer accommodation, marketing, could be suggested, or new markets found.

In conclusion, I can recommed Nova Scotia as a delightful holiday resort, and where a good deal may be learned in agricultural methods. The people are very kind and sociable, and willing to give information. And finally, I would here like to record my best thanks to my friends across the Atlantic.

Lightning Source UK Ltd.
Milton Keynes UK
UKHW021050061118
331795UK00009B/1175/P

9 780265 288832